The Science and Practice of Lithium Therapy

Gin S. Malhi • Marc Masson • Frank Bellivier
Editors

The Science and Practice of Lithium Therapy

Springer

Editors
Gin S. Malhi
Professor and Chair
Academic Department of Psychiatry
Sydney Medical School
The University of Sydney
Sydney, Australia

Marc Masson
Medical Director
Nightingale Hospitals
Paris-Clinique du Château de Garches, Consultant
Service Hospitalo-Universitaire
Centre Hospitalier Sainte-Anne
Paris, France

Frank Bellivier
Professor, Assistance Publique – Hôpitaux de Paris
INSERM UMR-S1144
Paris-Diderot University
Paris, France

Fondation FondaMental
Creteil, France

ISBN 978-3-319-45921-9 ISBN 978-3-319-45923-3 (eBook)
DOI 10.1007/978-3-319-45923-3

Library of Congress Control Number: 2016959283

Printed on acid-free paper

This Springer imprint is published by Springer Nature
The registered company is Springer International Publishing AG
The registered company address is Gewerbestrasse 11, 6330 Cham, Switzerland

Foreword

My lifelong love affair with lithium began 65 years ago, soon after graduating from the University of Sydney. I worked with some of the pioneers of lithium and, in particular, helped to make it a safer molecule to prescribe to patients. That initial wonder and excitement of a new treatment has never left me, and over the decades, I have seen countless people benefit from the effects of this seemingly magical element. The initial questions concerning efficacy and tolerability remain relevant today, and though much more is now known about the mechanisms of action of lithium, its precise effects and the reasons for its clinical specificity remain obscure.

Lithium's popularity has waxed and waned over the past half century. My own migration to North America allowed me to contribute to its introduction into clinical practice in the States in the 1970s. But the constant lure of things that are new and shiny has meant that lithium's usage has gradually suffered as more and more agents have been added to the pharmacological armamentarium used to treat what was known as 'manic-depressive illness' but which now goes by the name of bipolar disorder.

A key problem throughout this time has been that of identifying those patients most likely to do well, termed (although only after the fact) 'lithium responders'. Clinical indicators have been useful but applied perhaps more in research than in practice. Researchers have examined neurotransmitters, the stress axis, genes and brain structure, with a more recent focus on its functional architecture using emerging imaging techniques. As each biological layer has been exposed, enthusiasm has been renewed for identifying a neurobiological basis to lithium response. Along the way important clues have been found indicating that lithium responsivity does indeed have a biological basis but is, in part at least, heritable, suggesting that the pathophysiology of those with bipolar disorder that respond to lithium is perhaps different, but a definitive model is yet to emerge. This inability to clearly identify the context within which lithium excels has meant that clinically its use has been haphazard and vulnerable to therapeutic trends and clinical mythology.

Despite this, lithium has remained in pole position across international guidelines, not only for mood stability and prophylaxis, arguably its strongest actions, but also, remarkably, for acute mania and acute depression. In addition, it remains one of the few agents that can be utilised as an augmentation strategy across all mood disorders in conjunction with all manner of antidepressants and anticonvulsants. It is also one of the few agents that has profound discernible antisuicidal properties, and recently, there has been growing interest in its potential neuroprotective effects, opening the door to prevention and perhaps even neural regeneration.

The recent BALANCE study has restored confidence in its clinical use or, more accurately, provided evidence for its already existing administration in practice. Similarly, studies examining its tolerability profile have downgraded many of its side effects, whilst others examining its seemingly unique properties have upgraded its antisuicidality and potential neuroprotective roles.

Research into the effects of lithium has now once again become fashionable, reflecting perhaps a resurgence of interest in lithium clinically. And it is in these interesting times that this book by Malhi, Masson and Bellivier reviews and brings together the science and practice of lithium therapy. The book consists of two major parts that provide an accessible overview of interesting and emerging findings from research being conducted in relation to lithium and addresses the many clinical questions lithium therapy poses for clinicians worldwide.

The first part of the book discusses many intriguing aspects of this unique element that, despite its simplicity, creates unimaginable complexity in tissues throughout the body. It begins with a coruscating chapter that focuses on its molecular properties. The concluding chapters of this first part attempt to dissect the key question of lithium response from various neurobiological perspectives and to lay down a suitable foundation for the subsequent clinical part of the book that discusses more practical aspects of lithium therapy. Chapters in this part of the book critically appraise the efficacy of lithium, noting the many studies that still need to be undertaken and the opportunistic data that have been acquired because of lithium's repeated use as a comparator for newer molecules; it provides clinical insights for all practitioners and culminates in a balanced positioning of lithium use in the treatment of modern-day bipolar disorder.

I commend the editors on collating leading experts from all corners of the world and for the development of a text that will prove useful for clinicians and academics alike. It should be a standard text for anyone managing mood disorders or researching the field. The book is well written, erudite and presented beautifully with colourful illustrations and diagrams.

Many of the contributors constitute a who's who of lithium researchers worldwide – and drawing on a multitude of experts' contributions, the book brings the reader up to speed with exciting developments from around the globe and positions future endeavours authoritatively. I applaud the authors and the editors for undertaking and completing such a monumental and worthwhile task. It is my pleasure to introduce you to an old friend – described through the words of new friends.

Professor Samuel Gershon
Professor Emeritus of Psychiatry
School of Medicine
University of Pittsburgh
Pittsburgh, PA, USA

Preface

> In the fight against the two-headed dragon that surfaced from musings of the mind, the one champion that has reigned supreme for more than half a century is lithium. Surviving precariously from one decade to the next, this humble, silvery white element continues to demonstrate its silverback status amongst the burgeoning pharmacological armamentarium that is presently used to treat bipolar disorder.
>
> Numerous combatants belonging to the armies of antidepressants and anticonvulsants, flanked by their armigers brandishing pharmaceutical coats of arms, have successively thrown down the gauntlet to lithium, resulting in interregna marked by its displacement from prescription use. However, on each occasion, the shining knight, true to its mettle, has risen to the challenge and managed to reclaim its succession to the throne.
>
> Malhi (2010)[1]

Lithium has a long, distinguished and sometimes turbulent history as a therapeutic agent. In recent years its re-emergence as a mood stabiliser was effectuated by the publication of a seminal study in *The Lancet* in 2010.[2] An accomplished team of researchers from Oxford orchestrated BALANCE, and its compelling findings restored lithium – this element of blue blood – to its rightful throne.[3] As the first solid element to have formed in the known universe, lithium's use in the treatment of manic-depressive illness, the forerunner to modern-day bipolar disorder, has, as demonstrated by this book, a solid foundation. Its efficacy perhaps reflects its primordial origins and the fact that we have assumed form in its presence.

The origins of this book lie in a café in Paris in 2013, soon after a paper entitled *The Science and Practice of Lithium Therapy* was published in the *Australian and New Zealand Journal of Psychiatry*.[4] We three editors, already friends, met informally and discussed our joint vision for a book that would capture our passion for better understanding the mechanisms of lithium treatment. The idea gradually took

---

[1] Malhi G (2010) The king is dead, long live the king! The restoration of BALANCE *Bipolar Disorders* 12: 681–684.

[2] The BALANCE investigators and collaborators (2010) Lithium plus valproate combination therapy versus monotherapy for relapse prevention in bipolar I disorder (BALANCE): a randomised open-label trial. *The Lancet* Volume 375, Issue 9712, 385–395.

[3] Malhi GS (2014) Lithium: an element of blue blood? [French] *Ann Med Psychol* 172: 155–157.

[4] Malhi GS, Tanious M, Das P and Berk M (2012) The science and practice of lithium therapy. *Australian & New Zealand Journal of Psychiatry* 46: 192–211.

form of a single volume that would both coalesce the science related to this intriguing element and synthesise clinical experience distilled from practice. Together we scratched out an outline of what we considered to be important areas and then approached colleagues – all world-renowned experts – to assist with the task. It soon became apparent that research into the properties of lithium had been expanding at a magnificent rate and that to capture this faithfully would require many more chapters than we had originally envisaged. This task could only be achieved with the aid of a dedicated group of authors with a particular set of qualities: the necessary expertise, a willingness to participate and a capacity for fun and being easy to work with.

These qualities were to prove essential. Writing a book is inevitably challenging and time consuming – often much more so than expected. It is an activity that seems like a good idea right up to the point at which one has to begin. Fingers on keyboard, gazing blankly at the computer screen, one is forced to revisit the key question: is this book necessary and why? The editors must keep the answer to this question clearly in view as they undertake the formidable task of bringing together a wealth of information, harmonising it so that it is easy to assimilate, cross-linking it so that a narrative is maintained and ultimately giving form to a complete picture. During this process we often returned to the primary question of 'why'. Each time the resounding answer was lithium is still the best long-term therapy for the management of manic-depressive illness, and so the translation of existing knowledge into clinical practice is long overdue. This book serves that purpose. At the same time, future research concerning the mechanisms and clinical effects of this unique element is essential, and again this book addresses this need. These have been the drivers for its creation, which has been 3 years in gestation.

As editors, we would like to offer our sincere thanks to all the contributors, to whom we are genuinely indebted. They have each given up their expertise and time, spurred on by a shared belief in the benefits this book can bring. We also thank our friends and family for their forbearance in allowing us to devote our personal time to this project.

This book is intended for anyone and everyone working in the field of mood disorders and dealing with lithium therapy – doctors, patients and their families. We trust that it provides intriguing insights into the emerging science and growing clinical knowledge of lithium therapy and that it will suitably equip clinicians and academics worldwide to put this 'collective wisdom' into practice.

Sydney, NSW, Australia — Gin S. Malhi
Garches, France — Marc Masson
Paris, France — Frank Bellivier

Contents

Contributors

Martin Alda, MD, FRCPC Department of Psychiatry, Queen Elizabeth II Health Sciences Centre, Halifax, Nova Scotia, Canada

Christine Allwang, MD Clinic for Psychosomatic Medicine and Psychotherapy, Klinikum Rechts der Isar der Technischen Universitaet Muenchen, Munich, Germany

Jean-Michel Aubry, MD Division of Psychiatric Specialties, Department of Mental Health and Psychiatry, and Faculty of Medicine, University Hospitals of Geneva, University of Geneva, Geneva, Switzerland

Frank Bellivier, MD, PhD Department of Psychiatry, AP-HP GH Saint-Louis – Lariboisiere – F. Widal, Paris-Diderot University & INSERM UMR-S 1144, Paris, France

Chloé Benizri, MD, MSc INSERM U955, Team 15 "Psychiatry Genetics", Hôpitaux Universitaires Mondor, INSERM U955, Pôle de Psychiatrie, Créteil, France

Marc-Louis Bourgeois, MD, PsychoD Department of Psychiatry, IPSO CHS Charles Perrens Bordeaux, Bordeaux Cedex, France

Stephanie Brun, MD Centre Aliénor d'Aquitaine, Maternity Unit, Bordeaux University Teaching Hospital, Bordeaux, France

Lucie Chevillard, PhD INSERM UMR-S 1144, Response Variability to Psychotropics, Paris, France

Andrea Cipriani, MD, PhD Department of Psychiatry, University of Oxford, Warneford Hospital, Oxford, UK

Salvatore Cisternino, PhD Faculté de Pharmacie – Université Paris Descartes – Laboratoire de Pharmacocinétique, INSERM UMR-S 1144, Paris, France

Camille Couffignal, PharmD, PhD UMR 1137, Inserm and Paris-Diderot University, Paris, France

Aroldo A. Dargél, MD, PhD Institut Pasteur, Paris, France

Pritha Das, PhD Academic Department of Psychiatry, Sydney Medical School, The University of Sydney & CADE Clinic, Royal North Shore Hospital, Sydney, Australia

Alexandre Dayer, MD Division of Psychiatric Specialties, Department of Mental Health and Psychiatry, and Faculty of Medicine, University Hospitals of Geneva, University of Geneva, Geneva, Switzerland

Xavier Declèves, PharmD, PhD Department Cochin Hospital, AP-HP, Paris & Research Team, Paris, France

Souleiman El Balkhi, PhD Pharmacology, Toxicology and Pharmacovigilance Department, CHU Limoges Hospital, Limoges, France

Bruno Etain, MD, PhD Department of Psychiatry, AP-HP GH Saint-Louis – Lariboisiere – F. Widal, Paris-Diderot University & INSERM UMR-S 1144, Paris, France

Kristina Fritz, PhD Academic Department of Psychiatry, Sydney Medical School, The University of Sydney & CADE Clinic, Royal North Shore Hospital, Sydney, Australia

John R. Geddes, MD University Department of Psychiatry, Oxford University, Warneford Hospital, Oxford, UK

Pierre Alexis Geoffroy, MD Department of Psychiatry, AP-HP GH Saint-Louis – Lariboisiere – F. Widal, Paris-Diderot University & INSERM UMR-S 1144, Paris, France

Danielle Gessler Academic Department of Psychiatry, Sydney Medical School, The University of Sydney & CADE Clinic, Royal North Shore Hospital, Sydney, Australia

Guy M. Goodwin, DPhil FMedSci University Department of Psychiatry, University of Oxford, Oxford University, Warneford Hospital, Oxford, UK

Chantal Henry, MD, PhD Pôle de psychiatrie, Hôpital A. Chenevier, Créteil, France

Josselin Houenou, MD, PhD APHP, Hôpitaux Universitaires Mondor, Pôle de psychiatrie, DHU PePsy, INSERM U955, Team 15 "Psychiatry Genetics", Créteil, France

UNIACT Lab, Psychiatry Team, Neurospin Neuroimaging Center, Gif Sur Yvette, France

Pôle de Psychiatrie, Hôpitaux Universitaires Mondor, Créteil, France

Barbara Jakubowicz, MD APHP, Hôpitaux Universitaires Mondor, Pôle de psychiatrie, DHU PePsy, INSERM U955, Team 15 "Psychiatry Genetics", Créteil, France

UNIACT Lab, Psychiatry Team, Neurospin Neuroimaging Center, Gif Sur Yvette, France

Grünfeld Jean-Pierre, MD, PhD Department of Nephrology, Necker Hospital, Paris, France

Helen Jones Department of Psychiatry, University of Oxford, Warneford Hospital, Oxford, UK

Dominique Larcher Laboratoire de Réactivité et Chimie des Solides CNRS UMR 7314, Université de Picardie - Jules Verne, Amiens, France

Frédéric Limosin, MD, PhD Hôpital Corentin Celton, Service de Psychiatrie de l'adulte et du sujet âgé, groupe hospitalier, Hôpitaux Universitaires Paris Ouest, Assistance Publique–Hôpitaux de Paris, Issy les Moulineaux, France

Gin S. Malhi, BSc (Hons), MBChB, MD Academic Department of Psychiatry, Sydney Medical School, The University of Sydney, and CADE Clinic, Royal North Shore Hospital, Sydney, Australia

Cynthia Marie-Claire, PhD Variability of Response to Psychotropics, INSERM UMR-S 1144, Paris, France

Marc Masson, MD Medical Director, Nightingale Hospitals Paris-Clinique du Château de Garches; Consultant, Service Hospitalo-Universitaire, Centre Hospitalier Sainte-Anne, Paris, France

Thomas Mauras, MD Department of Psychiatry (Dr. Marcel), CH Sainte Anne, Paris, France

Bruno Mégarbane, MD, PhD Department of Medical and Toxicological Critical Care Medicine, Lariboisière Hospital – Paris Diderot University, Paris, France

Rebecca F. McKnight, BA (Oxon.), MA (Oxon.), BMBCh. Department of Psychiatry, Warneford Hospital, OxfordWarneford Lane, Oxford, UK

Nathalie Nanzer Child and Adolescent Department, Child and Adolescent Psychiatric Service, Guidance Infantile, Geneva University Hospital, Geneva, Switzerland

Tim Outhred, PhD Academic Department of Psychiatry, Sydney Medical School, The University of Sydney & CADE Clinic, Royal North Shore Hospital, Sydney, Australia

Nader Perroud, MD Division of Psychiatric Specialties, Department of Mental Health and Psychiatry, and Faculty of Medicine, University Hospitals of Geneva, University of Geneva, Geneva, Switzerland

Claudia Pisanu, PhD Department of Biomedical Sciences, University of Cagliari, Cagliari, Italy

Sami Richa, MD, PhD Faculty of Medicine, Department of Psychiatry, Hotel Dieu de France de Beyrouth, Saint-Joseph University, Alfred Naccache, Beirut, Lebanon

Nicolas Schaad, PhD Pharmacie Interhospitalière de la Côte, Dépts Neurosciences Fondamentales et, Anesthésiologie, Pharmacologie Clinique et Soins Intensifs (APSI), Faculté de Médecine, Université de Genève, Morges, Switzerland

Jean-Pierre Schuster, MD Hôpital Corentin Celton, Service de Psychiatrie de l'adulte et du sujet âgé, groupe hospitalier, Hôpitaux Universitaires Paris Ouest, Assistance Publique–Hôpitaux de Paris, Issy les Moulineaux, France

Jan Scott, MD, BS, PhD, FRC Psych. Academic Psychiatry, Institute of Neuroscience, Newcastle University, Newcastle, UK

Aude Servais, MD, PhD Department of Nephrology, Necker Hospital, Paris, France

Sarah Sportiche Department of Psychiatry, AP-HP GH Saint-Louis – Lariboisiere – F. Widal, Paris-Diderot University & INSERM UMR-S 1144, Paris, France

Alessio Squassina, PhD Department of Biomedical Sciences, University of Cagliari, Cagliari, Italy

Anne-Laure Sutter-Dallay, MD, PhD Perinatal Psychiatry Network, University Department of Adult Psychiatry, Centre Hospitalier Charles Perrens, Bordeaux University, Bordeaux, France

Jean-Marie Tarascon Collège de France, Paris, France

ALISTORE-European Research Institute, Amiens, France

Réseau sur le Stockage Electrochimique de l'Energie, Amiens, France

About the Editors

Gin S. Malhi, BSc (Hons), MBChB, FRCPsych, FRANZCP, MD, is Professor of Psychiatry and Chair in the Faculty of Medicine at the University of Sydney. He is Department Head at Royal North Shore Hospital where he is Director of the CADE Clinic, and Medical Director of the Mood Disorders Unit at Northside Clinic, Sydney. After graduating with degrees in Pharmacology and Medicine from the University of Manchester (UK), he completed his psychiatry training in Cambridge and then the Bethlem and Maudsley Hospitals in London. As a Clinical Lecturer at the Institute of Psychiatry (IOP) in London, he pursued research in depression and psychosurgery, and upon moving to Sydney (Australia) in 1999 he began research in neuroimaging and bipolar disorder at the University of New South Wales (UNSW). Since 2003, he has been a Chief Investigator on successive National Health Medical Research Council (NHMRC) Program Grants and during this time he has published more than 400 papers and 20 books/book chapters. After serving as Editor-in-Chief of *Acta Neuropsychiatrica*, he has been the Editor-in-Chief of the *Australian and New Zealand Journal of Psychiatry* since 2010 and is now also an Editor of *Bipolar Disorders*. In 2013, he received the Royal Australian and New Zealand College of Psychiatrists (RANZCP) Senior Academic Researcher Prize, and then in 2015 the RANZCP College Citation, and in the same year the Mogens Schou Award from the International Society of Bipolar Disorders for Education and Teaching. In 2016, in recognition of his many achievements, he received the Distinguished Professor Award from the University of Sydney.

Marc Masson, MD, is a Psychiatrist and the Medical Director of the Nightingale Hospital Paris-Clinique du Château de Garches and Consultant, Service Hospitalo-Universitaire, Centre Hospitalier Sainte-Anne, Paris, France. After graduating from Bordeaux University in France, he worked as a psychiatrist at Assistance Publique Hôpitaux de Paris (APHP). He has served as General Secretary of The Société Médico-Psychologique and Editor-in-Chief Assistant of The Annales Médico-Psychologiques and is now the Editor-in-Chief of L'Encéphale. His main areas of interest are bipolar disorders, psychopharmacology, neuropsychiatry and the history of psychiatry, and has published 4 books and 45 papers.

Frank Bellivier MD, PhD, is Professor of Adult Psychiatry at University Denis Diderot in Paris, and Head of the Departments of Psychiatry and Addiction at Saint-Louis – Lariboisière – F. Widal hospitals in Paris. He is also director of a research team on neuro-psychopharmacology of bipolar disorders and addictions at the Institut National de la Santé et de la Recherche Médicale (INSERM UMRS_1144, www.umrs1144.com). He received his medical degree at the University of Paris (René Descartes). There, he also completed his residency training and a PhD in Neurosciences, before undertaking a research fellowship at Pitié-Salpêtrière Hospital, Paris. He has led several large clinical programmes for patients with mood disorders, including the Early Onset Bipolar Disorders Program, funded by INSERM and the Assistance Publique Hôpitaux de Paris (APHP). Since 2007, he has led collaborative research programmes on genetic vulnerability factors in bipolar disorders and is currently conducting a prospective follow-up study on cognitive deficits and relapse risk factors in bipolar disorder. He is the recipient of several research prizes including the University of Paris 12 Research Prize (2000), the Jules Baillarger Prize (2002) and the NARSAD Independent Investigator Award (2004). His ongoing research program on biomarkers of lithium response is funded by the National Research Agency (ANR générique 2014), the Sorbonne Paris Cité University, the French Ministry of Research (PHRC) and the FondaMental/Dassault Fondations Research Prize (2013). He has published over 200 papers and is the author of many books and book chapters.

Part I

The Science of Lithium

The Lithium Ion

1

Dominique Larcher and Jean-Marie Tarascon

Abstract

For more than a century, lithium was considered an oddity and a laboratory curiosity of very limited use. Now, however, lithium is regarded as one of the most critical of the elements. It is strategically and economically crucial in many sectors of human activity, including nuclear, medicine, energy, lubricants, metallurgy, polymers and glass sectors, among others. Increasing interest from a variety of fields has given rise to substantial demand for this element and triggered fears, among a few alarmists, about resources and the possibility of supply shortages. The reasons behind the increasing interest in lithium are mainly rooted in its unique atomic, physical and chemical properties, and it is these that this chapter presents.

D. Larcher (✉)
Laboratoire de Réactivité et Chimie des Solides, CNRS UMR 7314, Université de Picardie Jules Verne, 33 rue Saint Leu, 80039 Amiens Cedex, France

ALISTORE-European Research Institute, 80039 Amiens, France

Réseau sur le Stockage Electrochimique de l'Energie (RS2E), FR CNRS, 3459 Amiens, France

Institut de Chimie de Picardie (ICP), CNRS FR3085, Université de Picardie Jules Verne, 80039 Amiens Cedex, France
e-mail: dominique.larcher@u-picardie.fr

J.-M. Tarascon
Collège de France, 11 place Marcelin Berthelot, 75231 Paris, Cedex 05, France

ALISTORE-European Research Institute, 80039 Amiens, France

Réseau sur le Stockage Electrochimique de l'Energie (RS2E) FR CNRS, 3459 Amiens, France

G.S. Malhi et al. (eds.), *The Science and Practice of Lithium Therapy*,
DOI 10.1007/978-3-319-45923-3_1

Key Points

- Lithium is the lightest metal and the lightest solid element.
- Lithium has the lowest density of all solid chemical elements, around half that of water.
- Li° is synthetic. In nature, lithium can only be found in an ionized state (Li^+) in salts or solutions.
- Its very small ionic radius causes bare Li^+ to interact strongly with water molecules, leading to large size for the hydrated lithium ion.
- The large energy of solvation of Li^+ explains the exceptionally low redox potential for $Li^+/Li°$, making Li° the most reducing of all elements.
- Combined size and charge effects explain solubility and solubility vs. temperature issues specific to lithium salts, as well as some other peculiar behaviours exhibited by many lithium salts.
- Until the 1920s, lithium was considered a laboratory curiosity, but it is now a strategic element used (30,000 t/year) in various domains: metallurgy, lubricants, glass and ceramics, electrochemical energy storage (batteries), nuclear energy, medicine/pharmacology, polymers, organic synthesis, air conditioners and others.
- In the Earth's crust, lithium is as abundant (by weight) as gallium or niobium, so more abundant than lead and tin.
- The diversity of applications and uses for lithium makes forecasting future demand and market evolution a difficult task, but known reserves (rocks, brines and salt lakes) are substantive (30–50 Mt) and recycling processes are already available.

1.1 Discovery, First Reports and Milestones

In 1800, in an iron ore mine on the island of Utö (Sweden), José Bonifácio de Andrada e Silva (1763–1838) discovered a new mineral, which he named petalite. At that time, Andrada e Silva was a Brazilian student completing his scientific education with an extensive study tour through many European countries (Portugal, France, Holland, Scandinavia, Hungary, Germany, Italy, etc.) while fostering potentially fruitful collaborations. Petalite samples were subsequently sent across Europe to many scientists (including Berzelius, Wollaston, Haüy and Ritter von Leonhard) for chemical and crystallographic analysis (Enghag 2004; Arfwedson 1818). In 1817, Johan August Arfwedson (1792–1841) first detected the presence of a new element in petalite ore (sheet silicate, $LiAlSi_4O_{10}$) and later on in other minerals such as spodumene (pyroxene, LiAlSi2O6) and lepidolite (mica, $KLi_2Al(Al, Si)_3O_{10}(F, OH)_2$) (Arfwedson 1818; de Blainville 1819). The name of this new element, 'lithium' (initially 'lithion' from the Greek λιθos = stone), is a tribute to its geological occurrence (unlike plant-based sodium and potassium) and was coined by the famous Swedish chemist Jöns Jakob Berzelius (1779–1848), who was at that time supervising Arfwedson's work (Arfwedson 1818). Only 1 year later (1818), Christian Gottlieb Gmelin (1792–1860) observed that lithium compounds impart a characteristic bright red/carmine colour to a flame. At that time, lithium was quickly recognized as the third member of the alkali family that already included sodium and potassium (Arfwedson 1818). This information fed into the first promising proposal for the classification of the chemical elements by Johann Wolfgang Döbereiner (1780–1849), who at that time was trying to complete his 'triad theory'. Döbereiner grouped elements with similar properties in triplets or triads (e.g. lithium/sodium/potassium, chlorine/bromine/iodine, calcium/strontium/barium) and made the fortuitous observation that the atomic weight (M) of the middle element matched the average of the other two, a rule that also held true for some of their other properties. For instance, $(M_{Li} + M_K)/2 = (6.94\ g/mol + 39.09\ g/mol)/2 = 23.015\ g/mol$, which is very close to $M_{Na} = 22.99\ g/mol$. However, with the increasing number of elements newly discovered throughout the nineteenth century, this primitive law was progressively found to be far from systematic and was ultimately discarded.

Early attempts to produce metallic lithium (Li°) by heating (reducing) a lithium-containing compound such as its oxide with iron or carbon failed. Based on the chemical similarities between lithium compounds and sodium/potassium compounds, melting-electrolysis was then suggested as the only way to isolate this metal. Sir Humphry Davy (1791–1829) had previously successfully used this method for sodium and potassium (in 1807) using electrolysers powered by Volta's cells. In the early 1820s, first attempts by Davy and William Thomas Brande (1788–1866) using the electrolysis of lithium oxide produced minute quantities of Li°. The first genuine production of the metal in relatively large amounts was performed by Robert Bunsen (1811–1899) and Augustus Matthiessen (1831–1870) in 1855, through the electrolysis of molten lithium chloride (Bunsen 1855). It is notable that, despite an unprecedented level of international scientific exchanges and partnerships, nearly 40 years elapsed between the first evidence of this element by J.A. Arfwedson (1817) and the reproducible isolation of it as a pure metal (1855) necessary for documented characterization.

Only 2 years later (17 June 1857), at the Science Faculty of the University of Paris, Louis Troost (1825–1911), aged 32, defended his PhD work entitled *Recherches sur le lithium et ses composés* (*Research on lithium and lithium compounds*) before a top-ranking committee that included, among others, Jean-Baptiste Dumas and Henri Etienne Sainte-Claire Deville (Troost 1857). This, probably the first doctoral defence focused on lithium compounds, and reporting on many physical and chemical properties of Li°, such as:

- Low density, low atomic weight, low melting point
- Highly reactive with water, nitrogen, phosphorus, chlorine, bromine, iodine at ambient temperature
- Reacts with glass, porcelain and alloys with many metals (platinum, gold, silver, potassium) below its melting point
- Reacts with sulphur to form water-soluble sulphides
- Not reactive with iron

Moreover, through this precise study of Li° and some of its compounds (e.g. lithium fluoride, lithium peroxide, lithium carbonate, lithium sulphate, lithium nitrate, lithium hydroxide), Troost also emphasized their peculiarity with respect to parent sodium and potassium compounds, from which they clearly depart, and the absence of evident systematic *isomorphism*. Clearly, similarities between lithium and magnesium further feed the so-called and well-known 'diagonal rule', as taught in first-year chemistry when studying the periodic table (Wallace and Cowell 1970; Verma et al. 2013).

Until the dawn of the twentieth century, lithium appeared to be an odd scientific curiosity and one that attracted little interest. In books published up until the 1920s, one can, for instance, read statements such as, 'The metal is expensive and has no commercial uses'. For a long time it was considered very rare, until Gustav Robert Kirchhoff (1824–1887) and Robert Bunsen (1811–1899) demonstrated through spectral analyses in the 1860s that it is actually one of the most widely distributed elements in nature, at least at trace levels (Kirchhoff and Bunsen 1860). We now know that lithium can be found in natural waters, living beings (plants, plankton, (in)vertebrate tissues, etc.), soils, rocks, salt deposits and brines. Industrially, the first large-scale high-purity production of Li° started in 1923 in Germany (Metallgesellschaft AG) via the electrolysis of a molten lithium chloride/potassium chloride eutectic mixture having a lower melting temperature (~380 °C) than that of pure lithium chloride (600 °C). This method relies on a process that was first proposed and carried out on a large scale by Antoine Nicolas Guntz (1859–1935) in 1893 in Nancy, France (Fig. 1.1).

Because of the present broad use of lithium in various sectors ranging from high power/energy electrochemical cells, the polymer and glass industries, metallurgy, nuclear activities and medicine (Schou 1957; Birch 1999), this chemical element now stands as essential for our society (Greenwood and Earnshaw 1984). This fully merits applying the powerful tools of modern chemistry and physics to provide a thorough description of its properties both in the metallic state (Li°) and, even more so, of its related ion (Li^+).

1.2 The Physico-chemical Properties of Metallic Lithium and Ionized Lithium

1.2.1 Metallic Lithium

As for other chemical elements, most of lithium's properties can now be qualitatively predicted and understood based on very basic considerations derived from quantum chemistry. These include the electronic structure, size of electronic shells and related monotonous evolution in size, electronegativity, ionization properties and binding properties along rows and columns of the periodic table of the elements. Within this predictive context, the alkali family (group I) is emblematic, as we will now illustrate by describing the fundamental properties of alkali metals, focusing on those of lithium, from which the physical and chemical characteristics of Li^+ will be deduced.

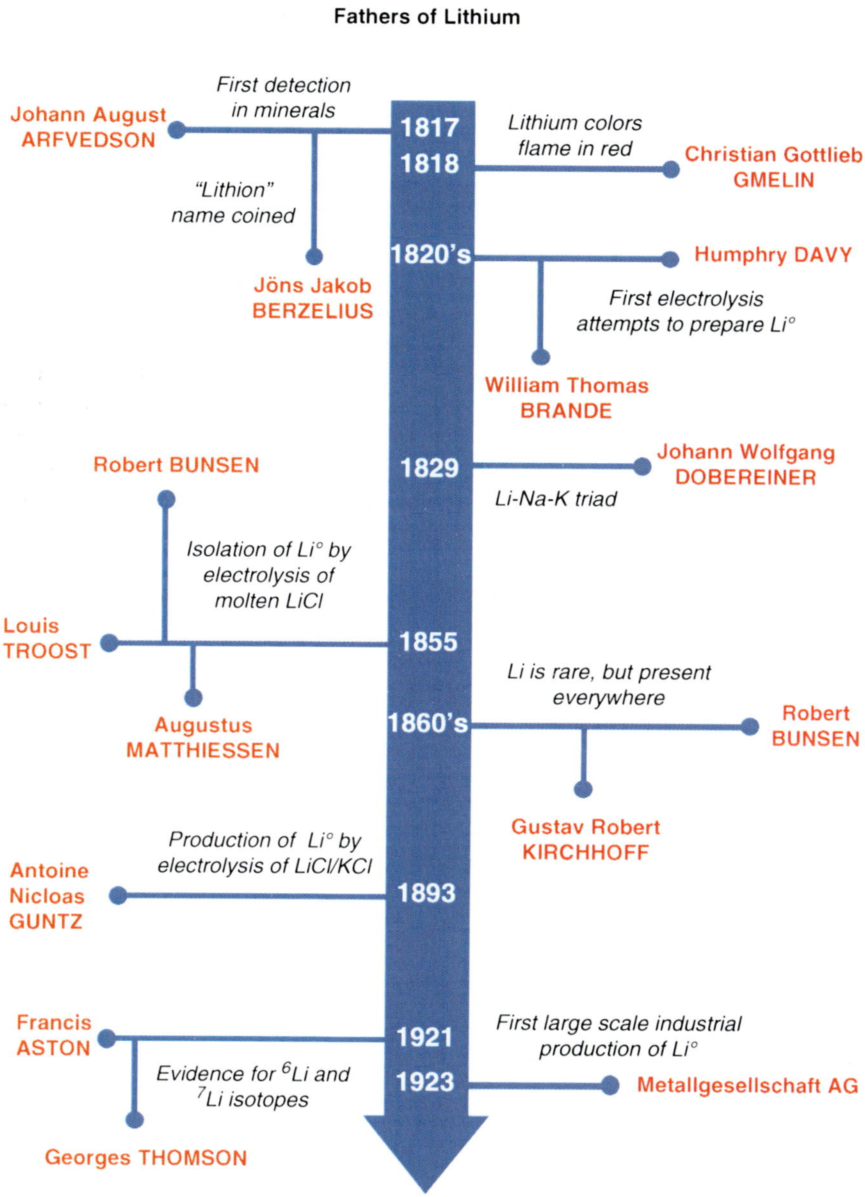

Fig. 1.1 Some key dates and names associated with the history of lithium

At neutral state, the electronic shell of lithium contains only one more electron ($Z=3$, $1s^22s^1$) than the nearest inert gas (helium, $Z=2$, $1\ s^2$). A similar situation holds for all the elements of the periodic table having this specific [inert gas]ns^1 electronic configuration and which form the so-called alkali metal family (Lithium:

Li, Sodium: Na, Potassium: K, Rubidium: Rb, Caesium: Cs and Francium: Fr) among which lithium is the lightest. Lithium has the lowest atomic weight of all solids/metals. In the following, the heaviest alkali metal (francium) will be neglected, as none of its known radioactive isotopes has a half-life exceeding 30 min. The size of alkali metals hosting only one outer valence electron (ns^1) increases with increasing n, the smaller and larger atomic size being for lithium (1.45 Å) and caesium (2.60 Å), respectively.

The combination of these low atomic weights and large atomic radii explains the very low densities of solid alkali metals all crystallizing in a body cubic-centred (bcc) system under ambient temperature and pressure conditions. Lithium ($\rho = 0.53$), sodium ($\rho = 0.97$) and potassium ($\rho = 0.86$) are the only solid elements less dense than water. Low density and low atomic weight logically impart lithium the highest gravimetric specific heat of all solids ($C = 3.6$ J·g^{-1}·K^{-1}). In its metallic state, the lonely outer s electron bonds weakly with neighbouring atoms, hence (i) the extreme softness of the alkali metals, which can easily be cut with a knife, their hardness ranging from 0.6 (lithium) to 0.2 (caesium) on the Mohs scale, and (ii) the low phase change (melting, vaporization, atomization/sublimation) temperatures and enthalpies (Table 1.1). Based on the increasing atomic size and resulting lower binding energy, these temperatures and enthalpies progressively drop when moving down from lithium to caesium (Table 1.1).

Table 1.1 Some primary selected comparative properties of the alkali elements

	Li	Na	K	Rb	Cs
Atomic number (Z)	3	11	19	37	55
Fundamental electronic configuration	[He] 2 s^1	[Ne] 3 s^1	[Ar] 4 s^1	[Kr] 5 s^1	[Xe] 6 s^1
Atomic weight (g·mole^{-1})	6.94	22.99	39.09	85.48	132.91
Density (solid, 25 °C, g·cm^{-3})	0.53	0.97	0.86	1.53	1.90
Hardness (Mohs scale)	0.6	0.4	0.5	0.3	0.2
Electronegativity (Pauling)	0.98	0.93	0.82	0.82	0.79
Melting temperature (°C)	180	97.9	63.5	39	28.4
Melting standard enthalpy (kJ·mole^{-1})	3	2.6	2.3	2.2	2.1
Boiling temperature (°C)	1336	883	757.5	700	670
Boiling standard enthalpy (kJ·mole^{-1})	146	97	80	72	68
Ionization enthalpy eV	5.39	5.14	4.34	4.18	3.8
Ionization enthalpy kJ·mole^{-1}	520	496	419	402	376
Standard redox M^+/M potential (E°, Volts, 25 °C)	−3.04	−2.714	−2.925	−2.98	−3.026
Flame colour	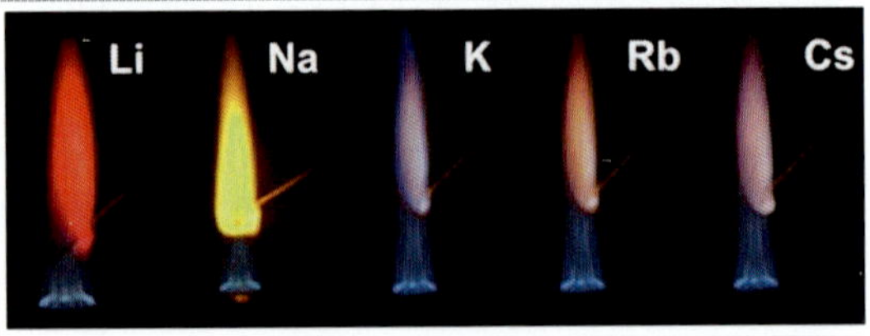				

1.2.2 Ionized Lithium (Li^+)

Because of their electronic structure, alkali elements have a clear tendency to lose their sole outer ns^1 electron, leading to monovalent M^+ ions having a much smaller size, since they reach the electronic structure of the nearest rare gas of the upper (n-1) row (Fig. 1.2). This confers to these monovalent and spherical cations both high stability and poor reactivity. Conversely, alkali metals are highly reactive, unstable and easily oxidized. Their ionization (from neutral M atom to M^+ cation) is accompanied by a severe decrease in size. Alkali metals are poorly electronegative (i.e. highly electropositive) as exemplified by potassium (0.82), rubidium (0.82) and caesium (0.79) having the lowest Pauling's electronegativity of all elements. This intrinsic electronic-driven instability explains why alkali metals can only be naturally found at M^+ ionized state, either in salts combined with other elements or dissolved in water. In the absence of any d electron in their electronic shell, alkali ions are noncoloured in aqueous solutions. Moreover, as they do not have unpaired electrons, alkali ions are diamagnetic, but they can be associated with a magnetic anion (e.g. superoxide). In nature, the chemistry of the alkali is mostly that of their monovalent ions. Note that alkali metals are among the few elements, along with fluorine and the alkaline earth elements, which exhibit only one known ionic formal charge. When moving down the alkali metal column in the periodic table, combined shielding and electronic shell size effects weaken the binding of the outermost electron to the nucleus, and the ionization energy (E_i)

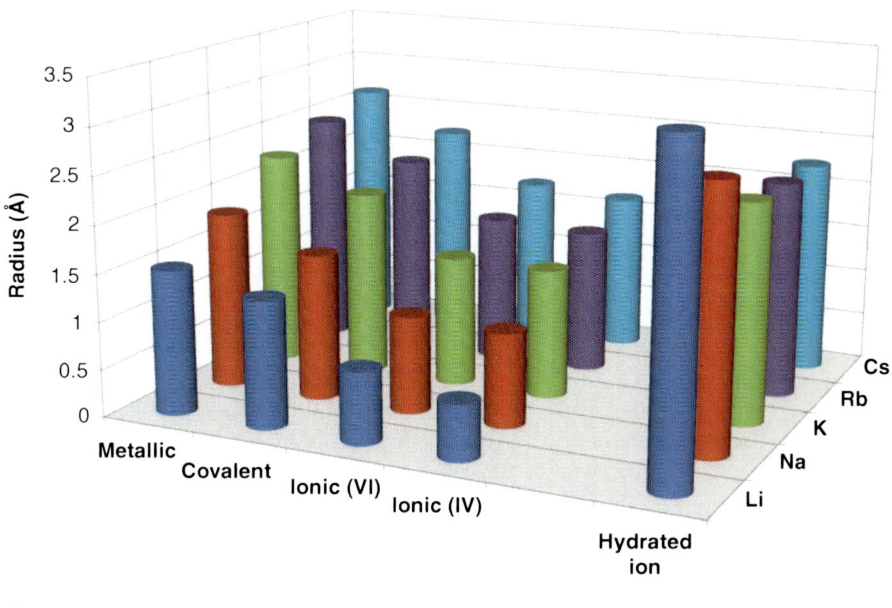

Fig. 1.2 Compared trends in metallic, covalent, ionic (tetrahedral and octahedral sites) and hydrodynamic radii for the alkali elements (Shannon 1976; Enghag 2004; Cotton et al. 1999)

continuously drops accordingly (Table 1.1). The five lowest reported E_i among all elements are owned by the alkali metals. Bottom alkali metals (i.e. large n or Z) have E_i values reaching the visible light energy range, so that they are used in photoelectric cells or as photosensors in automatic doors, escalators and other electronic devices (caesium). For lithium, ultraviolet light can trigger ionization. Aside from radiation, temperature can also cause ionization, with increasing efficiency when the Z value goes up. It is for this reason that caesium and rubidium can hardly be quantified through atomic absorption spectroscopy, a technique that requires a high level of atomization/excitation with the lowest possible ionization. On the way back to its fundamental ns^1 energy level, the ejected electron causes the emission of set of radiations with specific wavelengths, among which the most intense are the $^2S_{1/2} - {}^2P_{1/2}$ and $^2S_{1/2} - {}^2P_{3/2}$ doublet permitted transitions (Moore and Merrill 1968) well known to colour a hot flame. Lithium, sodium and potassium emit carmine/red ($\lambda = 670.7$ nm), yellow ($\lambda = 589$ nm) and violet/lilac colours, respectively. These specific radiations, which serve as atomic fingerprints, are classically used for qualitative as well as quantitative (flame absorption/emission spectrometry) analysis.

Light and heat trigger the ionization of alkali metals, but this energy conversion is localized and temporary. To be permanent, ionization has to result from the chemical reaction of M with reducible elements/compounds, leading to M^+-containing compounds or solutions. This is exemplified by the reaction of M reducing liquid water as follows:

$$M + H_2O \rightleftharpoons M^+ + OH^- + \tfrac{1}{2}H_2$$

Both the thermodynamics (increase in exothermicity) and kinetics of the reaction of alkali metals with water nicely mirror the drop in E_i when travelling down the column: lithium reacts slowly; sodium and potassium set fire; and caesium and rubidium explode. From lithium to caesium, the amount of heat released during this reaction increases, while the metal melting point goes down, so that only lithium does not melt in these conditions. When a flame is spotted, its colour perfectly matches that observed in photometric analysis (Table 1.1).

However, there is an apparent contradiction when comparing the trends in E_i and in the M^+/M standard redox $E°$ potentials (Table 1.1). For instance, lithium has the highest reducing power based on its $E°$ (−3.04 V vs. SHE), but it has the lowest ability among the alkali metals to lose its outermost 2 s^1 electron based on its E_i value (5.39 eV). As another example, sodium has the lowest reducing power based on its $E°$ (−2.71 V vs. SHE) while it is slightly less able than lithium to lose its 3 s^1 electron ($E_i = 5.14$ eV). Actually, the usual scale of standard redox potential is established in the presence of water; hence, the $E°$ values include the energy of solvation (hydration) of bare M^+ ion by the solvent molecules. This contrasts with E_i, which only reflects the energy gap between gaseous $M_{(g)}$ and $M^+_{(g)}$. For ease of comprehension, the formation of $M^+_{(aq)}$ from $M_{(s)}$ should be segmented into three steps:

(i) The atomization/sublimation of $M_{(s)}$: $M_{(s)} \rightarrow M_{(g)}$
(ii) The ionization of $M_{(g)}$: $M_{(g)} \rightarrow M^+_{(g)}$
(iii) The hydration of $M^+_{(g)}$: $M^+_{(g)} \rightarrow M^+_{(aq)}$

The variations in free standard energies (ΔG°) associated with each of these steps are well documented in the literature (Latimer et al. 1939; Hush 1948) and show monotonous evolutions as the atomic number (*Z*) or period number (*n*) increases, as expected (Fig. 1.3).

As previously rationalized, $\Delta G^\circ_{(i)}$ and $\Delta G^\circ_{(ii)}$ decrease as *n* increases, but the evolution in $\Delta G^\circ_{(iii)}$ deserves great attention. It shows a very large negative hydration energy for Li^+ (−502 kJ/mol) when compared to ionized sodium (−401 kJ/mol) and then continues to increase more gradually when *n* or *Z* increase further. Because of these large values and differences in $\Delta G^\circ_{(iii)}$, the sum of the free energies associated with these three steps does not follow a monotonous evolution (Fig. 1.3). Lithium is totally out of scale, because of its exceptional highly exothermic hydration by water, rendering it almost as reducing as caesium. This is shown in Fig. 1.4, which, incidentally, also illustrates the perfect proportionality between the overall total ΔG° (i + ii + iii) and E° (Nernst's law, $\Delta G^\circ = -n.\ F.\ \Delta E^\circ$) (Lide 1992). In the absence of any solvation, lithium would be the weakest reducing alkali element.

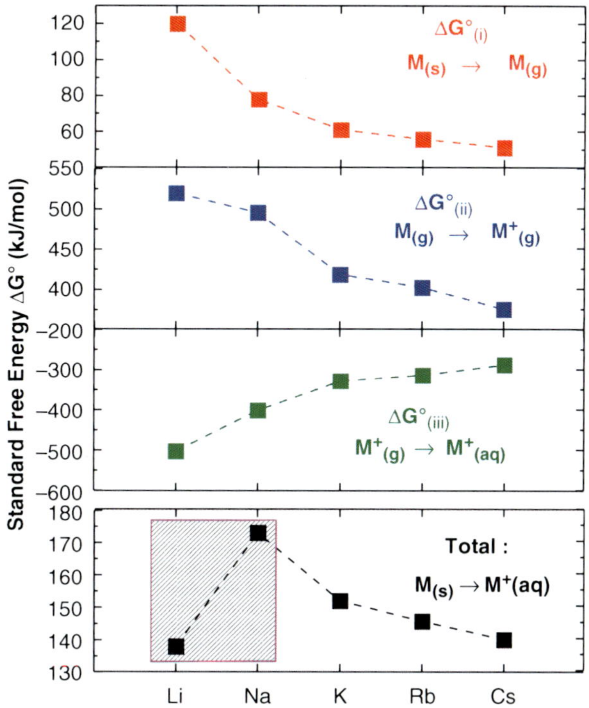

Fig. 1.3 Variations in standard free energy (ΔG°) for the sublimation (*red*), ionization (*blue*), solvation/hydration (*green*) and sum of the three (*black*) for each alkali element. Note the resulting out-of-trend behaviour of lithium due to its very exothermic hydration energy

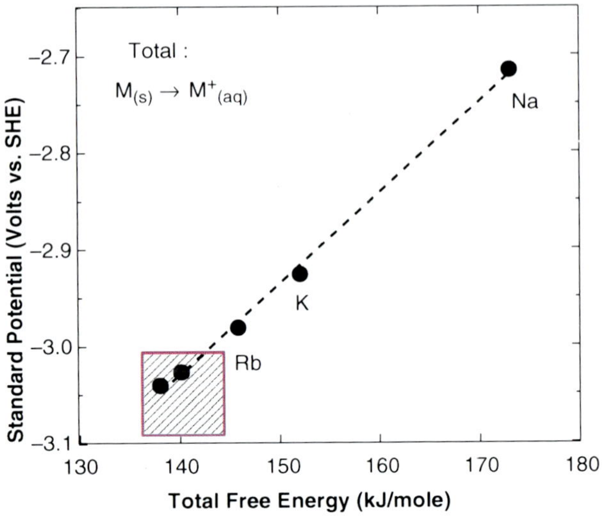

Fig. 1.4 Correlation between the total free energy for the $M_{(s)} \rightarrow M^+_{(aq)}$ reaction and the M^+/M standard redox potential. The very exothermic hydration reaction of Li^+ allows Li to be almost as reducing as Cs

In fact, the large exothermicity of Li^+ hydration can be easily accounted for by its very small size (0.6 Å in radius), resulting in its large *polarizing power* (= *charge density* = *ionic potential*) and its strong ability to interact with Lewis-base water (H_2O) molecules. The size of the central cavity created by the association of ligands/anions depends on their own size but also on their number/organization: a tetrahedron will delimit a smaller cavity than an octahedron. Li^+ is small enough to easily fit in the site at the centre of a tetrahedral association of polar water molecules, with which it will strongly interact. In $M(H_2O)_n^+$ clusters, the size of this cavity (r_{cavity}) hosting the charged M^+ ion can be directly computed from the Born equation (Born 1920) bridging r_{cavity} to the free energy of hydration $\Delta G°_{(iii)}$ as follows:

$$\Delta G°_{(iii)} = -\frac{N.z.e^2}{8\pi.\varepsilon_o.r_{cavity}}\left(1-\frac{1}{\varepsilon_r}\right)$$

This equation actually expresses the difference in energy required to charge the initially neutral cavity to a final *z* charge, in both vacuum (ε_o) and in a solvent having a relative dielectric constant ε_r. *N* is the Avogadro number (6.02 10^{23}), *e* is the elemental electronic charge (1.6 10^{-19} C) and the ε_r value is set to 80 for water. The resulting r_{cavity} values (Table 1.2) are closely related to the distance from the M^+ centre to the coordinated water molecules constituting the inner (or first, or primary) hydration shell (Latimer et al. 1939; Rashin and Honig 1985). As the size in bare ion increases, these r_{cavity} values also logically increase from Li^+ (1.38 Å) to ionized caesium (2.41 Å), and the small and roughly constant differences between these two radii (bare ion and cavity) suggest strong and similar interactions between M^+ and the first hydration shell molecules. Beyond the first solvation shell, the building of the second outer one follows the same rule. $Li^+(H_2O)_{\text{first shell}}$ having the lowest size/charge ratio, it will be more polarizing than the other alkali ions and will interact more with water molecules in its vicinity and so on. Finally, the smaller the bare

Table 1.2 Some selected properties related to the hydration properties of alkali ions (Lide 1992; Hush 1948; Rutgers and Hendrikx 1962; Hunt and Friedman 1983)

	Li	Na	K	Rb	Cs
Standard hydration free energy, $\Delta G°_{iii}$ (kJ·mole^{-1})	−502	−401	−328	−313	−287
R_{cavity} (Å)	1.38	1.73	2.11	2.21	2.41
$M^+_{(aq)}$ molar ionic conductivity, λ (Ω^{-1}·cm^2·mol^{-1}) at 25 °C	38.7	50.1	73.5	77.8	77.2
$M^+_{(aq)}$ ionic mobility, u (m^2·s^{-1}·V^{-1})	$4\cdot10^{-8}$	$5.2\cdot10^{-8}$	$7.6\cdot10^{-8}$	$8\cdot10^{-8}$	$8\cdot10^{-8}$
$M^+_{(aq)}$ hydrodynamic radius (Å)	3.40	2.76	2.32	2.28	2.28
Total $M^+_{(aq)}$ hydration number (n)	22–25	13–17	7–11	6–10	6–10

monovalent ions, the higher the number of molecules they interact with. There is thus an inverse relationship between the size of the bare ions and that of the fully hydrated species, also called the *hydrodynamic radius*. The $Li(H_2O)^{n+}$ entity is the largest among the alkali family, with a hydrodynamic radius estimated to be around 3.4 Å and n to around 25 molecules (Rutgers and Hendrikx 1962). This *hydration* vs. *size* issue of alkali ions is crucial, as it is directly involved in processes such as the ability of the biological cells to discriminate between different ions, surfactant effects affecting crystal growth (e.g. calcium carbonate) (Sims et al. 1995), in the efficiency of cationic exchanges or in the permeability of biological membranes (Volkov et al. 1997).

Having this in mind, one can also better understand some of the dynamic properties of these hydrated ions (Koneshan et al. 1998). In response to any driven force (electrostatic, osmotic), any ion will migrate through a solution dragging along its solvation shell. The larger that shell, the slower the ionic motion, as experimentally confirmed by measuring the ionic conductivity (Ω^{-1}·cm^2·mol^{-1}) and the ionic mobility (m^2·s^{-1}·V^{-1}), both steadily increasing from Li^+ to ionized caesium. $Li(H_2O)n^+$ has the lowest ability to move through a solution. Because of these large differences in hydration level, isotonic solutions of alkali salts can exhibit very different colligative properties (vapour pressure, freezing and boiling temperatures).

1.2.3 Some Peculiar Properties of Lithium Salts

Aside from the dissolved state, M^+ can only be found naturally in solid compound/salts (M^+X^-), with some of their properties also being driven by the ion size. Owing to the electronic structure of M^+, most of these salts are colourless and diamagnetic. Due to the small size of the lithium ion, the Li-X bond in anhydrous salts is much shorter and the Madelung lattice energy much larger than those reported for the parent (sodium ion, potassium ion, etc.) ones. A side effect of this, despite the very exothermic/favourable hydration enthalpy of Li^+ (−520 kJ·mole^{-1}), is the exceptionally low water solubility of some lithium salts (e.g. lithium carbonate, lithium fluoride, lithium phosphate) and the relatively low solubility of lithium hydroxide (Fig. 1.5) (Skolunov 1993; Lide 1992; Perry and Phillips 1995) as experimentally reported in the past by Arfwedson (1818) and later by Troost (1857). The high polarization power of Li^+ can confer to the Li-X bond a partial covalent character, the extent of which depends on the size vs. charge (i.e. polarizability) of the counter-anion. Also known as Fajans' rules (Casimir Fajans, 1887–1975), these qualitative empirical rules are well verified by the unexpected high solubility of lithium halides in low dielectric constant organic solvents (e.g. lithium chloride in acetone, $\varepsilon \sim 21$), thus behaving similarly to covalent lithium compounds (e.g. alkyls $Li\text{-}CH_3$, $Li\text{-}C_2H_5$).

The high polarization power of Li^+ and its resulting marked ability to interact with water also explain why most of its salts are naturally deliquescent (lithium chloride, lithium perchlorate) and/or crystallize as hydrated salts, the trihydrated form being very common (e.g. chlorine, bromine, iodine, nitrate, tetrafluoroborate, perchlorate). Other salts with lower hydration ratios are also classically reported, such as $LiCl{\cdot}2H_2O$, $Li_2SO_4{\cdot}H_2O$ and $LiOH{\cdot}H_2O$. In these hydrated salts, the lithium ion is tightly contained by the water molecules. Where these salts can be rendered

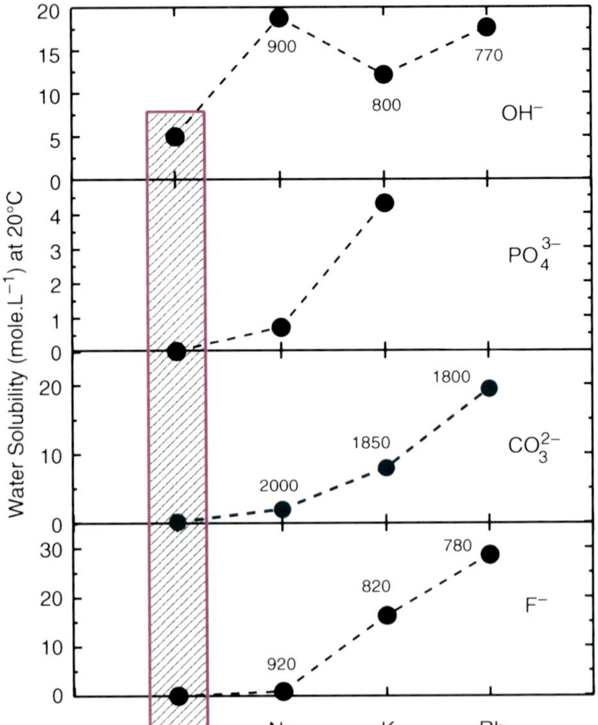

Fig. 1.5 Compared water solubility (20 °C) for alkali hydroxides, phosphates, fluorides and carbonates. Note the very low solubility for Li_2CO_3, Li_3PO_4 and LiF and the limited solubility of LiOH. Some lattice energies ($kJ{\cdot}mole^{-1}$) are indicated (Lide 1992)

anhydrous, they may then be used as drying/dehydration agents (e.g. lithium bromide and lithium chloride desiccants in air-conditioning systems). To a lesser extent, the aforementioned trend also holds for some sodium salts (Na_2SO_4). On another point, owing to the large exothermicity of the reaction of formation of $Li^+_{(aq)}$, the variation in standard enthalpy for the dissolution of lithium salts can be negative. Although counter-intuitive, this explains why their water solubility may increase when the temperature drops, according to Van't Hoff law and Le Chatelier principle. This property is used positively in the mining industry to separate lithium salts from other alkali compounds and is illustrated in Fig. 1.6 for Li_2CO_3 vs. Na_2CO_3.

In clays and layered crystal structures, the size of the interlayer space is governed by the size and charge of the intercalated species, so including their level of hydration. Lithium-containing interspaces are generally 3 Å larger than those containing other alkali ions, because of the much larger size of the hydrated lithium ion.

There are many other original aspects of the reactivity and formation of its salts that make lithium appear to be an oddity in the alkali family. Here are some additional examples:

- Upon heating, lithium hydroxide (LiOH) and lithium carbonate (Li_2CO_3) release water (H_2O) and carbon dioxide (CO_2), respectively, to form lithium oxide (Li_2O). Other MOH and M_2CO_3 (M = alkali) do not.
- Upon heating, lithium amide ($LiNH_2$) decomposes into Li_2NH and NH_3. Other alkali amides do not.
- Upon heating, lithium nitrate ($LiNO_3$) decomposes to lithium oxide (Li_2O); the other MNO_3 give nitrite MNO_2.

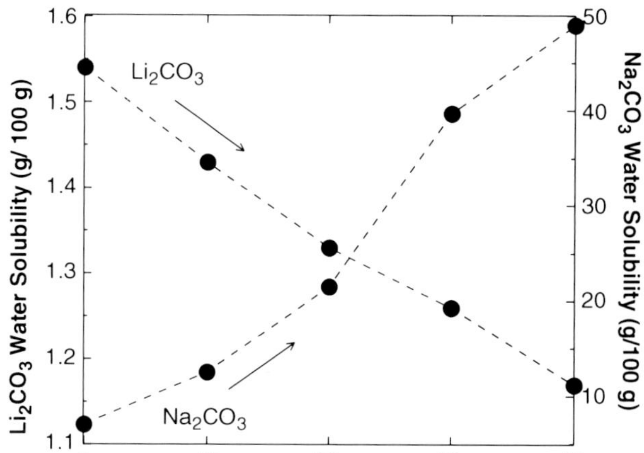

Fig. 1.6 Opposite effect of the temperature on the water solubility of anhydrous Li_2CO_3 and Na_2CO_3

- Lithium reacts with nitrogen gas to form lithium nitride (Li_3N). The others do not.
- Lithium bicarbonate ($LiHCO_3$) cannot be isolated in crystalline solid form. The other alkali can. This explains why Li_2CO_3 solubility is very dependent on the $CO_2/HCO_3^-/CO_3^{2-}$ content of the solution and why Li_2CO_3 is more soluble in carbonated aqueous solutions (e.g. buffers).
- When heated under a free supply of air, lithium mainly yields lithium oxide (Li_2O), while the other alkali metals give peroxide (sodium) or superoxide (potassium, rubidium, caesium) salts.
- Upon exposure to moisture, lithium chloride (LiCl) is deliquescent and crystallizes as $LiCl{\cdot}2H_2O$.
- Lithium directly reacts with carbon to form Li_2C_2 ionic carbide.

At this stage, it is worth recalling again, as already pointed out by the nineteenth-century chemists, that lithium shares most of its reactivity peculiarities with magnesium, an example of the validity of the 'diagonal relationship' rule previously mentioned. For instance, like lithium, magnesium reacts with oxygen to form magnesium oxide (MgO) and not peroxides as do the lower members of the alkali earth column (calcium peroxide, barium peroxide). Also, like lithium, magnesium reacts with nitrogen gas to form a nitride (Mg_3N_2). Such seemingly similar behaviour should not come as a total surprise, as both elements have very similar atomic and ionic sizes, but caution has to be exercised here to prevent any over-correlation, since Mg^{2+} (divalent) polarizability is consequently much higher than that of Li^+ (monovalent).

1.3 Occurrence, Uses, Resources and Main Industrial Products

Because of the wide panel of physical-chemical properties lithium offers, it is used as a metal (Li°), as ores/minerals, as part of numerous chemical compounds (e.g. lithium carbonate, Li_2CO_3; lithium hydroxide, LiOH; lithium chloride, LiCl; lithium bromide, LiBr; lithium oxide, Li_2O; lithium stearate and organo-lithium) and in many diverse sectors (Fig. 1.7).

For instance, lithium carbonate or lithium acetate is used for the treatment of manic depression, while the naturally occurring ^6Li isotope (7.5 %) (Aston and Thomson 1921; Thoennessen 2012) can be converted into tritium (^3H) for thermonuclear bombs and reactors. Industries use lithium for high-temperature lubricants (greases), to strengthen aluminium alloys, to produce shock-resistant ceramics, to lower the melting temperatures of glasses and to provide efficient dehumidification systems and for heat transfer applications. Lithium also has an impact on the field of synthetic chemistry as part of organic lithium reagents, which are extremely powerful bases and nucleophiles serving to prepare many chemicals. Last, and not least, lithium has become essential within the field of energy storage, with the emergence of lithium-ion battery technology.

The rate of lithium consumption within these different sectors is time dependent and a function of the transformational advances in research that enable new

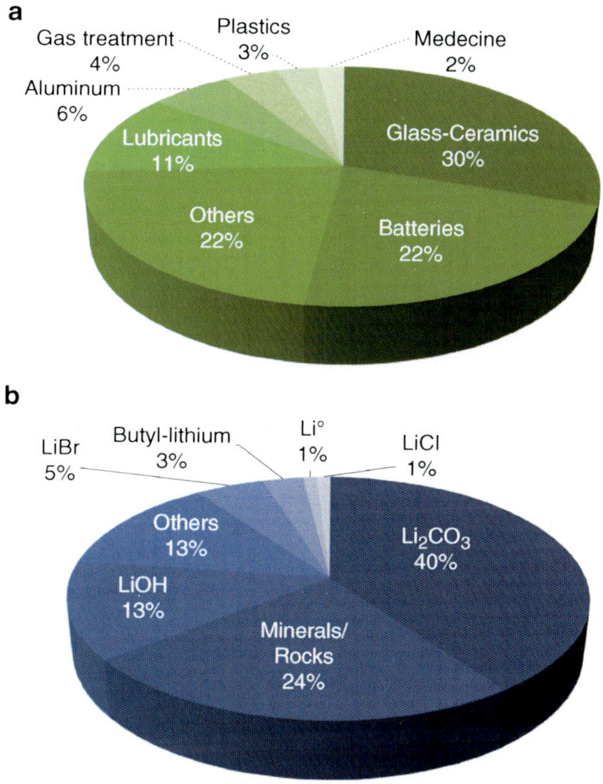

Fig. 1.7 (**a**) Repartition of the uses of lithium by industrial domains and (**b**) by chemicals (2011 data)

technologies to develop. Such evolutions can occasionally trigger severe concerns about global reserves of lithium, as witnessed a few years ago with the anticipation of a massive implementation of lithium-ion technology into electric vehicles (EVs) market. In the event, fears of a lithium shortage and spiralling costs proved too alarmist, but they also served some useful purposes: (i) they raised community awareness of the need for studies designed to estimate worldwide lithium reserves; (ii) they prompted investigation into the issue of lithium recycling; and (iii) they encouraged researchers to look for eventual possible alternatives to lithium. Addressing these issues necessitates a brief description of the natural occurrence of lithium and of industrial lithium-based products, to enable a comparison of present and forthcoming production and demands.

1.3.1 Occurrence and Abundance

In the Earth's upper crust, the concentration of lithium is about 20 mg/kg (=20 ppm) (ranked 33rd); this is roughly equivalent to that of gallium and niobium (Greenwood and Earnshaw 1984; Norton 1973). This also means that in the Earth's crust, lithium is more abundant than lead (14 ppm) and much more so than tin (3 ppm). In the oceans, lithium is present at a mean low concentration of 0.18 ppm (mg/kg). For the sake of comparison, a mean adult human body contains approximately 2 mg of lithium.

Yearly lithium metal production now approaches nearly 30 kt, with the major part coming from brines/salt lakes (60 %) and hard pegmatite rocks (26 %), while some clays (e.g. hectorite) and minor alternative deposits (e.g. geothermal, oil fields) account for the remainder (Mohr et al. 2012; Labbé and Daw 2012). There are numerous minerals, nearly 140, reported to contain lithium in amounts of up to 4 % by weight, with the most important being, in order of abundance: spodumene [$LiAlSi_2O_6$], lepidolite [$KLiAl(F,OH)_2Si_4O_{10}$], petalite [$LiAlSi_4O_{10}$], amblygonite [$(Li,Na)AlPO_4(F,OH)$], hectorite [$Na_{0.3}(Mg,Li)_3Si_4O_{10}(OH)_2$] and jadarite [$LiNaB_3SiO_7(OH)$]. The natural ore deposits generally contain from 0.5 % to 2 % lithium by weight, but their extraction is laborious and expensive. Some of them can be directly used in specific industries (e.g. spodumene and petalite are used in the glass industry) or are generally industrially transformed into carbonate or chloride.

In contrast, the recovery of lithium from brine lakes provides an easier prospect, as it is achieved using low-temperature processes. Continental (Bolivia, China, Chile, Argentina) and geothermal (California) brines and salt deposits are less concentrated (<0.15 wt%) than rocks, but their processing into commercial and industrial usable salts (e.g. carbonate) is less energy demanding, as it relies on the use of solar evaporation ponds. Usually, salty water is first pumped out of the lakes and then concentrated using solar energy into Li-Cl brine prior to being subsequently treated with a soda (sodium carbonate, Na_2CO_3) solution to precipitate lithium carbonate (Li_2CO_3). However, the quality of brine/deposit is generally bound to its magnesium content (magnesium/lithium ratio), which, if high, necessitates additional purification costs. As the industrial utilizations of lithium are relatively recent, its production is still in its infancy, and more deposits or lithium-bearing minerals are likely to be discovered in the future.

Often wrongly discarded in reports, numerous thermal spring waters contain non-negligible levels of dissolved lithium, of up to several mg/L. In France, one of these is amusingly named the ‘lithium source’ (Santenay) (Thomas-Caraman 1893) because of its exceptional lithium content (20 mg/L) due to large spodumene, petalite and lepidolite deposits in the surrounding hills. In the USA, some of these natural waters have even been bottled and sold under ‘lithia’-bearing brand names (e.g. Londonderry Lithia Water and Lithia Springs Water from Georgia); some are still on the market. Until recently (although no longer) some drinking water companies advertised the natural lithium content of their product on their bottle labels, such as *San Pellegrino* (Li^+: 0.18 mg/L). Still on the topic of beverages, we should note that back in the 1950s some soda manufacturers artificially added lithium salts to their drinks and advertised their claimed positive medical attributes (e.g. 7 UP Lithiated Lemon Soda).

Overall, if we include all the aforementioned sources of lithium, the worldwide reserve is estimated to be between 30 and 50 millions of tons, based on currently known deposits (Mohr et al. 2012; Labbé and Daw 2012). In the early 1970s, reserves were estimated to be just 1 million tons, which demonstrates the rapid and continuous growth in the known and available natural stock of lithium (Norton 1973) in recent years. Nevertheless, a legitimate question at this stage is how such a quantity compares with today's yearly consumption and future technology/markets trends. Thus the need to do a critical assessment of the demands.

1.3.2 Lithium Uses and Main Commercial Products

Various papers and books have nicely reviewed in great detail the various uses of lithium metal/alloys and Li^+ -bearing compounds (Ebensperger et al. 2005; Garrett 2004; Labbé and Daw 2012). Here we will limit ourselves to briefly reflect on a few application sectors that we judge relevant, so as to better sense their foreseen demands and the factors that may upset lithium production in the years to come. As the use of lithium salts in medical treatment is addressed in detail in later chapters, we will not deal with that in this section. Medical uses represent about 2 % of yearly global lithium consumption, mainly as lithium carbonate.

1.3.2.1 Glasses and Ceramics

For many years the glass and ceramic industries have widely used lithium-based compounds for manufacturing a broad variety of products including, among others, borosilicate glasses, containers and bottles as well as shock-resistant ceramics and porcelain tiles. The benefits of adding lithium to glass mainly reside in that element's capacity to lower the viscosity of the melt and reduce the melting temperature by a few hundred degrees. In ceramics, it helps to both lower the firing temperature and shorten the soaking time. So in both cases the addition of lithium salt reduces energy consumption and therefore lowers production costs. Another positive attribute is that the incorporation of lithium in such glasses/ceramics significantly reduces, or enables control of, their thermal expansion coefficient, making them resistant to thermal shocks, with applications ranging from cookware to spacecraft thermal shields. In addition, for both glass and ceramic processes, the ability to directly use both ores and lithium compounds, mainly lithium carbonate, as the source of lithium is very beneficial cost wise. As a result, today about 30 % of the yearly annual production of lithium is used within the glass-ceramic industrial sector, with demand expected to increase continuously at a steady rate of about 3 % over the next few decades.

1.3.2.2 Lubricants

The grease industry nowadays represents about 5 % of the total consumption of lubricants and also relies on considerable amounts of lithium hydroxide (1 kg of LiOH for 50–100 kg of grease) Garrett 2004. Aside from having a thickening effect, the addition of a lithium salt of fatty acid, resulting from the reaction of lithium hydroxide with a fatty acid (stearic or oleic), imparts some very useful properties to the grease. The previously mentioned specific properties of Li^+ bonding and resulting salt solubility issues enable the development of highly water-insoluble greases for applications in high-moisture environments. Additionally, these lithium-containing greases are stable over a large range of shear and temperatures. Such unique properties have resulted in lithium-containing greases conquering most (~70 %) of the highly demanding service applications, the others being served by either sodium- or calcium-based soaps. This industrial sector today consumes 11 % of the yearly production of lithium, with demand expected to grow at a steady 4 % per year over the next decades.

1.3.2.3 Aluminium Metallurgy

The aluminium industry also greatly benefits from lithium additives, both for manufacture and for alloying. For instance, adding lithium carbonate or lithium chloride to the electrolytic bath (alumina + Na_3AlF_6 cryolite) of an aluminium-making electrolytic cell lowers its melting temperature and increases its conductivity, so lowering the cell voltage and also reducing the production of hazardous hydrogen fluoride. This results in an increase in energy efficiency of the electrolytic cells and therefore lowers the production cost. Nevertheless, lithium demand in this sector is expected to diminish as lithium-free approaches are developed that reach similar efficiency without the cost of adding lithium. The intrinsic low atomic weight and low specific density of lithium carry another benefit for aluminium production, in that its addition (1–4 %) to aluminium enables the formation of aluminium-lithium alloys of lower density but with greater elastic modulus (stiffness) than pure aluminium. For example, the addition of 1 % lithium comes with a 3 % decrease in density and a 6 % increase in elastic modulus. Such alloys, which found early application in aircraft, were supplanted over time by less expensive and equally light fibre composites. Nonetheless, composites have their own limitations, and demand for light lithium-based alloys is bouncing back for satellite and space applications, for which weight is a crucial factor. (Lowering a satellite's weight by 1 kg enhances its orbital circulation by 30 days.) This interplay between lithium and the aluminium industries illustrates how research and technology advances can affect niche applications. Thus, it is difficult to establish reliable estimates of future demand.

1.3.2.4 Organic Synthesis, Desiccants and the Nuclear Industry

Lithium-bearing compounds also play a crucial role in some specific and strategic industrial sectors, although their overall yearly use here is limited to just a few percent of total consumption. For instance, highly hygroscopic lithium bromide and chloride are used in the dehumidification of air and other gases and can also be used to dry organic liquids, serving the role of desiccants. Within the field of chemistry, numerous organo-lithium compounds exist and can either act as a precursor for the design of complex pharmacologically active molecules or as a catalyst. Among them, the most well known is n-butyl-lithium (Li-C_4H_9), which is used as a stereospecific catalyst for the polymerization of butadiene, isoprene and styrene, aside from serving as a reducing agent to prepare lithium-based inorganic compounds.

In the 1950s, an important application of lithium was the conversion of its ^{6}Li isotope into tritium for hydrogen bombs [$^6Li + {}^1n \rightarrow {}^3H + {}^4He$]. This use could well re-emerge within the next century if the nuclear fusion International Thermonuclear

Experimental Reactor (ITER) program, which relies on the use of tritium (^3H) in controlled thermonuclear reactions, succeeds. Nevertheless, for reasons of uncertainty, none of the present models predicting the yearly consumption of lithium over the next decades take this scenario into account. In contrast, today's models do consider several scenarios based on various anticipated EV penetration rates in tomorrow's transportation market, because of the increasing importance of the lithium-ion battery in this sector.

1.3.2.5 Batteries and Energy Storage

Over the last few years, among all the industrial sectors discussed so far, battery technology has witnessed the greatest transformational changes with the arrival of primary lithium batteries in the 1970s and of lithium-ion batteries nearly 20 years later. Today, lithium-ion technology dominates the portable electronic market and is on the verge of capturing the EV market, while being a serious contender for grid/stationary applications. The lithium-ion cell uses lithium-based salts (e.g. lithium hexafluorophosphate, $LiPF_6$; lithium bis(trifluoromethanesulfonyl) imide salt, LiTFSI) in the electrolyte as well as lithium-based inorganic compounds (e.g. $LiCoO_2$, $LiFePO_4$, $LiMn_2O_4$) as the positive electrode. For the purpose of quantification, let us recall that, with lithium-ion technology, 0.15 kg of lithium is necessary to produce a 1 kWh battery. This means that 4 kg of lithium is needed to power today's EV cars, which rely on 25–30 kWh batteries (Chagnes et al. 2015). Solely based on the massive deployment of lithium-ion batteries in portable devices, the battery sector has today become the second largest consumer of lithium, with a 22 % share of the yearly lithium production that is still enjoying a 15 % yearly growth. Such growth has raised several concerns regarding lithium supply, bearing in mind that demand could further explode with the foreseen development of both EVs and grids, both large-volume applications. Numerous studies are presently focusing on forecasting lithium demands linked to the progressive EV market penetration. Various models exist with different assumptions regarding the degree of market penetration, car lifetime, increases in population and so on; hence, the estimates of consumption vary ranging up to as high as ~400 kt Li/year in 2050 (Chagnes et al. 2015). Regardless of this, all the reports are very optimistic and do not anticipate a lithium shortage for many years to come, based on the large lithium reserves and lithium recycling, which could represent a significant proportion (~50 %) of the supply in the future. We are not aware of any forecasting demand model regarding grid applications, perhaps not surprising as the implementation of batteries in this sector depends on national policies and basic decisions—such as centralized vs. localized power—that have not yet been made.

1.4 Summary

In this chapter we have sought to convey how lithium has now become a strategic component for society. The benefits of lithium are rooted in its unique chemical/physical properties, in terms of density, electropositive character, high reactivity and organic media enabling a very diversified chemistry. With a panoply of properties, lithium chemistry has, over the years, spread to numerous sectors. Future penetration will be time dependent and a function of transformational research advances that enable the development of new technologies.

Lithium usage is expected to further change rapidly with the foreseen deployment of EVs, which will boost the use of lithium-ion batteries and give dominance to the battery sector for yearly lithium consumption. We have referred to previous studies regarding lithium resource assessments as well as forecast supply and demand models to show that previous debates about lithium shortage were far too alarmist. Our annual consumption is presently 30 kt/year and could increase, depending upon various scenarios, to 400 kt/year (Vikström et al. 2013), but this is unlikely to threaten the current estimated world lithium reserves, which amount to 30–50 10^3 kt, especially given that numerous reserves have yet to be identified. Thus, at this point in time, lithium shortage is not an issue, but it may be a problem in the future. However, if a future shortage of lithium were to occur, some of the above applications, namely, grease and the glass-ceramic industries, could use sodium-based compounds as alternatives, with lower but still acceptable performances. Even for batteries, besides the development of the recycling of lithium (which is expected to account for 40% of the lithium supply by 2050) (Mohr et al. 2012), sodium ion is a viable alternative that is presently being studied.

Overall, lithium is a fascinating element about which there is much left to discover. Its underpinning rich chemistry is yet far from fully exploited and some lithium effects are still not fully understood.

Acknowledgements The authors would like to thank Romain Patrice, Joackim Bois, Rosa Palacin, Carine Lenfant, Raphaël Janot, Patrik Johansson and Mathieu Morcrette for fruitful discussions, sharing bibliographic records, advice and proofreading.

References

Arfwedson A (1818) Afhandlingar I Fysik, Kemi och Mineralogi VIII: 145, Stockholm

Aston FW, Thomson GP (1921) The constitution of lithium. Nature 106:827

Birch NJ (1999) Inorganic pharmacology of lithium. Chem Rev 99:2659–2682

Born M (1920) Volumen und Hydratationswiirme der Ionen. Z Phys 1(1):45–48

Bunsen R (1855) Darstellung des lithiums. Justus Liebigs Ann Chem 94(1):107–111

Chagnes A, Swiatowska J (Edit.) (2015) Lithium Process Chemistry, Resources, Extraction, Batteries, and Recycling, 1st Edition, Elsevier

Cotton FA, Wilkerson G, Murillo CA, Bochmann M (1999) Advanced inorganic chemistry. Wiley, New York

Ducrotay de Blainville HM (1819) in Journal de Physique, de Chimie, d'Histoire Naturelle et des Arts, Tome LXXXVIII, Paris, 164–169

Ebensperger A, Maxwell P, Moscoso C (2005) The lithium industry: its recent evolution and future prospects. Res Policy 30:218–231

Enghag P (2004) Encyclopedia of the elements: technical data – history – processing – applications. WILEY-VCH Verlag GmbH & Co KGaA, Weinheim

Garrett DE (2004) Handbook of lithium and natural calcium chloride, 1st edn. Academic, Cambridge, MA

Greenwood NN, Earnshaw A (1984) Chemistry of the elements. Pergamon Press, Oxford

Hunt JP, Friedman HL (1983) Aquo complexes of metal ions. Progress Inorg Chem 30:359–387

Hush NS (1948) The free energies of hydration of gaseous ions. Aust J Sci Res Ser A: Phys Sci 1:480

Kirchhoff G, Bunsen R (1860) Chemische Analyse durch Spectralbeobachtungen. Ann der Phys und Chem 110:161–189

Koneshan S, Rasaiah JC, Lynden-Bell RM, Lee SH (1998) Solvent structure, dynamics, and Ion mobility in aqueous solutions at 25°C. J Phys Chem B 102:4193–4204

Labbé JF, Daw G (2012) Panorama 2011 du marché du lithium, Rapport Public, BRGM/RP-61340-FR

Latimer WM, Pitzer KS, Slansky CM (1939) The free energy of hydration of gaseous ions, and the absolute potential of the normal calomel electrode. J Chem Phys 7:108

Lide DR (1992) CRC handbook of chemistry and physics, 73rd edn. CRC Press, Boca Raton

Mohr SH, Mudd GM, Giurco D (2012) Lithium resources and production: critical assessment and global projections. Minerals 2:65–84

Moore CE, Merrill PW (1968) Partial grotrian diagrams of astrophysical interest, national standard reference data series 23

Norton JJ (1973) Lithium, cesium, and rubidium—the rare alkali metals. In: Brobst DA, Pratt WP (ed) United States mineral resources in U.S. Geological Survey Professional Paper 820. pp 365–378

Perry DL, Phillips SL (1995) Handbook of inorganic compounds. CRC Press, Boca Raton

Rashin AA, Honig B (1985) Reevaluation of the born model of Ion hydration. J Phys Chem 89(26):5588–5593

Rutgers AJ, Hendrikx Y (1962) Ionic hydration. Trans Faraday Soc 58:2184–2191

Schou M (1957) Biology and pharmacology of the lithium ion. Pharmacol Rev 9(1):17

Shannon RD (1976) Revised effective ionic radii and systematic studies of interatomic distances in halides and chalcogenides. Acta Cryst A32:751–767

Sims SD, Didymus JM, Mann S (1995) Habit modification in synthetic crystals of aragonite and vaterite. J Chem Soc Chem Commun 10:1031–1032

Skolunov AV (1993) Solubility of alkali metal chlorides and hydroxides in water. Fibre Chem 25(6):463–467

Thoennessen M (2012) Discovery of isotopes with $Z \leq 10$. At Data Nucl Data Tables 98:43–62

Thomas-Caraman C (1893) Coup d'œil sur la Source Lithium de Santanay (*Côte d'Or*). O. Doin, Paris

Troost L (1857) Recherches sur le lithium et ses composés – PhD Thesis (June 17th, 1857, Paris). Ed. Mallet-Bachelier, Paris

Verma NK, Khanna SK, Kapila B (2013) Comprehensive chemistry, 11th edn. Laxmi Publications Pvt. Ltd, India

Vikström H, Davidsson S, Höök M (2013) Lithium availability and future production outlooks. Applied Energy 110:359–387

Volkov AG, Paula S, Deamer DW (1997) Two mechanisms of permeation of small neutral molecules and hydrated ions across phospholipid bilayers. Bioelectrochem Bioenerg 42:153–160

Wallace HG, Cowell VW (1970) Chemistry – an electronic and structural approach. W.E. Addison, John Murray, London

The Pharmacokinetics of Lithium

2

Camille Couffignal, Lucie Chevillard, Souleiman El Balkhi, Salvatore Cisternino, and Xavier Declèves

C. Couffignal
Inserm, U1137, Paris, France

University Paris Diderot, UMR-S 1137, Paris, France
e-mail: camille.couffignal@aphp.fr

L. Chevillard
Inserm, U1144, Paris, France

University Paris Descartes, UMR-S 1144, Paris, France

University Paris Diderot, UMR-S 1144, Paris, France
e-mail: luciechevillard@gmail.com

S. El Balkhi
Inserm, U1144, Paris, France

University Paris Descartes, UMR-S 1144, Paris, France

University Paris Diderot, UMR-S 1144, Paris, France

Pharmacology, Toxicology and Pharmacovigilance Department,
CHU Limoges Hospital, Limoges, France
e-mail: souleimanba.elbalkhi@chu-limoges.fr

S. Cisternino
Inserm, U1144, Paris, France

University Paris Descartes, UMR-S 1144, Paris, France

University Paris Diderot, UMR-S 1144, Paris, France

Département de Pharmacie, Assistance Publique – Hôpitaux de Paris, Jean Verdier Hospital, Paris, France
e-mail: salvatore.cisternino@aphp.fr

X. Declèves, PharmD, PhD (✉)
Inserm, U1144, Paris, France

University Paris Descartes, UMR-S 1144, Paris, France

University Paris Diderot, UMR-S 1144, Paris, France

Pharmacokinetics and Pharmacochemistry Department, Cochin Hospital, AP-HP, Paris & Research Team "Pathophysiological and Therapeutic Targets of the Blood-Brain Barrier", Inserm U1144, Variability to the Response of Psychotropic Drugs, Paris, France
e-mail: xavier.decleves@parisdescartes.fr

G.S. Malhi et al. (eds.), *The Science and Practice of Lithium Therapy*,
DOI 10.1007/978-3-319-45923-3_2

Abstract

This chapter summarizes the current knowledge about the pharmacokinetics of lithium in humans. It covers the various marketed pharmaceutical forms and salts of lithium, as well as the variability factors that influence its pharmacokinetics, such as age, body weight, pathophysiological modifications and drug-drug interactions. Pharmacokinetics parameters and their variability are important because lithium has a narrow therapeutic index. Therapeutic drug monitoring to determine lithium blood concentrations remains the key component of clinical surveillance. This chapter focuses special attention on the brain pharmacokinetics of lithium, which are not yet fully understood. A better comprehension of the brain disposition of lithium and associated variability factors may assist our future understanding of the pharmacokinetics-pharmacodynamics relationships of lithium.

Key Points

- The pharmacokinetics (PK) of lithium is linear within the dose regimen used in clinical practice.
- Lithium is orally well absorbed, mainly distributed in the total body water, not metabolized and almost completely eliminated unchanged by renal clearance.
- Lithium has an extremely narrow therapeutic window. To avoid adverse effects or, conversely, lack of drug response, it is necessary to routinely administer therapeutic drug monitoring (TDM).
- There are several variability factors of lithium pharmacokinetics (PK), such as age, body weight, renal function and drug-drug interactions. These also justify close TDM.
- The neuropharmacokinetics of lithium remains poorly understood. Further non-clinical and clinical experiments are needed to better understand brain lithium PK and its factors of variability.

2.1 Introduction

The benefits of lithium in the treatment of 'psychotic excitement' emerged at the beginning of the 1950s (Cade 1949). Today, lithium salts (carbonate or citrate) are still the first-choice medication for the treatment of bipolar disorder (Malhi et al. 2015). The clinical use of lithium remains tricky due to its extremely narrow therapeutic range: the concentrations required for therapeutic effects are close to concentrations that can cause toxicity. Indeed, a relatively minor increase in lithium serum concentrations may induce severe adverse effects, while their decrease may precipitate the worsening of psychiatric symptoms. Despite a narrow therapeutic index, the clinical profile of lithium determined by efficacy and tolerability has not yet been defined by clear pharmacokinetics-pharmacodynamics (PK-PD) relationships.

Therapeutic drug monitoring (TDM) to determine lithium serum concentrations remains the key component of clinical surveillance. To understand the mechanisms underlying modifications of lithium serum concentrations, it is important to consider the better known pharmacokinetic (PK) properties of lithium salts, as well as PK variability factors. This chapter summarizes our current knowledge of the PK of lithium for the various marketed pharmaceutical forms and salts, as well as those variability factors that influence its PK, such as age, body weight, pathophysiological modifications and drug-drug interactions (DDIs).

2.2 Bioanalysis

The TDM of lithium blood levels is performed by the determination of lithium in serum or plasma drawn 12 h after the last dose; there is universal consensus for this method among bipolar disorder experts (Grandjean and Aubry 2009a). Serum or plasma is separated from red blood cells (RBC) by density centrifugation, and lithium concentrations are measured using one of the methods cited hereafter, either directly or after an appropriate dilution. Tubes containing lithium heparinates (LiH) should absolutely be avoided when blood is drawn to avoid any overestimation of lithium serum concentrations. (Depending on the tube and the blood volumes, LiH can add up to 3 mmol/L of lithium.) As lithium moves rapidly into erythrocytes, serum should be separated quickly from the red blood cells, preferably within 1 h (Mitchell 2001). Routinely, the most frequently used analytical methods for plasma or serum lithium measurement include flame emission spectrometry (FES), atomic absorption spectroscopy (AAS) and ion-selective electrode (ISE). Other less commonly used techniques include capillary ion analysis (CIA), colourimetry, inductively coupled plasma optical emission spectroscopy (ICP-OES) and inductively coupled plasma mass spectrometry (ICP-MS). Elevated haematocrit values, as well as various drugs, including carbamazepine, quinidine, procainamide, N-acetylprocainamide, lidocaine and valproic acid, may interfere and introduce a bias in ISE determination of lithium concentration. The total lithium in the brain can also be measured in vivo using magnetic resonance spectroscopy (Gyulai et al. 1991); see Sect. 2.7. The cerebrospinal fluid (CSF) lithium concentration may also be determined using routine techniques of plasma/serum lithium determination.

The RBC lithium concentration can also be determined and is usually used to calculate the RBC to plasma lithium ratio as follows:

$$\text{RBC to plasma lithium ratio} = \frac{[\text{RBC lithium concentration}]}{[\text{plasma lithium concentration}]}$$

In some studies, the RBC lithium concentration is calculated indirectly using the following equation:

$$[\text{RBC lithium}] = \frac{[\text{whole blood lithium}] - ((1 - \text{hematocrit}) \times [\text{plasma lithium concentration}])}{\text{hematocrit}}$$

Intra-erythrocytes lithium could be measured directly after RBC separation and haemolysis using one of the previously mentioned methods. Usually, erythrocytes are isolated using density centrifugation and are haemolysed by adding de-ionized water. Proteins are precipitated with trichloroacetic acid and lithium concentrations are measured using FES, AAS or ISE. Internal standards (caesium or potassium) are used routinely to guarantee the quality of FES measurement. However, intra-erythrocyte potassium measurement is preferred by some clinicians to avoid haematocrit variation-related changes. Internal quality controls should cover the infra-therapeutic (0.2–0.4 mmol/L), the therapeutic (0.4–1 mmol/L) and the toxic (≥1.5 mmol/L) zones. Among the commercial quality controls are those of Bio-Rad™, Asqualab™, ProBioQual™, Fumouze™ and Randox™. It is also recommended, and in some countries obligatory, to participate in an external quality control survey.

2.3 Peripheral Pharmacokinetics

As described in Chap. 1, lithium is a naturally occurring alkali metal with a small molecular weight (6.941 g/mol) that does not exist in nature as an uncombined free metal form but as a single-charged cation or Li^+ associated with counter organic or inorganic ions to form different solid salts. These water-soluble salts give the physicochemical properties of weak bases and hydrophilic characteristics. For the treatment of bipolar disorder, lithium carbonate is the most frequently used lithium salt, but other salts such as acetate, citrate, gluconate and sulphate are also available. Furthermore, lithium carbonate is prescribed in two forms: an immediate-release and a sustained-release preparation. The relative low solubility of lithium carbonate in water, as compared with other salts, has been used to produce sustained-release forms of lithium by decreasing its absorption rate by the intestine. Lithium salts are only available as oral tablets or solution. It is the oral PK properties of lithium salts that are summarized in this chapter, as no intravenous forms of lithium have been available in the market. This chapter reviews the peripheral PK of lithium by considering its absorption, distribution, metabolism and excretion properties (ADME) and then examines several factors that are responsible for lithium PK variability.

2.3.1 Absorption

In clinical practice, lithium is administered by the oral route as tablets or capsules (carbonate) or solution (citrate). Solid pharmaceutical forms of lithium salts include immediate-release and controlled- or extended- or sustained-release forms, depending on the pharmaceutical firm that markets the drugs. Lithium citrate oral solution may be useful in patients unable to swallow capsules or tablets; 5 mL of a commercially available solution contains about 8 mEq of lithium and is approximately equivalent to 300 mg of lithium carbonate. The relatively high water solubility of lithium salts used in therapy permits rapid and total solubility of solid pharmaceutical forms containing lithium salts within the gastrointestinal tract fluids, leading to good oral absorption of lithium – in terms of the amount absorbed by the gut. Moreover, the rate of oral absorption of lithium salts depends on the rate of solubility of the different lithium salts and the pharmaceutical forms, i.e. immediate- or sustained-release forms, within the intestine.

Following oral ingestion, fast and effective absorption is observed within 1–6 h for the immediate-release form (later for the sustained-release form) through the upper gastrointestinal tract (Malhi and Tanious 2011). A complete dissociation of lithium salts occurs after their oral absorption, giving rise to lithium as a single cation in the circulating blood. Chloride and sulphate salts are rapidly and almost completely absorbed, while the less water-soluble carbonate salt is absorbed more slowly (Lehmann 1997). Lithium is mainly absorbed in the jejunum and ileum (Bettinger and Crismon 2006). Although the mechanisms involved are not well known, lithium transport across the intestine involves both active and passive mechanisms (Kersten et al. 1986). The bioavailability of all formulations of lithium is good, with a range of 80–100% (Hunter 1988; Price and Heninger 1994). Some gastrointestinal factors influence the oral absorption of lithium; for example, food may influence the rate and extent of absorption. Studies by Gai and colleagues (1999, 2000) using two different sustained-release formulations of lithium showed that the extent (bioavailability) of oral absorption was not modified under fasting or fed conditions. However, meals (normal diet, high fat and high fat/protein meals) significantly increased the maximal serum concentration (C_{max}) of lithium as compared with fasting conditions, suggesting that the rate of lithium oral absorption may be increased by meals, even if the observed time of maximum plasma concentration (T_{max}) was similar across fed and fasting conditions (Gai et al. 1999, 2000).

2.3.1.1 Immediate-Release Preparations

After oral administration of a single immediate-release dose of lithium citrate solution or carbonate tablets, T_{max} is observed at approximately 1–3 h and 2–6 h, respectively (Frye et al. 1998; Lee et al. 1998; Thornhill 1978; Ward et al. 1994). Several advantages have been found for citrate solution of lithium over the solid carbonate forms: it increases the *rate* of lithium absorption, avoids prolonged exposure of the small intestine to lithium and subsequently decreases the gastrointestinal side effects of lithium and decreases interindividual variability of lithium disposition (probably through reduced variability of the absorption of solid forms) (Guelen et al. 1992). Because of their better water solubility, chloride and sulphate in immediate-release solid lithium formulations also peak more quickly than do the immediate-release lithium carbonate forms (Altamura et al. 1977).

2.3.1.2 Sustained-Release Preparations

Sustained-release forms of lithium are prescribed much more frequently in clinical practice than immediate-release forms are, especially in European countries (Grandjean and Aubry 2009a). The first intent of a sustained-release form is to limit high stiff/peak lithium C_{max} that might otherwise reach concentrations in the toxic range and produce a subsequent reduced time in the narrow therapeutic range of lithium concentration (0.6–1.2 mmol/L) (Amdisen 1977; Cooper et al. 1978; Sproule 2002).

After oral administration of a single dose of sustained-release lithium carbonate, T_{max} is extended by between 4 and 12 h as compared with its immediate-release form. The absorption half-life of the immediate-release form is 0.78 ± 0.05 h; for the sustained-release form of lithium carbonate, it is 3.73 ± 0.37 h (Thornhill 1978). Cooper and colleagues (1978) have compared the bioavailability of a slow-release form of lithium with an immediate-release form administered as a single dose or after reaching steady-state conditions in normal volunteers and manic patients. The C_{max} of the slow-release form was delayed (T_{max} = 4.6 ± 0.6 h) compared to the immediate-release form (T_{max} = 1.5 ± 0.2 h). A lowered C_{max} was also observed for the slow-release form (C_{max} = 0.33 ± 0.06 mmol/L) as compared to the immediate-release form (C_{max} = 0.66 ± 0.10 mmol/L). In contrast, lithium absorption, in terms of systemic exposure (area under the curve, AUC), was not statistically significant as the $AUC_{0\text{-}48h}$ were similar between the two forms (7.65 ± 1.24 and 6.53 ± 1.38 mmol.h/L for the immediate- and slow-release forms, respectively), indicating equal bioavailability (Cooper et al. 1978).

In a similarly designed study in healthy subjects, the sustained-release preparation was associated with a marked reduction of the C_{max}/C_{min} ratio, which represents fluctuation of lithium serum concentrations at steady state, as comparison with an immediate-release form. But as stated above, lithium serum systemic exposure did not differ between the two forms, with an $AUC_{0\text{-}96h}$ of 0.214 ± 0.107 and 0.252 ± 0.097 mmol.h/L for the immediate- and sustained-release forms, respectively (Castrogiovanni 2002). Other studies confirmed the seminal study of Cooper and colleagues showing that, after repeated doses, oral exposure at steady state was similar to that observed after a single administration of an immediate-release form of lithium but with a blunted C_{max} and an increased T_{max} with similar AUC (Hunter 1988; Perry et al. 1981). Importantly, the sustained-release form exhibited higher plasma concentrations than the immediate-release form and reduced interindividual variability by as much as 50% in bipolar disorder patients (Cooper et al. 1978).

2.3.2 Distribution

Because of their physicochemical properties and small molecular size, lithium ions largely distribute into extracellular and intracellular body water spaces, as do sodium ions. Among all the models tested, the best model to describe lithium serum PK is a two-compartment model with first-order absorption rate constant (ElDesoky et al. 2008; Findling et al. 2010; Sproule et al. 2000; Taright et al. 1994). However, as shown in Table 2.1, some studies have also used a one-compartment model (Hoegberg et al. 2012; Yukawa et al. 1993) or a non-compartmental analysis. However, a lack of robustness of the analytical methods used, and the number of subjects and time points, did not always permit accurate determination of the number of compartments. Lithium, like sodium ion, is water soluble and its binding to plasma proteins is negligible (Clarke et al. 2004; Price and Heninger 1994). Unlike

Table 2.1 Main pharmacokinetic parameters of lithium in some studies from the literature in patients with bipolar disorder (mean ± standard error, mean [minimum-maximum] or (% coefficient of variation))

Study	Subject	Salt	PK analysis	Age (years)	BW (kg)	CrCL (L/h)	Dose frequency	Dose (mmol or mg)	C_{max} (mmol/L)	C_{min}, ss (mmol/L)	T_{max} (h)	$AUC_{0-\infty}$ (mmol. min/L)	$t_{1/2}$ (h)	CL/F (L/h)	Vd/F (L/kg)
Single dose															
Standard release															
Lee et al. (1998)	8	Li_2CO_3	CA (2C)	26.9 ± 8.5	59.7 ± 11.6	–		900 mg	0.97 ± 0.17	–	1.59 ± 0.78	722.6 ± 262.7	16.3 ± 7.18	1.13 ± 0.39	1.43 (0.39)
Frye et al. (1998)	13	–	CA	–	–	–		600 mg	0.72 ± 0.15	–	1.50 ± 0.91	611.4 ± 173.4	–	–	–
Findling et al. (2010)	39	Li_2CO_3	CA (2C)	11.9 ± 2.5	51.6 ± 16.9	7.96 ± 2.67		600–900 mg	–	–	–	–	27.3 (41.0)	1.34 (51.6)	39.7[a](8.84)
At steady state															
Standard release															
Chapron et al. (1982)	6	Li_2CO_3	–	81 [73–88]	68 ± 14	3.74	Bid	0.15 mmol/kg/day	–	–	–	–	28.5 ± 4.9	0.83	z
Hardy et al. (1987)	9	Li_2CO_3	–	73 [67–80]	56 ± 10	4.18	Od	0.21 mmol/kg/day	–	–	–	–	26.9 ± 5.5	0.94	0.64 (0.16)
Cooper et al. (1978)	12	Li_2CO_3	NCA	Adult	Non obese	–	Bid	24.3–81.0 mmol/day		0.74 ± 0.15					

(continued)

Table 2.1 (continued)

Study	Subject	Salt	PK analysis	Age (years)	BW (kg)	CrCL (L/h)	Dose frequency	Dose (mmol or mg)	C_{max} (mmol/L)	C_{min}, ss (mmol/L)	T_{max} (h)	$AUC_{0-\infty}$ (mmol. min/L)	$t_{1/2}$ (h)	CL/F (L/h)	Vd/F (L/kg)
Sustained release															
Cooper et al. (1978)	12	Li_2CO_3	NCA	Adult	Non obese	–	Bid	24.3–81.0 mmol/day	–	0.84 ± 0.17					
ElDesoky et al. (2008)	50	Li_2CO_3	CA (2C)	33 ± 10	67 ± 3.6	6.3	Bid	–	–	–	–	–	–	0.51	15.2[a]
Hoegberg et al. (2012)	11	$Li_3C_6H_5O_7$	CA (1C)	52 [21–80]	72 [45–125]		Bid	Adjusted dose							0.74 (0.19)
Potkin et al. (2002)	10	Li_2CO_3	CA	32.8 ± 1.9	87.6 ± 3.3	–	Bid	16.2–64.8/day	1.11	0.55 [0.32–0.77]	1.07 [0.55–4.07]	554 [408–750]/12 h	–	–	–

1C one-compartment, *2C* two-compartment, *CA* compartmental analysis, *Li2CO3* lithium carbonate, *Li3C6H5O7* lithium citrate, *NCA* non-compartmental analysis, *Bid* twice per day, *Od* one per day, *BW* body weight

[a]In litres

other organic drugs used in mental disorders, lithium has a slow diffusion from extracellular to intracellular compartments due to its low passive diffusion permeability across cell membranes. The means of the central volume of distribution of lithium calculated in several studies were from 0.8 to 1.2 L/kg with an initial distribution into the total extracellular body fluids (Ward et al. 1994). Therefore, lithium has a small volume of distribution, slightly higher than the total body water volume, which makes it susceptible to large variations of total body water volumes that are frequently observed following some pathophysiological changes. For example, with ageing, alteration of body composition such as increased body fat, decreased fat-free mass and total body water will decrease the volume of distribution of lithium, decreasing the time to reach the steady state with repeated doses of lithium together with a reduced elimination half-life and an increase of the C_{max}/C_{min} ratio. Oedemas or, conversely, dehydration syndrome may also change the percentage of total body water and consequently the volume of distribution of lithium. The distribution of lithium into the tissue from the blood occurs variably at differing rates: for example, high and low uptake rates occur in the liver/kidneys and brain/muscles, respectively. The kidney, bone and thyroid lithium concentrations are twice those simultaneously measured in serum (Amdisen 1977; Kusalic and Engelsmann 1999; Wilting et al. 2007) with a T_{max} from 2 to 4 h later than that observed in the serum. Lithium concentrations in the liver and muscles are lower than those in serum and other tissues (kidney, thyroid, bone). Erythrocyte concentrations of lithium correlate in most patients with plasma levels but are subject to large interindividual variability. In early studies, lithium concentration in RBC was proposed as a better surrogate of brain lithium concentration than serum lithium concentration. Today, this statement remains poorly substantiated, and lithium RBC concentrations are not consistently used in clinical practice (Lee et al. 1975; Schreiner et al. 1979; White et al. 1979).

2.3.3 Metabolism

Lithium is not subject to first-pass hepatic metabolism (Bauer et al. 2006). A small amount (2 %) of lithium is eliminated in the bile within 24 h of its oral administration (Terhaag et al. 1978). Thus, the PK of lithium is not influenced by any changes in hepatic biotransformation or hepatic blood flow.

2.3.4 Excretion

Lithium is freely filtered by the renal glomerulus with a filtration rate similar to that of other small molecules circulating freely in the plasma that are unbound to proteins. Once filtered, about 75–80 % of lithium is actively reabsorbed in the proximal tubules, while a small proportion can also be reabsorbed in loop of Henle and by the epithelial channel which involves sodium (eNaC) in the early distal tubules (Hayslett and Kashgarian 1979) under sodium-restricted conditions (Thomsen and Shirley 2006; Trepiccione and Christensen 2010). The rate of lithium renal elimination is linearly correlated with its serum levels (Bauer et al. 2006). The terminal elimination half-life ($t_{1/2}$) of lithium in the plasma ranges from 16 to 30 h in healthy volunteers and patients with normal renal function (Arancibia et al. 1986; Cooper et al. 1978; Granneman et al. 1996; Hunter 1988; Lee et al. 1998; Luisier et al. 1987; Thornhill 1978). This elimination phase can be observed as early as 4–6 h after the T_{max}. The lithium elimination half-life during long-term lithium treatment or in patients with impaired renal function increases and can reach 50 h. A linear relationship between the dose and the AUC of serum lithium levels versus time has been determined following several oral doses of lithium (Amdisen 1977); this linear relationship is an indicator of any saturation of the different PK phases (absorption, distribution, elimination) of lithium at usual therapeutic doses (12–36 mmol/24 h). However, this does not argue against the involvement of carrier-mediated processes in absorption, distribution and excretion, as sufficient concentrations to saturate and kinetically reveal these transport systems could not easily be reached in vivo.

Lithium is mainly eliminated by the kidneys as a free ion. Its clearance ranges from 0.6 to 2.4 L/h (Ward et al. 1994) with large interindividual variability (Amdisen 1977; Frye et al. 1998; Perry et al. 1981). This variability has been identified in PK population studies (Taright et al. 1994; Yukawa et al. 1993), interindividual lithium clearance variability being from 25 % to 38 % in bipolar patients with normal renal function. Lithium total body clearance has been reported as variable depending on several pathophysiological states, such as age, total body weight and renal function (ElDesoky et al. 2008). It has been clearly demonstrated that the effect of weight parameters (total body weight or lean body weight) and creatinine renal clearance accurately predict steady-state lithium concentrations by decreasing interindividual variability (Sproule et al. 2000). Renal lithium clearance ranges from 10 to 40 mL/min and decreases with age (linearly with age-related decrease in renal glomerular filtration), low sodium intake, volume depletion in dehydration or with age, cardiac failure, concomitant use of drugs affecting renal glomerular filtration such as non-steroidal anti-inflammatory agents (see Sect. 2.5.1.3), kidney disease and any event increasing proximal tubular reabsorption of sodium and water (i.e. dehydration, decreased salt intake, extra-renal salt loss, diuretics) and increased with pregnancy and total body weight.

2.4 Factors Influencing the Pharmacokinetics of Lithium

Intra-subject variability in lithium PK is low in patients who receive lithium and achieve constant normal renal clearance during administration of the medication. However, several physiological and co-medication factors can contribute to high interindividual variability in lithium PK. This latter variability necessitates individualization of lithium dosage regimens and close monitoring of lithium serum concentrations to avoid no response or, conversely, severe adverse effects of lithium therapy. This section summarizes the major factors that influence lithium PK.

2.4.1 Renal Function

In the case of significant impaired renal function, plasma lithium concentrations increase quickly and may trigger acute toxicity symptoms. A preliminary check of renal function is thus highly recommended before initiation of lithium treatment, followed by regular checks throughout the course of treatment (see Sect. 2.8). Lithium by itself causes a modest decline in renal function, thus increasing its elimination half-life and steady-state levels. End-stage renal disease (ESRD) is a very rare complication of long-term lithium treatment affecting only 1 % of patients over 15 years of treatment (Tredget et al. 2010). No study has evaluated the variability of the PK parameters for patients with ESRD. A known renal failure after lithium initiation is considered a contraindication, especially when a sodium-poor diet is also prescribed (Grandjean and Aubry 2009a).

2.4.2 Body Weight

Reiss and colleagues (1994) found that, in obese patients, the steady-state volume of distribution in litres was not significantly different from that in normal weight individuals (44.5 ± 9.8 L versus 41.3 ± 8.4 L in obese and nonobese subjects, respectively), which is consistent with the similarly calculated ideal body weight (IBW), and thus significantly less as expressed in litres/kg (0.42 ± 0.09 L/kg and 0.66 ± 0.16 L/kg for obese and nonobese patients, respectively). Total lithium clearance was greater in obese patients (33.9 ± 7.0 mL/min in obese patients versus 23.0 ± 6.2 mL/min in nonobese subjects). The authors thus demonstrated that the volume of distribution of lithium was significantly correlated with ideal body weight and fat-free mass but not with total body weight. Lithium clearance was significantly correlated with total body weight but not with creatinine clearance. The major conclusion was the necessity to increase the maintenance dose in obese patients to achieve similar lithium steady-state levels as those observed in nonobese patients, because of a higher lithium clearance in obese patients. However, the starting dose should be the same as that used in nonobese patients since the volume of distribution of lithium in obese patients is more related to the ideal body weight than the total body weight.

2.4.3 Circadian Rhythm

Circadian rhythm has a specific influence on the PK of lithium. A study in rats by Olesen and Thomsen (1985) reported a significant decrease in serum lithium concentration during the early night portion of the light-dark cycle, when rats are most active, associated with a large increase in renal clearance of lithium. In mice, the toxicity of acute intraperitoneal injection of lithium chloride (LiCl) (940 mg/kg) increased from 20 % during the night to 70 % during the day (Hawkins et al. 1978).

Similarly, Shito and colleagues (1992) reported time-dependent changes in both lithium clearance and toxicity. Chapter 6 is devoted entirely to this topic.

2.4.4 Age

Elderly bipolar disorder patients require a lower dose of lithium to achieve the same steady-state levels observed in young adults with normal renal clearance (see Sect. 2.6.2). In contrast, child bipolar disorder patients have a shorter elimination half-life and greater clearance compared to adults (see Sect. 2.7.1).

2.4.5 Cardiovascular Disease

Associated with age, hypertension and congestive heart failure may decrease the glomerular filtration rate and accordingly further reduce lithium renal excretion (Sproule et al. 2000), indicating that a lower dose of lithium may be required in these patients.

2.4.6 Pregnancy and Lactation

Briefly, lithium clearance increases by 30–50 % in pregnant women due to an increase in its renal glomerular filtration rate, especially during the last months of pregnancy. There is a rapid return to the pre-pregnancy state immediately after delivery (Grandjean and Aubry 2009b; Schou 1990; Viguera et al. 2002). Therefore, TDM must be carried out more frequently for pregnant women. Lithium also diffuses freely from maternal plasma to breast milk (Schou 1990; Yoshida et al. 1999) with serum concentrations in infants ranging from 10 % to 50 % (and possibly up to 80 %) of those in maternal plasma (Viguera et al. 2002). All studies in pregnancy and breastfeeding populations are necessarily retrospective, with the safety of the infant-mother couple being the first concern. (See Chap. 18.)

2.5 Drug-Drug Interactions (DDIs)

Drug-drug interactions involving lithium arise from pharmacodynamic and/or PK interactions. In this section, we will focus on DDI of PK origin that affect lithium serum concentrations. As lithium is absorbed well by the gastrointestinal tract, not metabolized, not bound to plasma proteins and almost entirely eliminated from the body through renal excretion, the major DDIs of PK origin are due to modifications of lithium renal glomerular filtration and/or tubular reabsorption. Competition between sodium and lithium for their tubular reabsorption processes is also a critical key element for lithium renal excretion, and drug-induced sodium depletion is one of the main mechanisms modifying lithium renal clearance.

2.5.1 Drugs That Decrease Lithium Clearance

2.5.1.1 Thiazide and Thiazide-Like Diuretics

Thiazide and thiazide-like diuretics (chlorothiazide, hydrochlorothiazide, bendroflumethiazide, chlorthalidone, indapamide, methyclothiazide, metolazone, polythiazide) are inhibitors of sodium reabsorption in the distal tubules. The mechanism by which lithium serum concentrations are potentially increased with co-medication of these diuretics is directly linked to a compensatory up-regulation of sodium (and thus lithium) reabsorption in the proximal tubules. This has been clearly demonstrated for chlorothiazide, which reduces the renal clearance of lithium by 40–70 %, depending on the diuretic doses. Co-medication of lithium (300 mg of lithium carbonate) alone and with chlorothiazide (500 mg/day for 1 week) was studied in healthy volunteers. Chlorothiazide increased lithium plasma concentration by 26.2 % and decreased renal lithium clearance by 26.5 % (Poust et al. 1976). Subsequently, it was reported in healthy volunteers (Crabtree et al. 1991) that hydrochlorothiazide (25 mg twice daily) in combination with lithium for 5 days increased average lithium serum concentrations by 23 %. Therefore, DDI of lithium with thiazide diuretics has been extended to other thiazide and thiazide-like diuretics even in the absence of ad hoc clinical studies. As a general rule, clinicians reduce (usually halve) the dose of lithium once a thiazide diuretic is added to the medication regimen, and serum lithium concentrations should therefore be more closely monitored. Another clinical approach may simply be to avoid these diuretics in patients with cardiac disorders, as so many alternatives are now available on the market for this indication.

2.5.1.2 Inhibitors of Angiotensin-Converting Enzyme (ACE) and Angiotensin-Receptor Antagonists

Inhibitors of ACE are drugs widely used in the treatment of hypertension and heart failure and as protective agents against renal failure diabetes mellitus. The effect of ACE inhibitors on lithium renal clearance is variable and reports on toxicity have been less common than with thiazide diuretics. However, in the early 1980s, several case reports evidenced an increase in lithium serum concentrations together with lithium toxicity once co-medication with ACE inhibitors (enalapril, captopril, lisinopril) was initiated in the following days or weeks. This has been reviewed thoroughly by Lehmann and Eberhard, while the molecular mechanism of the DDI remains poorly understood (Lehmann and Ritz 1995). A decrease in aldosterone plasma levels observed in patients treated with lithium may induce sodium depletion and thus lithium retention. DDI between enalapril and lithium has been studied in the only prospective study analysing the effect of enalapril for 10 days on lithium serum concentrations in nine healthy volunteers. However, only one patient exhibited a 31 % increase in lithium serum concentrations when enalapril was added, whereas there was no statistical difference for the other eight patients (DasGupta et al. 1992). Another longitudinal case-control study of 20 hypertensive patients previously stabilized on lithium therapy evaluated the effect of ACE inhibitors (lisinopril, enalapril and captopril) on steady-state lithium concentrations and lithium toxicity signs. After initiation of the ACE inhibitor, steady-state lithium concentrations increased by 36.1 %, and four patients presented symptoms suggestive of lithium toxicity (Finley et al. 1996). The clinical significance of this DDI between lithium and ACE has been reported in a population-based study that included 413 patients admitted to hospital at least once for lithium toxicity. A dramatically increased risk of lithium toxicity was observed within a month of introducing an ACE inhibitor (Juurlink et al. 2004). Several case reports have also described potential DDI between angiotensin-2 receptor antagonists (losartan, valsartan, telmisartan).

2.5.1.3 Nonsteroidal Anti-inflammatory Drugs (NSAIDs)

As for ACE inhibitors, the mechanism of action of this DDI remains poorly understood. The possible main mechanism by which NSAIDs increase serum lithium concentrations is directly linked to the inhibition of cyclooxygenase (COX) in the kidney reducing the plasma levels of prostaglandin E2 (PGE2). Reduction in PGE2 levels by NSAIDs decreases the renal blood flow and subsequently the glomerular filtration rates of lithium; it may concomitantly mediate sodium retention as well as that of lithium. Furthermore, micropuncture experiments showed that dosing rats with indomethacin increased the reabsorption of lithium in loop of Henle; this latter effect might be another mechanism by which NSAIDs lower lithium renal excretion (Boer et al. 1993). The effect of NSAIDs on lithium clearance was reported for the first time at the end of the 1970s, where indomethacin increased plasma lithium concentrations by 59 % in three psychiatric patients and by 30 % in four volunteers (Frölich et al. 1979). Diclofenac produced a similar effect, decreasing lithium renal clearance by 23 % and increasing lithium plasma levels by 26 % in women treated with lithium (Reimann and Frölich 1981) – in contrast to aspirin and sulindac, which had no effect (Ragheb and Powell 1986; Reimann et al. 1983). Other NSAIDs have also been investigated for their effect on serum lithium concentrations. Several case reports and clinical studies have reported that naproxen (Ragheb and Powell 1986), piroxicam (Kerry et al. 1983) and selective inhibitors of COX-2 such as celecoxib (Slørdal et al. 2003) and rofecoxib (now withdrawn from the market) (Lundmark et al. 2002) increased serum lithium concentrations.

2.5.2 Drugs That Increase Lithium Clearance

2.5.2.1 Acetazolamide

Acetazolamide decreased sodium, bicarbonate and chloride reabsorption in the renal tubules following inhibition of carbonic anhydrase. The renal excretion of lithium was thus also increased from 19.6 % to 31.4 % in healthy volunteers co-treated without and with acetazolamide, respectively (Colussi et al. 1989). In humans the effect of acetazolamide has been attributed to its actions within the proximal renal tubules where it inhibits both sodium and lithium reabsorption as demonstrated in a micropuncture study in rats (Fransen et al. 1993).

2.5.2.2 Osmotic Diuretics

The increase in lithium renal excretion by osmotic diuresis was reported early in the 1970s in dogs and humans (Obek 1972; Thomsen and Schou 1968). A possible mechanism of action involves the creation of a gradient in the renal tubules that facilitates water and sodium and lithium ion excretion. This DDI has been clearly reported by Noormohamed and Lant (1995), showing that lithium renal excretion can be increased by the infusion of mannitol.

2.5.2.3 Methylxanthine Diuretics

Several studies have reported that methylxanthines (aminophylline, theophylline, caffeine) can increase lithium excretion rate by altering sodium disposition in the kidneys. The ability of theophylline to increase lithium renal excretion was first documented by Perry and colleagues (1984). Despite considerable inter-subject variability, they showed a mean increase of 30 % in the renal excretion of lithium with the addition of theophylline. The results of this first study were later confirmed in another study enrolling ten normal subjects who received lithium twice daily for 1 week and were then infused with either normal saline or theophylline. The authors showed that theophylline infusions increased lithium clearance by 51 % (Holstad et al. 1988), suggesting that this procedure may even be used to increase lithium elimination in lithium intoxication; but the high incidence of adverse effects of this medication compromised its use in this indication, and haemodialysis remains today the established procedure to enhance lithium elimination (Scharman 1997). The effect of methylxanthine caffeine has also been studied in a few studies. The first study reported that, with a single dose of lithium, caffeine evoked a possible increase in lithium elimination (Thomsen and Schou 1968). As chronic caffeine consumption may increase lithium renal elimination, abrupt caffeine withdrawal may rapidly decrease lithium blood levels. Despite high interindividual variability, a clinical study reported that lithium blood levels increased significantly (by 24 %) in 73 % of the patients included in the study when caffeine was eliminated from the diet. The clinical significance of this interaction remains unclear, and there are no case reports supporting the idea that caffeine withdrawal may precipitate worsening of psychiatric symptoms in bipolar patients treated with lithium.

2.5.3 Drugs with Variable Effects on Lithium Clearance

2.5.3.1 Loop and Potassium-Sparing Diuretics

Since lithium reabsorption mainly occurs in the proximal tubules, it is likely that diuretics affecting sodium and water exchange in the loop of Henle and in the distal tubules may have only small effects on the proximal reabsorption of lithium. However, small amounts of lithium are also reabsorbed in the loop of Henle, suggesting that loop diuretics such as furosemide may also decrease lithium renal clearance (Atherton et al. 1991). A micropuncture study in rats revealed means of fractional lithium excretion of 0.23 and 0.40 in the control- and furosemide-treated animals, respectively. Furthermore, no significant lithium reabsorption occurs in nephron segments beyond the loop (Shirley et al. 1992). Furosemide was the first loop diuretic demonstrated to be weakly involved in DDI with lithium. A statistically non-significant decrease in lithium renal excretion of 11 % has also been reported in six human volunteers co-medicated with lithium and furosemide compared to subjects medicated with lithium alone (Thomsen and Schou 1968). Conversely, chronic treatment with furosemide in hypertensive patients showed a 5 % decrease in lithium serum concentrations, suggesting that the effect of furosemide on lithium renal excretion is not immediately apparent (Saffer and Coppen 1983). However, initiation of a loop diuretic has been associated with increased risk of lithium toxicity in a large clinical study with more than 10,000 elderly patients. The relative risk of lithium toxicity was 5.5 within 1 month after the initiation of loop diuretic medication (Juurlink et al. 2004). In conclusion, because of conflicting reports on the effect of loop diuretics on lithium renal excretion, the potential clinical significance of this DDI is not clear but warrants careful lithium TDM.

The effect of potassium-sparing diuretics is no clearer than that of loop diuretics. These diuretics, acting exclusively in the distal tubules and collecting tubes, may have a weak effect on proximal reabsorption of lithium. Amiloride is the first-choice treatment of lithium-induced diabetes insipidus and has been evaluated for its capacity to alter lithium renal excretion. In a micropuncture study conducted in rats, amiloride did not modify lithium renal clearance (Walter et al. 1995). The absence of the effects of amiloride on lithium renal excretion was confirmed using sodium-depleted meals in human volunteers co-treated with lithium and amiloride (Bruun et al. 1989). The possible presence of lithium transport beyond the proximal tubule was also examined by measuring lithium excretion after the administration of triamterene, another potassium-sparing diuretic, exclusively acting in the cortical collecting tubule. Triamterene (100 mg orally) increased lithium renal clearance slightly, by 30 %, in eight healthy volunteers. Similarly, two studies also evidenced the small effect of spironolactone on the renal excretion of lithium such that spironolactone increased, the renal excretion of lithium, but the change was not statistically significant (Baer et al. 1971; Thomsen and Schou 1968).

2.6 Specific Populations

2.6.1 Children and Adolescents

Several years after lithium was introduced as a treatment for adult patients with bipolar disorder, it was proposed as a treatment for affective disorders in children and adolescents (Tueth et al. 1998). Lithium PK in children was first reported by Vitiello and colleagues (1988). The PK in both serum and saliva of a single oral dose of lithium carbonate 300 mg was investigated in nine children aged from 9 to 12 years. Lithium PK was best described by a bi-exponential profile, as already reported in adults. The mean elimination half-life was 17.9 h in the serum and similar in the saliva. Lithium levels were 2.8 times higher in the saliva than in the serum, with a saliva/serum lithium concentration correlation coefficient of 0.93. The PK of lithium in children thus had the same features as in adults, with a trend towards a shorter elimination half-life due to a higher total clearance as expressed per kilogram. This shorter elimination half-life is mostly due to a higher lithium renal excretion clearance. A more recent population PK study with many covariates (Findling et al. 2010) evaluated the PK of lithium in children and adolescents (39 subjects from 7 to 17 years old) suffering bipolar I disorder after administration of 600 or 900 mg lithium carbonate as immediate-release capsules. The PK of lithium was also best described using a two-compartment model with a lag time and first-order absorption rate constant. Lithium clearances did not vary systematically with age group, dose, sex or creatinine clearance. A positive correlation was observed between clearance and volume of distribution parameters and total body weight, fat-free mass and BMI. In conclusion, this study showed that children have a shorter elimination half-life and greater clearance compared with adults; thus, steady-state lithium serum concentrations will be achieved in children more rapidly and at lower levels than in adults. Therefore, in children, the dose required may be higher in milligrams per kilogram than that used in adults and might be adapted according to body weight: for children <20 kg and >20 kg, the FDA recommended dose is 300 mg and 600 mg or 900 mg, respectively, with an adapted dose regimen based on steady-state plasma levels as in bipolar adults. The therapeutic range of serum lithium concentrations is equal to that in adult patients, i.e. 0.9 ± 0.3 mmol/L (Geller et al. 1998; Kowatch et al. 2000).

2.6.2 Elderly Patients

For all prescriptions for elderly patients, it is necessary to know the impact of age on PK. In the case of lithium, the patient response may be modified by three age-related clinical conditions (Sproule et al. 2000): physiological changes, comorbidities and co-medications. Major changes between a younger adult population (<50 years) and elderly patients were observed regarding PK parameters of lithium:

- A lower dose administered to elderly patients provides a serum concentration ranging from equivalent to that in adults to values increased by 20–40 % (aged >70 versus <50 years) (Hewick et al. 1977; Vestergaard and Schou 1984). This is confirmed by the ratio of lithium dose to lithium serum concentration in elderly patients, which was lower than that in the youngest group (mean age 70.1 ± 4 years with a ratio of 0.37 ± 0.11 versus two younger age groups with mean ages of 34.9 ± 7.2 and 53.0 ± 5.6 with a ratio 0.58 ± 0.15) (Greil et al. 1985). In contrast, oral absorption of lithium was not altered in elderly patients in terms of both rate and extent.
- Volume of distribution was modified following several physiological changes such as increase in body fat together with a decrease in total body water and dehydration. Due to the hydrophilic property of lithium, this causes a decrease in the volume of distribution of lithium as previously reported (Lehmann and

Merten 1974), where an age-related reduction in volume of distribution of 23 % was found in patients aged from 52 to 65 years as compared to younger patients aged from 23 to 28 years.

- Renal lithium clearance also decreased linearly with age-related decreases in creatinine clearance with a 60 % reduction in older patients aged from 52 to 65 years compared to younger patients aged from 23 to 28 (Lehmann and Merten 1974), together with a prolonged half-life (Chapron et al. 1982).

Some key studies are summarized in Table 2.1. The impact of co-medications is well explored and all drugs that altered the renal function in elderly patients might be involved in DDI with lithium (Shulman and Herrmann 1999; Young et al. 2004). (see Sect. 2.5).

2.7 Neuropharmacokinetics

Lithium is a widely used first-line treatment for bipolar disorder that is thought to have neuroprotective properties; however, its therapeutic mechanisms of action are still largely unknown, although it is presumed to act on intracellular targets. Interestingly, even at adequate serum levels, about half of patients do not respond to lithium therapy (Soares et al. 2001), indicating that the correlation between serum and brain lithium concentrations and clinical effect might be weak or nonexistent (Grandjean and Aubry 2009a; Kato et al. 1994, 1996; Komoroski 2005; Riedl et al. 1997; Sachs et al. 1995). In this section, we don't consider the variability factors from a pharmacodynamic point of view that may also explain discrepancies between serum lithium levels and CNS effects.

Since the magnitude of the pharmacologic effect and side effects of a drug depends on the concentration at receptor sites in the target tissue, the brain lithium concentration is of great interest, although, to date, lithium PK in the brain is not well understood. Nevertheless, it differs from that in other organs and tissues (Komoroski et al. 1997).

In dynamic studies, a rise of brain lithium concentration occurred after a 24-h delay compared to the rise in serum, red blood cells, muscle or CSF, suggesting that the rate of lithium transfer from blood to CSF across the blood-CSF barrier (BCSFB) is faster than that occurring at the blood-brain barrier (BBB) (Hillert et al. 2012; Komoroski et al. 1990; Renshaw et al. 1986). The brain concentrations and mean brain-to-serum ratio of lithium for maintenance treatment in bipolar patients are around 0.2–0.3 mEq/L and 0.80 ± 0.19, respectively (Grandjean and Aubry 2009a; Komoroski 2005). The half-life for lithium is 28 h in the human brain, longer than the serum elimination half-life (16 h), suggesting that lithium is released more slowly from the brain into the blood (Girard et al. 2001).

Lithium distribution, first determined by endogenous measurements in the brain tissue from human autopsies or animals, is heterogeneous: some reports are of relatively small variations of 20–30 %, while others found large differences among different regions, without brain phosphorus and intracellular pH influence (Kato et al. 1993). Human samples have shown levels to be consistently highest in the midbrain and occipital cortex and typically lowest in the pons and in the white matter located at the centrum semiovale (Francis and Traill 1970; Spirtes 1976; Wittrig et al. 1970). Similarly, rat studies have shown both insignificant (Ho et al. 1970) and significant (Ebadi et al. 1974; Edelfors 1975; Mukherjee et al. 1976; Nelson et al. 1980) differences in brain lithium regional distribution following acute and chronic administration. Some researchers conducted short-term experiments that indicated high lithium concentrations in the basal ganglia, neocortex and putamen and low levels in the pons, medulla, cerebellum, spinal cord and white matter structures. Intermediate levels were found in the thalamus, hypothalamus, septal nuclei, dentate gyrus, hippocampus and substantia grisea centralis (Ebadi et al. 1974; Mukherjee et al. 1976; Nelson et al. 1980). Moreover, monkeys given lithium orally for 3–6 weeks once daily (Spirtes 1976) had the highest concentrations in the anterior thalamus and head of the caudate nucleus; the lowest levels were in the spinal cord, cerebellum and cortical tips (frontal and temporal poles) and intermediate levels in the same structures as previously demonstrated in rat studies. These previous results were confirmed using ^{7}Li magnetic resonance spectroscopy (Renshaw and Wicklund 1988). Furthermore, lithium distribution appears to differ after chronic administration compared to acute administration (Ramaprasad 2004; Sandner et al. 1994). Thus, chronic lithium administration may culminate in brain tissue accumulation as shown by lithium levels in brain homogenate, which are significantly higher than in CSF or dialysates after 24 h and 3 weeks in rats (Hillert et al. 2012). Such accumulation has been speculated to cause switching from depression into mania (Kato et al. 1992). This phenomenon of lithium accumulation has been ascribed to lithium's binding to intracellular proteins or other structures, as well as to saturation/ion competition or fewer efflux transport mechanisms at the plasma membrane.

As for many other physiological inorganic ions (e.g. sodium, potassium), the intrinsic plasma membrane permeability of lithium is very low. However, depending on the type of cell, the expression of some specific proteins embedded in the plasma membrane is capable of transporting inorganic ions across the membrane providing specific but rate-limiting ion movement. Such channel- and/or carrier-mediated systems involved in lithium membrane permeability still need to be investigated in brain cells. Because of shared physico-chemical features with sodium, lithium is thought to utilize sodium/organic or inorganic co-transporters – observed in a primary culture of cortical neuronal mouse cells (Komoroski 2005; Wada et al. 2006). The molecular mechanisms involved in lithium transport in the body, and particularly in brain cells, need to be fully elucidated; however, the importance of such molecular transport mechanisms has been recently underscored by a clinical pharmacogenomics study that shows SLC4A10, a sodium/carbonate (NBC, SLC4 family) transporter expressed at least at the BCSFB, is involved in the lithium response (Rybakowski 2013). Although to date no data have been published on the role of the sodium/proton transporter (NHE, SLC9 family), this may also be involved in lithium brain movement by analogy with lithium renal excretion.

2.8 Therapeutic Drug Monitoring (TDM) for Lithium

2.8.1 The Justification for TDM

The oral dose of lithium necessary to obtain recommended serum concentrations can vary from 450 to 1,300 mg/day, depending on the individual (Grandjean and Aubry 2009a). This implies that an appropriate dose has to be determined for each patient, and there must be regular TDM of lithium blood levels. Close monitoring of lithium concentrations helps to achieve this individualization of treatment, and the administration of an adequate dose regimen ensures prophylaxis of psychiatric symptoms while minimizing the risks of relapse and toxicity. The wide interindividual variation in renal clearance of lithium and response to treatment complicates its therapeutic use by clinicians. In addition, a great number of factors contribute to interindividual variations of lithium PK parameters (see Sect. 2.4). Serum lithium concentration should thus be checked, and the lithium dose should be adjusted to obtain the desired range after initiation of lithium therapy, after any change in lithium dose, when renal function has been altered and/or when there has been intercurrent disease or any change in co-medication that may affect lithium PK. The lithium level should be determined when a steady state has been reached, usually 5–7 days after initiation, or dose adjustment, as its terminal plasma half-life is close to 1 day (Yatham et al. 2005).

2.8.2 The Administration of Lithium

2.8.2.1 The Influence of Pharmaceutical Form on PK

Lithium salts are available in a variety of forms (carbonate, citrate, chloride, acetate and sulphate) (see Chap. 1). These salts are present in either immediate- or sustained-release pharmaceutical formulations (see Sects. 2.3.1.1 and 2.3.1.2). To avoid high C_{max} and potential adverse effects, the sustained-release formulations are the mostly widely used; especially in Europe (Mitchell 2001). The nature of lithium salts and formulations influences the PK parameters such as T_{max}, C_{max} and absorption half-lives. However, under steady-state concentrations, C_{min} measured 12 h after the last administration of lithium and AUC are similar whatever the form of lithium. It is worth noting that in the majority of PK studies, immediate-release lithium carbonate has been administered (Thornhill 1978).

2.8.2.2 Initiation of Lithium Therapy

Lithium has a relatively narrow therapeutic index that requires finding a suitable balance between achieving effectiveness and avoiding adverse effects. The most widely used method to initiate lithium therapy is the empirical titration method (Sienaert et al. 2013); here, lithium is started at low daily divided doses, and serum levels are determined once steady-state concentrations have been reached, usually 3–7 days after initiation of treatment. The daily dose is then adjusted with gradual increments to reach the desired serum level.

2.8.3 Optimal Dosage and Administration Schedule

The administration schedule can also influence the efficacy and tolerability of lithium therapy and it should be chosen carefully. The classic schedule is to prescribe lithium twice daily or, less commonly, three times daily. A twice-a-day regimen is well established with the sustained-release preparations, and there is little justification for a three-times-daily regimen, especially because midday dose compliance is usually poor. A multiple-dose schedule is thought to produce a broader and lower peak in plasma lithium concentration with fewer adverse effects. For this reason, current recommendations advise that lithium should be administered in multiple divided doses. However, this regimen also produces a lower trough concentration, which could, in turn, make breakthrough symptoms more likely. The initial recommended daily dose is usually 450–900 mg, depending on age and body weight. The lithium dose should be adjusted after 1 week to obtain the desired serum concentration. The usual maintenance daily dose is 925–1,300 mg for patients aged <40 years, 740–925 mg for those aged 40–60 years and 550–740 mg for patients >60 years. To limit a rapid increase in lithium blood concentrations, an alternative schedule could be adopted by starting with a lower dose (200–300 mg/day) and then adjusting upwards week by week. This may necessitate waiting several weeks before the desired blood concentration is obtained, so delaying the therapeutic benefit. However, the use of atypical antipsychotics in the acute management of severe manic and mixed episodes has rendered this time lapse before achieving an adequate therapeutic lithium level less important (Sienaert et al. 2013). Recently, the single daily dose schedule has been advised by an increasing number of specialists and is proposed as a viable option for administration. Since lithium exerts its therapeutic effects within the CNS, the brain-tissue concentrations of lithium determine its efficacy and tolerability. Yet Jensen and colleagues (1992) reported that no relationship was found between brain lithium concentration at 12 h and the form of lithium dosing schedule. Furthermore, a single daily dose regimen is associated with less polyuria, a reduction in permanent renal damage and better compliance. In addition, because lithium clearance is lower overnight, it has been suggested that this dose regimen could be reduced up to 25 % (Malhi and Tanious 2011). However, less frequent administration regimens have been tested and shown not to maintain prophylactic efficacy against recurrent episodes of bipolar disorder, with reports of higher relapse rates and shorter time to relapse (Jensen et al. 1996a, b).

2.8.3.1 Optimal Plasma Lithium Concentrations

For initiation of lithium therapy de novo, available worldwide guidelines indicate that a target plasma concentration of 0.6–0.8 mmol/L seems to be a reasonable compromise. The French HAS guidelines (http://www.has-sante.fr/) recommend achieving and maintaining concentrations of around 0.5–0.8 mmol/L for immediate-release forms and 0.8–1.2 mmol/L for sustained-release forms, with close monitoring of adverse effects. In comparison, the UK National Institute for Health and Care Excellence (NICE) advocates a relatively narrow range for plasma concentrations (0.6–0.8 mmol/L) (Goodwin and Goldstein 2003), whereas the American Psychiatric

Association (APA) recommends a gradual titration to achieve 0.5–1.2 mmol/L (APA 2002). The Canadian Network for Mood and Anxiety Treatments (CANMAT) advises a higher range (0.8–0.9 mmol/L). The reason for these slight disparities in the optimal lithium concentration is a difference in emphasis between each of the preceding guidelines: either on the prophylaxis and absence of adverse effects and toxicity that low concentrations (0.4–0.6 mmol/L) are reported to offer (Severus et al. 2008) or on the necessity to avoid manic episodes that concentrations above 0.8 mmol./L can ensure (Malhi et al. 2012).

2.8.4 Maintaining Lithium Treatment

Once steady state has been achieved, the recommended interval for routine serum checking varies from 6 to 12 weeks to 6 months in stable patients. Recommended lithium concentrations at this stage (when lithium is used for prophylaxis) are far from well established. Recent studies suggest that titration of lithium plasma levels should be guided according to the symptomatic profiles of patients. CANMAT advocates lithium plasma levels of 0.6–0.8 mmol/L for maintaining lithium therapy, while APA guidelines suggest that lower levels of 0.4–0.6 mmol/L may be sufficient and that higher concentrations (0.8–1.0 mmol/L) are only necessary to control mania-prone patients. On the other hand, there is a consensus that 0.4 mmol/L is the minimum plasma level to ensure prophylaxis with low risk of relapse. This was confirmed by Severus and colleagues, who have shown that depression-prone patients are likely to benefit from prophylactic lithium levels at 0.4–0.8 mmol/L, whereas those predisposed to mania need higher levels at 0.6–1.0 mmol/L (Kleindienst et al. 2007; Severus et al. 2008).

2.8.4.1 The 'Lithiumeter'

On the basis of clinical guideline recommendations and recent studies, Malhi and colleagues (2011) have coined the term 'lithiumeter' and suggested a visual scale for gauging optimal lithium plasma levels for the treatment of bipolar disorder (see Appendix 3).

2.8.4.2 Lithium Dosage Prediction and Pharmacokinetic-Based Therapeutic Adjustment

Initial lithium dosage is most often calculated via empirical methods and then titrated on the basis of serum lithium determinations to achieve appropriate concentrations. However, this method often requires the hospitalization of patients for 3–7 days, the necessary delay to reach steady state. To shorten the process of dose titration, several authors have developed PK methods for predicting individual dosage requirements with varying degrees of success. As lithium is exclusively excreted by the kidneys, it is possible to estimate the daily dose of lithium required to obtain a desired blood concentration by using the renal clearance of lithium. However, Sienart and colleagues (2013) have critically reviewed a great number of a priori methods that propose a model for initiating lithium based upon the patient's characteristics and 'test-dose' based methods. They state that, although the empirical titration method is not quick, 'faster' methods remain understudied and cannot be recommended for clinical use.

2.9 Summary

Since lithium was first marketed 70 years ago, its peripheral PK, together with variability factors, has been extensively investigated leading to a degree of consensus in guidelines for its practical use throughout the world. Lithium is absorbed relatively well with few variability factors altering its absorption. Lithium is mainly distributed within body fluids and eventually eliminated by renal excretion. Most of the variability factors affecting lithium PK are those modifying its volume of distribution (age, body weight and pathophysiological situations) and its renal clearance (age, body weight and DDI). These factors of variability have to be fully understood by clinicians to avoid non-response or, conversely, adverse effects and severe toxicity. Systematic and routine TDM is therefore necessary to maintain lithium serum concentrations within the narrow therapeutic window.

While the peripheral PK of lithium has been extensively studied, lithium PK in the CNS, the compartment in which its actions take effect, remains poorly investigated. New imaging techniques such as MRI may assist in finding its brain PK parameters. Furthermore, the mechanisms by which lithium is able to cross the brain barriers, including the BBB and the BCSFB, are still unknown. Further study of the channels and/or transporters implicated in the membrane transfer of lithium through these barriers is needed, similar to the studies that have been conducted in the kidneys to understand renal clearance and variability factors. Finally, we have yet to build an understanding of how lithium exhibits a different PK in the brain compared with blood PK, and how these brain concentrations are linked to its effect through PK-PD relationships. Taken together, such novel insights will undoubtedly improve our understanding of the pharmacology of lithium and may enable us to diagnose (or predict) which patients will, or will not, respond to lithium therapy.

References

Altamura AC, Gomeni R, Sacchetti E, Smeraldi E (1977) Plasma and intracellular kinetics of lithium after oral administration of various lithium salts. Eur J Clin Pharmacol 12:59–63

Amdisen A (1977) Serum level monitoring and clinical pharmacokinetics of lithium. Clin Pharmacokinet 2:73–92

American Psychiatric Association. Practice guideline for the treatment of patients with bipolar disorder (revision). Am J Psychiatry. 2002;159(4 Suppl):1–50

Arancibia A, Corvalan F, Mella F, Concha L (1986) Absorption and disposition kinetics of lithium carbonate following administration of conventional and controlled release formulations. Int J Clin Pharmacol 24:240–245

Atherton JC, Doyle A, Gee A, Green R, Gingell S, Nicholls F, Pempkowiak L, Plange-Rhule J (1991) Lithium clearance: modification by the loop of Henle in man. J Physiol 437:377–391

Baer L, Platman SR, Kassir S, Fieve RR (1971) Mechanisms of renal lithium handling and their relationship to mineralocorticoids: a dissociation between sodium and lithium ions. J Psychiatr Res 8:91–105

Bauer M, Grof P, Müller-Oerlinghausen B (2006) Lithium in neuropsychiatry: the comprehensive guide. Informa Healthcare, Abingdon, Distributed in North and South America by Taylor & Francis, Abingdon, Oxon; Boca Raton, FL

Bettinger TL, Crismon ML (2006) Lithium. In: Burton ME, Shaw LM, Schentag JJ, Evans WE (eds) Applied pharmacokinetics and pharmacodynamics: principles of therapeutic drug monitoring, 4th edn. Lippincott Williams & Wilkins, Philadelphia, pp 798–812

Boer WH, Fransen R, Boer P, de Roos R, Koomans HA (1993) Prostaglandin synthesis inhibition stimulates lithium reabsorption in Henle's loop in rats. Kidney Int 43:301–306

Bruun NE, Skøtt P, Lønborg-Jensen H, Giese J (1989) Unchanged lithium clearance during acute amiloride treatment in sodium-depleted man. Scand J Clin Lab Invest 49:259–263

Cade JFJ (1949) Lithium salts in the treatment of psychotic excitement. Med J Aust 2:349–352

Castrogiovanni P (2002) A novel slow-release formulation of lithium carbonate (Carbolithium Once-A-Day) vs. standard Carbolithium: a comparative pharmacokinetic study. Clin Ter 153:107–115

Chapron DJ, Cameron IR, White LB, Merrall P (1982) Observations on lithium disposition in the elderly. J Am Geriatr Soc 30:651–655

Clarke WB, Guscott R, Lindstrom RM (2004) Binding of lithium and boron to human plasma proteins II. Biol Trace Elem Res 97:117–124

Colussi G, Rombolà G, Surian M, De Ferrari ME, Airaghi C, Benazzi E, Malberti F, Minetti L (1989) Effects of acute administration of acetazolamide and furosemide on lithium clearance in humans. Nephrol Dial Transplant Off Publ Eur Dial Transpl Assoc Eur Ren Assoc 4:707–712

Cooper TB, Simpson GM, Lee JH, Bergner PE (1978) Evaluation of a slow-release lithium carbonate formulation. Am J Psychiatry 135:917–922

Crabtree BL, Mack JE, Johnson CD, Amyx BC (1991) Comparison of the effects of hydrochlorothiazide and furosemide on lithium disposition. Am J Psychiatry 148:1060–1063

DasGupta K, Jefferson JW, Kobak KA, Greist JH (1992) The effect of enalapril on serum lithium levels in healthy men. J Clin Psychiatry 53:398–400

Ebadi MS, Simmons VJ, Hendrickson MJ, Lacy PS (1974) Pharmacokinetics of lithium and its regional distribution in rat brain. Eur J Pharmacol 27:324–329

Edelfors S (1975) Distribution of sodium, potassium and lithium in the brain of lithium-treated rats. Acta Pharmacol Toxicol (Copenh) 37:387–392

ElDesoky ES, Kumar V, Alorainy MS, Hamdi MM, Derendorf H (2008) Estimation of lithium clearance from routine clinical data in Egyptian bipolar patients. A population pharmacokinetic approach. Int J Clin Pharmacol Ther 46:617–626

Findling RL, Landersdorfer CB, Kafantaris V, Pavuluri M, McNamara NK, McClellan J, Frazier JA, Sikich L, Kowatch R, Lingler J, Faber J, Taylor-Zapata P, Jusko WJ (2010) First-dose pharmacokinetics of lithium carbonate in children and adolescents. J Clin Psychopharmacol 30:404–410

Finley PR, O'Brien JG, Coleman RW (1996) Lithium and angiotensin-converting enzyme inhibitors: evaluation of a potential interaction. J Clin Psychopharmacol 16:68–71

Francis RI, Traill MA (1970) Lithium distribution in the brains of two manic patients. Lancet 2:523–524

Fransen R, Boer WH, Boer P, Dorhout Mees EJ, Koomans HA (1993) Effects of furosemide or acetazolamide infusion on renal handling of lithium: a micropuncture study in rats. Am J Physiol 264:R129–134

Frölich JC, Leftwich R, Ragheb M, Oates JA, Reimann I, Buchanan D (1979) Indomethacin increases plasma lithium. Br Med J 1:1115–1116

Frye MA, Kimbrell TA, Dunn RT, Piscitelli S, Grothe D, Vanderham E, Corá-Locatelli G, Post RM, Ketter TA (1998) Gabapentin does not alter single-dose lithium pharmacokinetics. J Clin Psychopharmacol 18:461–464

Gai MN, Ferj S, García E, Seitz C, Thielemann AM, Andonaegui MT (1999) Evaluation of the in vitro and in vivo performance of two sustained-release lithium carbonate matrix tablets. Effect of different diets on the bioavailability. Drug Dev Ind Pharm 25:131–140

Gai MN, Thielemann AM, Arancibia A (2000) Effect of three different diets on the bioavailability of a sustained release lithium carbonate matrix tablet. Int J Clin Pharmacol Ther 38:320–326

Geller B, Cooper TB, Sun K, Zimerman B, Frazier J, Williams M, Heath J (1998) Double-blind and placebo-controlled study of lithium for adolescent bipolar disorders with secondary substance dependency. J Am Acad Child Adolesc Psychiatry 37:171–178

Girard F, Suhara T, Sassa T, Okubo Y, Obata T, Ikehira H, Sudo Y, Koga M, Yoshioka H, Yoshida K (2001) 7Li 2D CSI of human brain on a clinical scanner. Magma N Y N 13:1–7

Goodwin FK, Goldstein MA (2003) Optimizing lithium treatment in bipolar disorder: a review of the literature and clinical recommendations. J Psychiatr Pract 9:333–343

Grandjean EM, Aubry J-M (2009a) Lithium: updated human knowledge using an evidence-based approach. Part II: clinical pharmacology and therapeutic monitoring. CNS Drugs 23:331–349

Grandjean EM, Aubry J-M (2009b) Lithium: updated human knowledge using an evidence-based approach: part III: clinical safety. CNS Drugs 23:397–418

Granneman GR, Schneck DW, Cavanaugh JH, Witt GF (1996) Pharmacokinetic interactions and side effects resulting from concomitant administration of lithium and divalproex sodium. J Clin Psychiatry 57:204–206

Greil W, Stoltzenburg MC, Mairhofer ML, Haag M (1985) Lithium dosage in the elderly. A study with matched age groups. J Affect Disord 9:1–4

Guelen PJ, Janssen TJ, De Witte TC, Vree TB, Benson K (1992) Bioavailability of lithium from lithium citrate syrup versus conventional lithium carbonate tablets. Biopharm Drug Dispos 13:503–511

Gyulai L, Wicklund SW, Greenstein R, Bauer MS, Ciccione P, Whybrow PC, Zimmerman J, Kovachich G, Alves W (1991) Measurement of tissue lithium concentration by lithium magnetic resonance spectroscopy in patients with bipolar disorder. Biol Psychiatry 29:1161–1170

Hardy BG, Shulman KI, Mackenzie SE, Kutcher SP, Silverberg JD (1987) Pharmacokinetics of lithium in the elderly. J Clin Psychopharmacol 7:153–158

Hawkins R, Kripke DF, Janowsky DS (1978) Circadian rhythm of lithium toxicity in mice. Psychopharmacology (Berl) 56:113–114

Hayslett JP, Kashgarian M (1979) A micropuncture study of the renal handling of lithium. Pflüg Arch Eur J Physiol 380:159–163

Hewick DS, Newbury P, Hopwood S, Naylor G, Moody J (1977) Age as a factor affecting lithium therapy. Br J Clin Pharmacol 4:201–205

Hillert M, Zimmermann M, Klein J (2012) Uptake of lithium into rat brain after acute and chronic administration. Neurosci Lett 521:62–66

Ho AK, Gershon S, Pinckney L (1970) The effects of acute and prolonged lithium treatment on the distribution of electrolytes, potassium and sodium. Arch Int Pharmacodyn Thérapie 186:54–65

Hoegberg LCG, Jürgens G, Zederkof VW, Holgersson B, Andersson JE, Dalhoff KP, Larsen EB, Angelo HR (2012) A computerised sampling strategy for therapeutic drug monitoring of lithium provides precise estimates and significantly reduces dose-finding time. Basic Clin Pharmacol Toxicol 110:259–263

Holstad SG, Perry PJ, Kathol RG, Carson RW, Krummel SJ (1988) The effects of intravenous theophylline infusion versus intravenous sodium bicarbonate infusion on lithium clearance in normal subjects. Psychiatry Res 25:203–211

Hunter R (1988) Steady-state pharmacokinetics of lithium carbonate in healthy subjects. Br J Clin Pharmacol 25:375–380

Jensen HV, Hemmingsen L, Holm J, Christensen EM, Aggernaes H (1992) Urinary excretion of albumin and retinol-binding protein in lithium-treated patients: a longitudinal study. Acta Psychiatr Scand 85:480–483

Jensen HV, Davidsen K, Toftegaard L, Mellerup ET, Plenge P, Aggernaes H, Bjørum N (1996a) Double-blind comparison of the side-effect profiles of daily versus alternate-day dosing schedules in lithium maintenance treatment of manic-depressive disorder. J Affect Disord 36:89–93

Jensen HV, Plenge P, Stensgaard A, Mellerup ET, Thomsen C, Aggernaes H, Henriksen O (1996b) Twelve-hour brain lithium concentration in lithium maintenance treatment of manic-depressive disorder: daily versus alternate-day dosing schedule. Psychopharmacology (Berl) 124:275–278

Juurlink DN, Mamdani MM, Kopp A, Rochon PA, Shulman KI, Redelmeier DA (2004) Drug-induced lithium toxicity in the elderly: a population-based study. J Am Geriatr Soc 52:794–798

Kato T, Takahashi S, Inubushi T (1992) Brain lithium concentration by 7Li- and 1H-magnetic resonance spectroscopy in bipolar disorder. Psychiatry Res 45:53–63

Kato T, Takahashi S, Shioiri T, Inubushi T (1993) Alterations in brain phosphorous metabolism in bipolar disorder detected by in vivo 31P and 7Li magnetic resonance spectroscopy. J Affect Disord 27:53–59

Kato T, Inubushi T, Takahashi S (1994) Relationship of lithium concentrations in the brain measured by lithium-7 magnetic resonance spectroscopy to treatment response in mania. J Clin Psychopharmacol 14:330–335

Kato T, Fujii K, Shioiri T, Inubushi T, Takahashi S (1996) Lithium side effects in relation to brain lithium concentration measured by lithium-7 magnetic resonance spectroscopy. Prog Neuropsychopharmacol Biol Psychiatry 20:87–97

Kerry RJ, Owen G, Michaelson S (1983) Possible toxic interaction between lithium and piroxicam. Lancet 1:418–419

Kersten L, Fleck C, Bräunlich H (1986) Evidence for an intestinal lithium excretion in rats. Exp Pathol 29:55–63

Kleindienst N, Severus WE, Greil W (2007) Are serum lithium levels related to the polarity of recurrence in bipolar disorders? Evidence from a multicenter trial. Int Clin Psychopharmacol 22:125–131

Komoroski RA (2005) Biomedical applications of 7Li NMR. NMR Biomed 18:67–73

Komoroski RA, Newton JE, Walker E, Cardwell D, Jagannathan NR, Ramaprasad S, Sprigg J (1990) In vivo NMR spectroscopy of lithium-7 in humans. Magn Reson Med Off J Soc Magn Reson Med Soc Magn Reson Med 15:347–356

Komoroski RA, Pearce JM, Newton JE (1997) The distribution of lithium in rat brain and muscle in vivo by 7Li NMR imaging. Magn Reson Med Off J Soc Magn Reson Med Soc Magn Reson Med 38:275–278

Kowatch RA, Suppes T, Carmody TJ, Bucci JP, Hume JH, Kromelis M, Emslie GJ, Weinberg WA, Rush AJ (2000) Effect size of lithium, divalproex sodium, and carbamazepine in children and adolescents with bipolar disorder. J Am Acad Child Adolesc Psychiatry 39:713–720

Kusalic M, Engelsmann F (1999) Effect of lithium maintenance therapy on thyroid and parathyroid function. J Psychiatry Neurosci JPN 24:227–233

Lee CR, Hill SE, Dimitrakoudi M, Jenner FA, Pollitt RJ (1975) The relationship of plasma to erythrocyte lithium levels in patients taking lithium carbonate. Br J Psychiatry J Ment Sci 127:596–598

Lee CF, Yang YY, Hu OY (1998) Single dose pharmacokinetic study of lithium in Taiwanese/Chinese bipolar patients. Aust N Z J Psychiatry 32:133–136

Lehmann K (1997) Pharmakokinetik von Lithiurnsalzen. In: Müller-Oerlinghausen PDB, Berghöfer A, Greil PDW (eds) Die Lithiumtherapie. Springer, Berlin, pp 148–160

Lehmann K, Merten K (1974) Elimination of lithium in dependence on age in healthy subjects and patients with renal insufficiency. Int J Clin Pharmacol 10:292–298

Lehmann K, Ritz E (1995) Angiotensin-converting enzyme inhibitors may cause renal dysfunction in patients on long-term lithium treatment. Am J Kidney Dis Off J Natl Kidney Found 25:82–87

Luisier PA, Schulz P, Dick P (1987) The pharmacokinetics of lithium in normal humans: expected and unexpected observations in view of basic kinetic principles. Pharmacopsychiatry 20:232–234

Lundmark J, Gunnarsson T, Bengtsson F (2002) A possible interaction between lithium and rofecoxib. Br J Clin Pharmacol 53:403–404

Malhi GS, Tanious M (2011) Optimal frequency of lithium administration in the treatment of bipolar disorder: clinical and dosing considerations. CNS Drugs 25:289–298
Malhi GS, Tanious M, Gershon S (2011) The lithiumeter: a measured approach. Bipolar Disord 13(3):219–26. doi:10.1111/j.1399-5618.2011.00918.x
Malhi GS, Tanious M, Das P, Berk M (2012) The science and practice of lithium therapy. Aust N Z J Psychiatry 46:92–211
Malhi GS, Bassett D, Boyce P, Bryant R, Fitzgerald PB, Fritz K, Hopwood M, Lyndon B, Mulder R, Murray G, Porter R, Singh AB (2015) Royal Australian and New Zealand College of Psychiatrists clinical practice guidelines for mood disorders. Aust N Z J Psychiatry 49:1087–1206
Mitchell PB (2001) Therapeutic drug monitoring of psychotropic medications. Br J Clin Pharmacol 52(Suppl 1):45S–54S
Mukherjee BP, Bailey PT, Pradhan SN (1976) Temporal and regional differences in brain concentrations of lithium in rats. Psychopharmacology (Berl) 48:119–121
Nelson SC, Herman MM, Bensch KG, Barchas JD (1980) Localization and quantitation of lithium in rat tissue following intraperitoneal injections of lithium chloride. II. Brain. J Pharmacol Exp Ther 212:11–15
Noormohamed FH, Lant AF (1995) Renal handling of lithium and the effects of mannitol and arginine vasopressin in man. Clin Sci (Lond) 89(1):27–36
Obek A (1972) Mannitol infusions and lithium excretion. Br Med J 4:550
Olesen OV, Thomsen K (1985) Diurnal variations in serum lithium and renal lithium clearance in rats given lithium as a single small dose or as multiple high doses. Acta Pharmacol Toxicol (Copenh) 57:171–175
Perry PJ, Dunner FJ, Hahn RL, Tsuang MT, Berg MJ (1981) Lithium kinetics in single daily dosing. Acta Psychiatr Scand 64:281–294
Perry PJ, Calloway RA, Cook BL, Smith RE (1984) Theophylline precipitated alterations of lithium clearance. Acta Psychiatr Scand 69:528–537
Potkin SG, Thyrum PT, Bera R, Carreon D, Alva G, Kalali AH, Yeh C (2002) Open-label study of the effect of combination quetiapine/lithium therapy on lithium pharmacokinetics and tolerability. Clin Ther 24:1809–1823
Poust RI, Mallinger AG, Mallinger J, Himmelhoch JM, Neil JF, Hanin I (1976) Effect of chlorothiazide on the pharmacokinetics of lithium in plasma and erythrocytes. Psychopharmacol Commun 2:273–284
Price LH, Heninger GR (1994) Lithium in the treatment of mood disorders. N Engl J Med 331:591–598
Ragheb M, Powell AL (1986) Lithium interaction with sulindac and naproxen. J Clin Psychopharmacol 6:150–154
Ramaprasad S (2004) Lithium spectroscopic imaging of rat brain at therapeutic doses. Magn Reson Imaging 22:727–734
Reimann IW, Frölich JC (1981) Effects of diclofenac on lithium kinetics. Clin Pharmacol Ther 30:348–352
Reimann IW, Diener U, Frölich JC (1983) Indomethacin but not aspirin increases plasma lithium ion levels. Arch Gen Psychiatry 40:283–286
Reiss RA, Haas CE, Karki SD, Gumbiner B, Welle SL, Carson SW (1994) Lithium pharmacokinetics in the obese. Clin Pharmacol Ther 55:392–398
Renshaw PF, Wicklund S (1988) In vivo measurement of lithium in humans by nuclear magnetic resonance spectroscopy. Biol Psychiatry 23:465–475
Renshaw PF, Haselgrove JC, Bolinger L, Chance B, Leigh JS Jr (1986) Relaxation and imaging of lithium in vivo. Magn Reson Imaging 4:193–198
Riedl U, Barocka A, Kolem H, Demling J, Kaschka WP, Schelp R, Stemmler M, Ebert D (1997) Duration of lithium treatment and brain lithium concentration in patients with unipolar and schizoaffective disorder – a study with magnetic resonance spectroscopy. Biol Psychiatry 41:844–850

Rybakowski JK (2013) Genetic influences on response to mood stabilizers in bipolar disorder: current status of knowledge. CNS Drugs 27:165–173

Sachs GS, Renshaw PF, Lafer B, Stoll AL, Guimarães AR, Rosenbaum JF, Gonzalez RG (1995) Variability of brain lithium levels during maintenance treatment: a magnetic resonance spectroscopy study. Biol Psychiatry 38:422–428

Saffer D, Coppen A (1983) Frusemide: a safe diuretic during lithium therapy? J Affect Disord 5:289–292

Sandner G, Di Scala G, Oberling P, Abbe JC, Stampfler A, Sens JC (1994) Distribution of lithium in the rat brain after a single administration known to elicit aversive effects. Neurosci Lett 166:1–4

Scharman EJ (1997) Methods used to decrease lithium absorption or enhance elimination. J Toxicol Clin Toxicol 35:601–608

Schou M (1990) Lithium treatment during pregnancy, delivery, and lactation: an update. J Clin Psychiatry 51:410–413

Schreiner HC, Dunner DL, Meltzer HL, Fieve RR (1979) The relationship of the lithium erythrocyte: plasma ratio to plasma lithium level. Biol Psychiatry 14:207–213

Severus WE, Kleindienst N, Seemüller F, Frangou S, Möller HJ, Greil W (2008) What is the optimal serum lithium level in the long-term treatment of bipolar disorder – a review? Bipolar Disord 10:231–237

Shirley DG, Walter SJ, Sampson B (1992) A micropuncture study of renal lithium reabsorption: effects of amiloride and furosemide. Am J Physiol 263:F1128–1133

Shito T, Ando T, Okamoto T, Nakano S (1992) Chronopharmacokinetics and chronotoxicity of lithium in mice eating normal and low-sodium diets. Chronobiol Int 9:114–123

Shulman KI, Herrmann N (1999) The nature and management of mania in old age. Psychiatr Clin North Am 22:649–665, ix

Sienaert P, Geeraerts I, Wyckaert S (2013) How to initiate lithium therapy: a systematic review of dose estimation and level prediction methods. J Affect Disord 146:15–33

Slørdal L, Samstad S, Bathen J, Spigset O (2003) A life-threatening interaction between lithium and celecoxib. Br J Clin Pharmacol 55:413–414

Soares JC, Boada F, Spencer S, Mallinger AG, Dippold CS, Wells KF, Frank E, Keshavan MS, Gershon S, Kupfer DJ (2001) Brain lithium concentrations in bipolar disorder patients: preliminary (7)Li magnetic resonance studies at 3 T. Biol Psychiatry 49:437–443

Spirtes MA (1976) Lithium levels in monkey and human brain after chronic, therapeutic, oral dosage. Pharmacol Biochem Behav 5:143–147

Sproule B (2002) Lithium in bipolar disorder: can drug concentrations predict therapeutic effect? Clin Pharmacokinet 41:639–660

Sproule BA, Hardy BG, Shulman KI (2000) Differential pharmacokinetics of lithium in elderly patients. Drugs Aging 16:165–177

Taright N, Mentré F, Mallet A, Jouvent R (1994) Nonparametric estimation of population characteristics of the kinetics of lithium from observational and experimental data: individualization of chronic dosing regimen using a new Bayesian approach. Ther Drug Monit 16:258–269

Terhaag B, Scherber A, Schaps P, Winkler H (1978) The distribution of lithium into cerebrospinal fluid, brain tissue and bile in man. Int J Clin Pharmacol Biopharm 16:333–335

Thomsen K, Schou M (1968) Renal lithium excretion in man. Am J Physiol 215:823–827

Thomsen K, Shirley DG (2006) A hypothesis linking sodium and lithium reabsorption in the distal nephron. Nephrol Dial Transplant Off Publ Eur Dial Transpl Assoc Eur Ren Assoc 21:869–880

Thornhill DP (1978) Pharmacokinetics of ordinary and sustained-release lithium carbonate in manic patients after acute dosage. Eur J Clin Pharmacol 14:267–271

Tredget J, Kirov A, Kirov G (2010) Effects of chronic lithium treatment on renal function. J Affect Disord 126:436–440

Trepiccione F, Christensen BM (2010) Lithium-induced nephrogenic diabetes insipidus: new clinical and experimental findings. J Nephrol 23(Suppl 16):S43–48

Tueth MJ, Murphy TK, Evans DL (1998) Special considerations: use of lithium in children, adolescents, and elderly populations. J Clin Psychiatry 59(Suppl 6):66–73

Vestergaard P, Schou M (1984) The effect of age on lithium dosage requirements. Pharmacopsychiatry 17:199–201

Viguera AC, Cohen LS, Baldessarini RJ, Nonacs R (2002) Managing bipolar disorder during pregnancy: weighing the risks and benefits. Can J Psychiatry Rev Can Psychiatr 47:426–436

Vitiello B, Behar D, Malone R, Delaney MA, Ryan PJ, Simpson GM (1988) Pharmacokinetics of lithium carbonate in children. J Clin Psychopharmacol 8:355–359

Wada M, Shimada A, Fujita T (2006) Functional characterization of Na+ -coupled citrate transporter NaC2/NaCT expressed in primary cultures of neurons from mouse cerebral cortex. Brain Res 1081:92–100

Walter SJ, Sampson B, Shirley DG (1995) A micropuncture study of renal tubular lithium reabsorption in sodium-depleted rats. J Physiol 483(Pt 2):473–479

Ward ME, Musa MN, Bailey L (1994) Clinical pharmacokinetics of lithium. J Clin Pharmacol 34:280–285

White K, Cohen J, Boyd J, Nelson R (1979) Relationship between plasma, RBC, and CSF lithium concentrations in human subjects. Int Pharmacopsychiatry 14:185–189

Wilting I, de Vries F, Thio BMKS, Cooper C, Heerdink ER, Leufkens HGM, Nolen WA, Egberts ACG, van Staa TP (2007) Lithium use and the risk of fractures. Bone 40:1252–1258

Wittrig J, Anthony EJ, Lucarno HE (1970) An ashing technique for endogenous lithium in human brain and other biological tissues. Dis Nerv Syst 31:408–411

Yatham LN, Kennedy SH, O'Donovan C, Parikh S, MacQueen G, McIntyre R, Sharma V, Silverstone P, Alda M, Baruch P, Beaulieu S, Daigneault A, Milev R, Canadian Network for Mood and Anxiety Treatments (2005) Canadian Network for Mood and Anxiety Treatments (CANMAT) guidelines for the management of patients with bipolar disorder: consensus and controversies. Bipolar Disord 7(Suppl 3):5–69

Yoshida K, Smith B, Kumar R (1999) Psychotropic drugs in mothers' milk: a comprehensive review of assay methods, pharmacokinetics and of safety of breast-feeding. J Psychopharmacol Oxf Engl 13:64–80

Young RC, Gyulai L, Mulsant BH, Flint A, Beyer JL, Shulman KI, Reynolds CF (2004) Pharmacotherapy of bipolar disorder in old age: review and recommendations. Am J Geriatr Psychiatry Off J Am Assoc Geriatr Psychiatry 12:342–357

Yukawa E, Nomiyama N, Higuchi S, Aoyama T (1993) Lithium population pharmacokinetics from routine clinical data: role of patient characteristics for estimating dosing regimens. Ther Drug Monit 15:75–82

Lithium: Neurotransmission and Cellular Mechanism Pathways Underlying Neuroprogression in Bipolar Disorder

3

Gin S. Malhi, Tim Outhred, and Pritha Das

Abstract

Lithium has neuroprotective effects that are thought to underpin its clinical effectiveness as a mood stabilizer in the treatment of bipolar disorder. Lithium has a multitude of potentially therapeutic effects on excitotoxic neurotransmission pathways that cause cellular damage and decreased resilience in bipolar disorder. However, isolating the mechanisms that underpin lithium's therapeutic properties has remained difficult. This chapter reviews the therapeutic effects of lithium on neurotransmission and cellular signal transduction mechanisms underlying neuroprogression in bipolar disorder. In doing so, key glycogen synthase kinase-3β (GSK-3β)-mediated mechanisms are highlighted, as they appear pivotal to therapeutic response. These actions are considered in the context of lithium's effects on bipolar disorder neurocircuitry, mood dysregulation and neurocognitive dysfunction. Employing the integrative framework developed within this chapter, the underlying mechanisms of lithium treatment can be investigated. This endeavour is imperative for future development of pharmaceutical agents that are more specific and better tolerated than lithium and could be beneficial to untreated and unresponsive patient populations. Future investigation of the specific therapeutic mechanisms of lithium is likely to provide insights into the pathophysiology of bipolar disorder, and given GSK-3β-mediated mechanisms are implicated in neurodegenerative diseases, the identification of lithium's therapeutic actions and novel treatment targets may be useful for the treatment of other neuroprogressive diseases, such as Alzheimer's and Parkinson's disease.

G.S. Malhi (✉) • T. Outhred • P. Das
Academic Department of Psychiatry, Northern Sydney Local Health District, St Leonards, NSW 2065, Australia

Sydney Medical School Northern, The University of Sydney, Sydney, NSW 2006, Australia

CADE Clinic, Royal North Shore Hospital, Northern Sydney Local Health District, Level 3, Main Hospital Building, St Leonards, NSW 2065, Australia
e-mail: gin.malhi@sydney.edu.au

G.S. Malhi et al. (eds.), *The Science and Practice of Lithium Therapy*,
DOI 10.1007/978-3-319-45923-3_3

Key Points

- Bipolar disorder is associated with progressive worsening of cellular function and deterioration of clinical presentation, which is known as neuroprogression.
- Lithium therapy is thought to slow down and stop, and possibly in some cases even repair, this deterioration; however, isolating its specific therapeutic effects from its numerous actions remains a challenge.
- Among all potential actions, glycogen synthase kinase-3β (GSK-3β)-mediated pathways appear to initiate diverse protective and reparative actions across neurotransmission and cellular signal transduction mechanisms.
- This chapter provides a framework in which the key actions of lithium can be placed within a context of therapeutic changes to the neurocircuitry involved in mood regulation and neurocognition.
- Investigating the therapeutic mechanisms of lithium continues to be a worthy pursuit given that novel treatment targets for bipolar disorder and other neuroprogressive disorders are likely to be revealed.

3.1 Introduction

Bipolar disorder is associated with great personal and societal burden (Merikangas et al. 2007; Bonnín et al. 2012) and the greatest risk of suicide of any psychiatric disorder (Nock et al. 2009). Oxidative stress, inflammation and neurotrophin expression are thought to contribute to the neuroprogressive changes in bipolar disorder that lead to functional deterioration (Berk 2009; Berk et al. 2011b; Fries et al. 2012). Lithium, known for its mood-stabilizing (Goodwin and Malhi 2007), anti-suicidal (Lewitzka et al. 2015) and neuroprotective actions (Malhi et al. 2013; Alda 2015), is effective in treating acute bipolar depression and mania (Malhi et al. 2011b, 2012) and remains the most effective first-line treatment for prophylaxis (Goodwin and Malhi 2007; Malhi et al. 2011a). The mechanisms underpinning lithium's therapeutic effects are unknown, but a wide range of potential targets have been found at cellular and intracellular levels (Malhi et al. 2013; Alda 2015). Understanding the mechanisms that underpin the therapeutic effects of lithium may help explain the pathophysiology of bipolar disorder and also facilitate the development of new pharmaceuticals with more specific actions, greater tolerability and fewer associated risks than lithium (Malhi et al. 2012). Finally, a deeper understanding of the effects of lithium on neuroprogression may have application to, and afford new treatment targets across, many disorders, with effects on neurodegenerative disorders already proving promising (Chuang and Manji 2007; Marmol 2008; Mauer et al. 2014).

This chapter reviews many of the key actions of lithium and provides an integrated framework for examining its therapeutic effects in bipolar disorder (see Fig. 3.1). To achieve this, we first review the evidence that suggests that neuroprogression occurs in bipolar disorder—specifically that it produces progressive

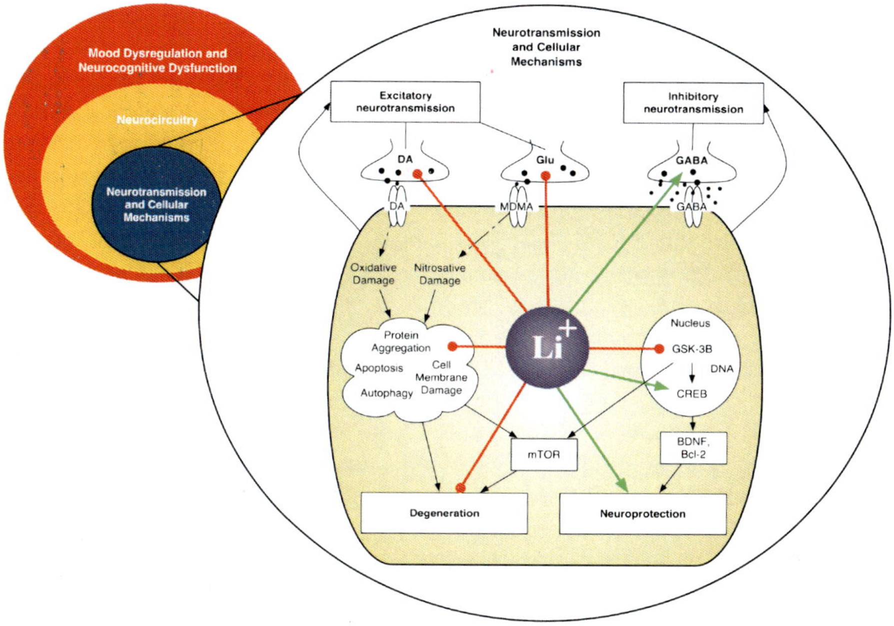

Fig. 3.1 A framework for understanding the impact of lithium treatment for bipolar disorder. Lithium inhibits (*red arrows*) and facilitates (*green arrows*) a number of neurotransmitter and cellular signal transduction mechanisms involved in degeneration and neuroprotection. In terms of key mechanisms, lithium inhibits neuroprogression through the excitotoxic damage-GSK-3β-mTOR and excitotoxic damage-GSK-3β-CREB-neurotrophin encoding pathways involved in neural degeneration and neuroprotection, where excitotoxic damage has occurred through oxidative and nitrosative damage pathways. These mechanisms are framed within the wider context of the therapeutic effects of lithium on the neurocircuitry dysfunction and mood dysregulation and neurocognitive dysfunction characteristic of bipolar disorder. This simplified framework is based on a previous review (see Malhi et al. 2013; 2016a)

changes within interacting neurotransmission and cellular signal transduction pathways. The evidence of how lithium acts on these systems is also examined in detail, with particular emphasis on glycogen synthase kinase-3β (GSK-3β)-mediated mechanisms because of their pivotal roles in both neuroprogression and the actions of lithium. Finally, discussion turns to how these neuroprogressive changes can alter neural circuitry functioning implicated in bipolar disorder and potentially how lithium produces clinical mood stability and restores neurocognitive function.

3.2 Neuroprogression, Bipolar Disorder and Lithium Treatment

Proponents of neuroprogression perspectives (e.g. Frank et al. 2015) view bipolar disorder as an illness that has deteriorating course, with progressive changes across multiple systems. Lithium has multiple potential actions that counteract neuroprogression and lead to therapeutic benefits. The process of neuroprogression and the effects of lithium treatment on this process can be examined at multiple levels of 'magnification', ranging from microscopic changes in cellular functions to macroscopic changes that manifest clinically (Fries et al. 2012; Malhi et al. 2013; see Fig. 3.2). Many interacting neurotransmitter and cellular signal transduction processes underlie the pathophysiology of bipolar disorder and its responses to therapy, but their influences are distal from bipolar disorder's characteristic mood dysregulation, neurocognitive dysfunction and clinical presentation. Therefore, it is important to consider intermediate levels, such as changes to neurocircuitry, that may occur with neuroprogression of the illness and in response to therapy (Nock et al. 2009; Berk et al. 2011b; Fries et al. 2012; Malhi et al. 2013). It is likely that through a series of interconnected pathways involving changes in neurotransmission and cellular and neurocircuitry mechanisms, lithium exerts its therapeutic effects as a mood stabilizer and serves as a prophylactic treatment (Berk 2009; Berk et al. 2011b; Fries et al. 2012; Malhi et al. 2013).

Clinically, treatment resistance in bipolar disorder is correlated to the number of episodes the individual has experienced—in other words the total illness burden and therefore lithium treatment response is likely to be a function of the level of

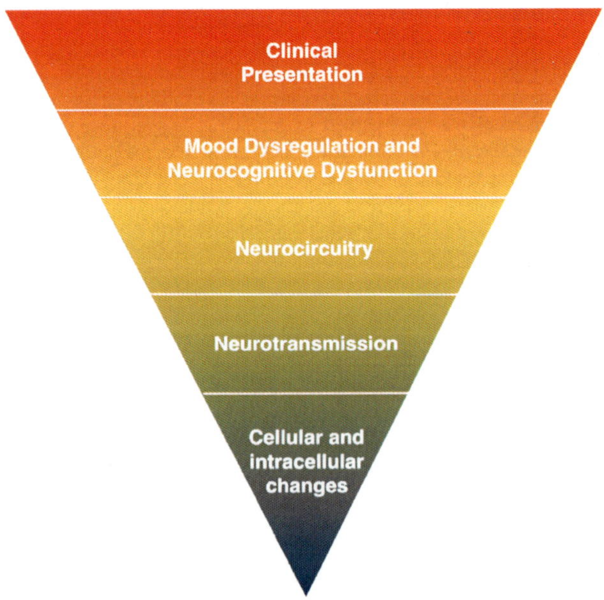

Fig. 3.2 Neuroprogression and the effects of lithium treatment in bipolar disorder can be understood from multiple levels of understanding, from cellular and intracellular changes to clinical presentation. To understand how neuroprogression and therapeutic changes at the cellular level result in changes in clinical presentation, the more intermediate changes at the neurotransmission, neurocircuitry and mood dysregulation and neurocognitive dysfunction levels are considered. It is important to consider these intermediate changes as they are more proximal mechanisms to both cellular and intracellular changes and clinical presentation (Malhi et al. 2013)

underlying neuroprogression that has already occurred (Berk et al. 2011a; Malhi et al. 2011b, 2012). Interestingly, lithium attenuates neural degradation and augments neuroprotection pathways by directly acting on deficits in neurotransmission and cellular signal transduction processes (Alda 2015), and these effects ultimately lead to changes in the structure and function of neurocircuitry implicated in mood dysregulation and neurocognitive dysfunction in bipolar disorder (Goodwin and Malhi 2007; Tsaltas et al. 2009; Malhi et al. 2011a, 2013; see Fig. 3.1).

Key neurotransmitters implicated in neuroprogressive changes in bipolar disorder include dopamine, glutamate and γ-aminobutyric acid (GABA; Berk 2009; Berk et al. 2011b; Malhi et al. 2013). Dopamine is an excitatory neurotransmitter thought to be involved in reward systems and cyclical mood dysregulation (Post et al. 1980; Staunton et al. 1982; Jacobs and Silverstone 1986; Berk et al. 2007; Cousins et al. 2010; Malhi et al. 2012). Glutamate is also an excitatory neurotransmitter but when in excess can produce toxic effects, and notably brain concentrations of glutamate have been found to be elevated during mania (Michael et al. 2003; Chuang and Manji 2007; Marmol 2008; Mauer et al. 2014). In contrast, GABA is an inhibitory neurotransmitter that acts to balance and regulate both dopamine and glutamate neurotransmission, and hence it is thought to contribute to mood stabilization (Brambilla et al. 2003; Kato 2008; Ng et al. 2009; Berk 2009; Ghasemi and Dehpour 2011; Berk et al. 2011b). In bipolar disorder, over-excitation of dopaminergic and glutamatergic pathways, along with diminished inhibition from underactive GABAergic pathways, may be the cause of oxidative and nitrosative damage in postsynaptic cells (Berk et al. 2011b). Countering these actions in bipolar disorder, lithium inhibits excitotoxic degradation and facilitates the availability of neuroprotective factors (Chen and Chuang 1999; Marmol 2008; Chiu and Chuang 2010; Malhi et al. 2013), stabilizing perhaps not only acute episodes of illness but also limiting further neuroprogression of the illness. The process of neuroprogression that leads to degenerative cellular changes and decreases in neuroprotective factors and the opposing actions of lithium are discussed in greater detail in the following sections.

3.3 Neurotransmission

Bipolar disorder is associated with changes in a wide range of neurotransmitters that eventuate in neural dysfunction. Lithium acts both pre- and postsynaptically to modulate neurotransmission (see Fig. 3.1), and though multiple interactions between neurotransmission pathways are implicated, the modulation and regulation of dopamine (Post et al. 1980; Staunton et al. 1982; Jacobs and Silverstone 1986; Berk et al. 2007; Cousins et al. 2010), glutamate (Dixon and Hokin 1998; Berk 1999; Brunello and Tascedda 2003; Ongür et al. 2008; Ghasemi and Dehpour 2011) and GABA (Ahluwalia et al. 1981; Vargas et al. 1998; Brambilla et al. 2003; Brunello and Tascedda 2003; Kato 2008; Ng et al. 2009; Ghasemi and Dehpour 2011) are thought to be key to the therapeutic effects of lithium (Malhi et al. 2013).

3.3.1 Dopamine

Dopamine agonists and antagonists produce robust manic and anti-manic behavioural effects (Berk et al. 2011b). An increase in dopaminergic activity in bipolar disorder produces oxidative damage to postsynaptic cells, because this increases the concentration of potentially damaging dopamine metabolites (Berk 2009; Berk et al. 2011b; Malhi et al. 2013). Increased dopamine requires increased metabolism via monoamine oxidase with the resulting production of hydrogen peroxide (H_2O_2) and dihydroxyphenylacetic acid (Maker et al. 1981; Berman and Hastings 1999). However, the metabolism of dopamine can also occur through non-enzymatic hydroxylation with iron (Fe^{2+}) and water (H_2O_2) leading to the formation of free radicals, which in turn induces oxidation of DNA, proteins and lipids (Graham et al. 1978). In animal models, lithium decreases extracellular dopamine, which in turn decreases reactivity to harmful stimuli (Gambarana et al. 1999; Ichikawa et al. 2005). When lithium treatment is withdrawn, dopamine levels increase and remain elevated for several days (Ferrie et al. 2005). Thus, regulation of dopamine may explain both treatment response and relapse in patients as lithium levels move into and out of the therapeutic window (Malhi et al. 2011b).

The action of lithium on dopaminergic pathways is not only evident in normalizing presynaptic neurotransmission but also affects postsynaptic actions. G-protein-coupled dopamine receptors stimulate cellular signal transduction mechanisms (discussed in the next section), which initiate a cascade of processes that modulate dopamine neurotransmission. Chronic lithium administration alters the function of G-protein active and inactive subunits (Manji and Lenox 2000a), modulating the stimulation of the transduction mechanisms. The aforementioned oxidation pathway activates the production of GSK-3β (Stokes et al. 1999; Obata 2002; Grima et al. 2003), which is a component of the cell survival-promoting signalling pathway (discussed in the next section; Gould et al. 2006). Along with dopaminergic-mediated oxidative damage leading to compromised anti-oxidation, glutamatergic-mediated nitrosative damage induces membrane damage, protein aggregation and apoptosis initiation (Andreazza et al. 2009; Ghasemi and Dehpour 2011).

3.3.2 Glutamate

In bipolar disorder, increased glutamatergic neurotransmission in bipolar disorder increases cellular damage via the activation of the *N*-methyl-D-aspartic acid (NMDA) receptor (Berk et al. 2000; Plein and Berk 2000; Zuo et al. 2007; Zou et al. 2010; Diazgranados et al. 2010), which facilities calcium (Ca^{2+}) influx (Nonaka et al. 1998; Bown et al. 2003) and consequent nitric oxide (NO) production (another highly reactive free radical). This leads to nitrosative damage to DNA, proteins and lipids (Andreazza et al. 2009; Ghasemi and Dehpour 2011). Glutamate is usually cleared by astrocytes, where it is converted to glutamine

(Choi et al. 1987; Berk et al. 2000, 2011b; Kato 2007), but its clearance may be reduced by the effects of inflammatory cytokines on astrocytes (Zou et al. 2010). Bipolar disorder patients have elevated intracellular Ca^{2+} mediated in part by increase glutamate receptor activity. Raised levels of glutamate have been associated with neuronal death and are thus also implicated in the pathogenesis of neurodegenerative diseases (Zou et al. 2010) including Alzheimer's disease (Robinson 2000) and Parkinson's disease (Zipp et al. 1998; Zoghbi and Orr 2000; Orr et al. 2002).

Lithium is known to act on glutamatergic pathways and this may partly explain its long-term mood-stabilizing properties (Dixon and Hokin 1998; Brunello and Tascedda 2003; Ghasemi and Dehpour 2011). Specifically, lithium selectively competes with magnesium at binding sites on NMDA glutamate receptors (Tsapakis and Travis 2002), leading to acute stimulation, which in turn increases the availability of glutamate in the postsynaptic neuron (Hokin et al. 1996). However, with chronic lithium administration, glutamate transmission is restored by directly downregulating the NMDA receptor and increasing glutamate reuptake. Given that lithium downregulates glutamate neurotransmission at the NMDA receptor, it is able to ameliorate the damaging excitotoxic effects of Ca2+ influx via the aforementioned NO-nitrosative pathway (Nonaka et al. 1998; Bown et al. 2003). Interestingly, other NMDA antagonists also have rapid effects on mood (Diazgranados et al. 2010).

3.3.3 GABA

The final neurotransmitter pathway that plays a significant role in the neuroprogression of bipolar disorder is the GABAergic pathway. Patients with bipolar disorder have decreased GABAergic neurotransmission, decreasing inhibitive signalling (Lenox et al. 1998; Shiah and Yatham 1998; Brambilla et al. 2003; Kato 2008; Ng et al. 2009). Low GABA levels result in an increase in excitatory dopaminergic and glutamatergic neurotransmission that, as previously discussed, can lead to excitotoxicity (Rajkowska 2002). This again contributes to activity within the oxidative and nitrosative pathways that results in apoptosis and cell loss (Rajkowska 2002).

Lithium modifies GABAergic neurotransmission, and increases in GABAergic transmission occur in relation to lithium's long-term mood-stabilizing effects. Specifically, lithium increases the level of GABA in the plasma and cerebrospinal fluid (Ahluwalia et al. 1981; Vargas et al. 1998; Brunello and Tascedda 2003). Given that increasing GABA levels decreases excitatory neurotransmission, lithium's effects on GABA also promote the release of neuroprotective proteins (Chuang et al. 2002), and an increase in GABA, in response to lithium, reduces the level of glutamate, which further downregulates NMDA receptor activity (Ghasemi and Dehpour 2011). Hence, long-term mood stabilization effects of lithium stem from a combination of its effects on GABA and the excitatory dopaminergic and glutamatergic neurotransmitter systems.

3.4 Cellular Signal Transduction Mechanisms

Cellular signal transduction mechanisms are involved in the process of neural degradation and neuroprotection in bipolar disorder (see Fig. 3.1). However, these processes themselves are also adversely impacted by the illness. The actions of lithium on several interacting cellular signal transduction mechanisms are thought to contribute to its therapeutic effects (Malhi et al. 2013; Alda 2015).

Centred on GSK-3β, there are some key therapeutic mechanisms involving neural degradation and neuroprotection processes that curb neuroprogression. GSK-3β is an enzyme that regulates glycogen synthesis and has direct involvement in the genetic transcription of neuroprotective factors (Ikonomov and Manji 1999; Berk 2009; Machado-Vieira et al. 2009; Berk et al. 2011b). It is a downstream target of monoaminergic systems and is thus implicated in mood regulation and neurocognition (Einat et al. 2006). During mania, GSK-3β is activated via dopamine overactivity (Beaulieu et al. 2004), and interestingly manipulations of GSK-3β can cause hyperactivity in animal models (Prickaerts et al. 2006). In the context of bipolar disorder and the effects of lithium, two key GSK-3β-mediated pathways will be discussed in the following sections (see Fig. 3.1).

3.4.1 The Excitotoxic Damage-GSK-3β-mTOR Cell Survival Pathway

GSK-3β is a component of the cell survival signalling pathway, which plays a critical role in multiple processes, including metabolism, proliferation, differentiation, cell structure, axogenesis and synaptogenesis (Ikonomov and Manji 1999; Gould et al. 2006; Berk 2009; Machado-Vieira et al. 2009; Berk et al. 2011b). Oxidative and nitrosative damage-mediated excitotoxicity activates the production of GSK-3β (Stokes et al. 1999; Obata 2002; Grima et al. 2003). However, lithium directly inhibits GSK-3β activity (Klein and Melton 1996; Chalecka-Franaszek and Chuang 1999; Manji and Lenox 2000b) and enhances inhibition of GSK-3β activity through other cellular process (Chalecka-Franaszek and Chuang 1999; Grimes and Jope 2001). GSK-3β inhibition activates the mammalian target of rapamycin (mTOR), which is an inhibitor of the autophagy process (Sarkar et al. 2008). Therefore, lithium inhibition of GSK-3β increases mTOR activity, which decreases neural degradation and increases neuroprotection through the facilitation of the Akt pathway (Tajes et al. 2009). These actions of lithium, therefore, inhibit autophagy and are likely to be protective for neurocognitive dysfunction. In addition to being involved in the pathophysiology of bipolar disorder, the excitotoxic damage-GSK-3β-mTOR pathway is implicated in the pathophysiology of neurodegenerative disorders (Parlato and Liss 2014) such as Alzheimer's (Hooper et al. 2008; Pei and Hugon 2008) and Parkinson's (Liu et al. 2013; Jiang et al. 2013; Li et al. 2014), with GSK-3β inhibition and mTOR being potential therapeutic targets. In addition to the impact of GSK-3β on the mTOR pathway, the inhibition of GSK-3β leads to modulation of neurotrophin expression.

3.4.2 The Excitotoxic Damage-GSK-3β-CREB-Neurotrophin Encoding Pathway

Neuroprogression in bipolar disorder occurs because of the excitotoxic consequences of recurrent affective episodes and reductions in cellular resiliency with time (Manji and Chen 2002; Rajkowska 2002). The excitotoxic damage-GSK-3β-CREB-neurotrophin encoding pathway is implicated in the neuroprogression of bipolar disorder and its effective treatment with lithium (Berk et al. 2011b; Malhi et al. 2013). GSK-3β inhibits the transcription factors β-catenin and cyclic AMP (cAMP) response element binding protein (CREB), by phosphorylation. This results in a decrease in the transcription of important genes encoding for neurotrophins (Gould et al. 2006; Böer et al. 2008; Berk et al. 2011b). Neurotrophins such as brain-derived neurotrophic factor (BDNF) and B-cell lymphoma 2 (Bcl-2) play a key role in neuronal survival and proliferation, and these are implicated in the neuroprogressive pathways of bipolar disorder (Kim et al. 2010; Berk et al. 2011b). BDNF is decreased in acute episodes of mania and depression (Cunha et al. 2006; Kapczinski et al. 2008; de Oliveira et al. 2009), and decreases in the BDNF levels correlate with increasing severity of clinical episodes (Cunha et al. 2006; Machado-Vieira et al. 2007; Fernandes et al. 2009). BDNF levels decrease during acute episodes, with further decrements as the disorder advances (Kapczinski et al. 2008), thereby perpetuating neuroprogression (Post 2007). However, in animals it has been shown that after 5 days of lithium administration, BDNF levels increase (Hashimoto et al. 2002) and, in patients, this delay is thought to reflect the time required for BDNF to reach neuroprotective levels and achieve mood-stabilizing and neurocognitive effects (Einat et al. 2006). This is supported by the finding that lithium nonresponders have lower BDNF levels than both lithium responders and healthy controls, suggesting further that a specific therapeutic effect of lithium is to facilitate BDNF (Rybakowski and Suwalska 2010). This is also in keeping with the finding that patients who responded to treatment during a manic episode had increased BDNF after the resolution of the episode (Tramontina et al. 2009). Both lithium's direct and indirect effects on GSK-3β facilitate BDNF expression through CREB activation (Tramontina et al. 2009), with lithium's additional actions mediated via cAMP (Lien et al. 2008; Manji et al. 2000; Chang et al. 2009).

In addition to BDNF, Bcl-2 is a neuroprotective protein that regulates cellular apoptotic pathways, preventing degradation (Manji et al. 2000) and, like BDNF, decreases in Bcl-2 are associated with mania (Lien et al. 2008). Interestingly, increased CREB increases Bcl-2 expression (Lien et al. 2008; Manji et al. 2000; Chang et al. 2009), thereby providing another action that facilitates neuroprotection. Similarly, chronic lithium therapy increases Bcl-2 expression (Chen and Chuang 1999), which reduces apoptosis (Ghribi et al. 2002). Therefore, increased BDNF and Bcl-2 expression with lithium treatment is thought to be a key mechanism by which it provides neuroprotection against glutamate-induced excitotoxicity (Hashimoto et al. 2002). However, the system is complicated and multifaceted given that reduced activation of the G-protein coupled dopamine receptor and the NMDA receptor regulate the second messenger excitotoxic damage-GSK-3β-mTOR and excitotoxic damage-GSK-3β-CREB-neurotrophin encoding pathways (Jope 1999; Manji and Lenox 2000a; Gould et al. 2002; Brunello and Tascedda 2003; Montezinho et al. 2007; Marmol 2008).

Hence, though the various potential cellular signal transduction mechanisms underpinning the actions of lithium on neural degradation and neuroprotection have been reasonably well characterized (see Malhi et al. 2016a), the precise pathways that are important in producing specific therapeutic outcomes remain unknown (Malhi et al. 2013). The manner in which these pathways possibly converge and produce specific therapeutic outcomes is also unclear. However, it appears that GSK-3β is fundamental to the therapeutic actions of lithium in terms of inhibiting neuroprogression in bipolar disorder (Freland and Beaulieu 2012; Ikonomov and Manji 1999; Machado-Vieira et al. 2009; Berk et al. 2011b; Malhi et al. 2013). However, the manner in which GSK-3β inhibition translates to clinical outcomes, or can be an effective sole treatment target, is unknown (Eldar-Finkelman 2002). Therefore, it is perhaps useful to consider the effects of lithium on neurocircuitry and how they relate to neuroprogression in bipolar disorder as an intermediate level of understanding between cellular changes and clinical outcomes.

3.5 Neurocircuitry

In addition to considering neurotransmitter and cellular signal transduction mechanisms that putatively underlie neuroprogression in bipolar disorder, it is important to also consider the structure and function of brain neurocircuitry, given that this is necessarily more proximal to clinical dysfunction of neurocognition and the capacity to regulate mood (see Fig. 3.2; Berk et al. 2011b; Fries et al. 2012; Malhi et al. 2013, 2015).

Bipolar disorder is associated with structural and functional abnormalities involving structures in the fronto-limbic network, which have implications for deficits in neurocognition and mood dysregulation (Emsell and McDonald 2009; Strakowski et al. 2012; Phillips et al. 2015; Malhi et al. 2015).

3.5.1 Structure

Bipolar disorder patients have reduced grey matter in the prefrontal cortex (Brooks et al. 2009) and anterior limbic regions (Bora et al. 2010). The reductions are widespread and include areas of the dorsomedial and ventromedial prefrontal cortex, the anterior cingulate cortex, insula and hippocampus. The rate of grey matter volume reduction is more rapid in bipolar disorder patients compared with controls (Brambilla et al. 2001; Lyoo et al. 2006) and different in some key regions such as the hippocampus relative to major depressive disorder (Wise et al. 2016). Grey matter loss is attributed to episodic excitotoxicity (Yksel and Öngür 2010) and subsequent oxidative and nitrosative damage (Rajkowska 2002), which is reflected in reduced neuronal densities (Ongür et al. 1998; Rajkowska et al. 2001). These findings are consistent with the neurotransmission and cellular signal transduction findings that suggest bipolar disorder is a neuroprogressive disorder (Berk et al. 2011b; Fries et al. 2012). Prefrontal and limbic structural abnormalities in bipolar disorder are thought to be related to executive control, emotion processing and mood regulation difficulties and other neurocognitive function abnormalities observed clinically (Bora et al. 2009, 2010; Berk et al. 2011b).

Effective lithium treatment normalizes the fronto-limbic network structural deficits implicated in bipolar disorder. Lithium-treated patients, compared with non-treated patients or healthy controls, show increases in total (Sassi et al. 2002; Bearden et al. 2007; Kempton et al. 2008) and regional prefrontal (Adler et al. 2005; Bearden et al. 2007) and limbic grey matter volume (Yucel et al. 2008; Foland et al. 2008; Usher et al. 2010; Bora et al. 2010; Hallahan et al. 2011; Hajek et al. 2012; van Erp et al. 2012), with increases in prefrontal grey matter volume shown to be specifically associated with lithium response (Moore et al. 2009). This is supported by the finding in rats that chronic administration of lithium promotes proliferation of neurons in the hippocampus (Chen et al. 2000). These findings clearly highlight the potential neuroprotective effects of lithium (Kempton et al. 2008).

3.5.2 Function

Patients with bipolar disorder display prefrontal cortex hypoactivity and limbic hyperactivity during emotional and cognitive tasks, and these findings correlate with trait and state emotional lability and mood disturbances (see Strakowski et al. 2012). In addition, bipolar disorder is characterized by dysfunctional connectivity among ventral prefrontal networks and limbic brain regions, particularly the amygdala (Chen et al. 2011; Houenou et al. 2011; Townsend and Altshuler 2012; Blond et al. 2012; Strakowski et al. 2012), indicating both difficulty in regulating mood and a dysfunction of emotion processing. Impaired prefrontal cortex regulation subsequently leads to a loss of neurological emotional homeostasis, emotional lability and mood disturbances (Strakowski et al. 2012). It is posited that a disruption of frontal regulatory networks allows for extreme mood states, switching among mood states and mixed states. These abnormalities have been conceptualized as dysfunction within oscillatory mechanisms, which perhaps worsen over time, and result in the many manifestations of the illness (Schneider et al. 2012; Malhi et al. 2015).

In comparison to the impacts of neuroprogression on the neurocircuitry of bipolar disorder, the impacts of lithium therapy are relatively understudied. In functional neuroimaging studies, lithium administration is associated with prophylactic effects on neurocircuitry after 14 days of treatment, acting on frontal regions in the euthymic phase of bipolar disorder with little impact during the depressed phase (Silverstone et al. 2005; Bell et al. 2005a, b). There appears to be normalization of the cognitive and emotional neural networks implicated in bipolar disorder with treatment when patients present in the depressed or manic phases of illness (Silverstone et al. 2005; Bell et al. 2005a, b). Thus, the impacts of lithium on neural processing are broad, yet the observed effects are dependent on both mood state and the task undertaken (Silverstone et al. 2005; Bell et al. 2005a, b). However, this line of research is preliminary given the relatively small sample sizes recruited by most studies to date. In terms of neurocognitive assessment-based findings, we have found that the benefits of lithium for neurocognitive functions—such as executive function, attention, memory and learning and processing and psychomotor speed—are mixed and need to be clarified and replicated in longitudinal research (for review, see Malhi et al. 2016b).

Nevertheless, there are promising findings from these relatively new lines of research that should encourage future work. Such research may provide further insights into the impacts of lithium on the neurocircuitry affected by neuroprogression (Malhi et al. 2013) and afford clinically useful markers for predicting treatment response and treatment monitoring (see Outhred et al. 2014).

The neuroprotective effects of lithium in bipolar disorder patients have been proposed on the basis of preservation of grey matter volume and functional changes within fronto-limbic neurocircuitry. However, the mechanisms linking these changes to neurotransmission and cellular signal transduction changes, and the significance of these for mood dysregulation, are not fully understood. Given the broad effects of lithium on neurotransmission, cellular signal transduction mechanisms and brain neurocircuitry, determining the specific pathways leading to beneficial therapeutic outcomes with lithium through changes in neural functioning poses the next challenge for research in this domain.

3.6 The Impact of Lithium on Mechanisms Underlying Neuroprogression in Bipolar Disorder

Given that lithium interacts with multiple targets across a number of neurotransmitter systems, more sophisticated and integrated investigations are required to unravel the complexity of its actions on neurotransmission. As such, it is difficult to pinpoint the specific mechanisms that are involved in producing the therapeutic effects of lithium for bipolar disorder. With further elucidation of these processes, it is envisaged that the potential targets for lithium and the development of novel medications for bipolar disorder, as well as other neurodegenerative disorders, will be realized (Chuang 2005; Chuang and Manji 2007; Marmol 2008). Regardless, it is likely that the actions of lithium on modulating dopaminergic, glutamatergic and GABAergic neurotransmission, which in turn impact cellular signal transduction pathways, are involved in inhibiting neurodegeneration and promoting neuroprotection, key benefits in themselves.

Based on extant research, the impacts of lithium on neurotransmission and cellular signal transduction mechanisms underlying neuroprogression can perhaps be best understood in the context of its effects on the neurocircuitry involved in mood dysregulation and neurocognitive dysfunction that characterize bipolar disorder. In a simplified manner, Fig. 3.1 depicts the actions of lithium in terms of inhibition and facilitation of several neurotransmitters and cellular signal transduction mechanisms involved in neurodegeneration and neuroprotection. Key actions appear to be the attenuation of excitotoxic dopaminergic and glutamatergic neurotransmission and facilitation of inhibitory GABAergic neurotransmission and the consequent and direct modulation of excitotoxic damage-GSK-3β-mTOR and excitotoxic damage-GSK-3β-CREB-neurotrophin encoding pathways. These mechanisms are thought to underpin the therapeutic effects of lithium that ultimately ameliorate the neurocircuitry dysfunction and mood dysregulation and neurocognitive functions that occur in bipolar disorder. Presently, much more is known about the changes to the neurotransmitter and cellular signal transduction pathways with lithium treatment than is known about how these changes then translate to clinical outcomes (Malhi et al. 2013). By examining the specific effects of integrating them within a framework of therapeutic changes that rectify mood dysregulation and neurocognitive dysfunction, future research may be guided towards a more complete depiction of the pathophysiology of bipolar disorder and its modification with lithium therapy.

3.7 Future Research

Given that lithium has broad effects on multiple processes that are potentially therapeutic (Malhi et al. 2013; 2016a), the task for researchers is to isolate those processes that lead to therapeutic response. By investigating each level and every process within a level (see Fig. 3.2), future research should be able to correlate neurotransmission, cellular signal transduction pathway and neurocircuitry changes to therapeutic improvement, specifically restoration of neurocognition. In doing so, the aetiology of bipolar disorder and its lifelong neuroprogression is likely to be better understood.

For lithium treatment, researchers need to associate changes in markers of neurotransmission and cellular changes with changes in the structure and function of key neurocircuits (Malhi et al. 2016a). Research into the precise effects of lithium treatment on neurocircuitry underpinning mood dysregulation and neurocognition is currently lacking. Future research along these lines may also provide valuable insights into the anti-suicidal and neuroprotective properties of lithium along with its mood-stabilizing effect and prophylaxis. Determining the specific pathways leading to therapeutic outcome in patients may also open up avenues for the development of more specific treatment targets. In doing so, the development of new pharmaceuticals that focus more specifically on neuroprogressive processes may be facilitated by employing reverse engineering (Malhi et al. 2013). Drugs that selectively target key therapeutic processes may have greater tolerability, reduced risk and wider application.

Experimental medicine models could be developed so that novel agents can be tested for likely efficacy if they induce neurotransmission, cellular changes and neurocircuitry changes underpinning treatment response. These models may be employed early in the pharmaceutical development pipeline, so that potentially effective agents can be screened for further testing. This would be advantageous for pharmaceutical development in terms of time and cost. Given GSK-3β's direct involvement in the putative neuroprogression of bipolar disorder and its pivotal role in therapeutic actions of lithium, its inhibition is an important potential pharmaceutical target that may provide greater selectivity. Similarly, because the excitotoxic damage-GSK-3β-mTOR and excitotoxic damage-GSK-3β-neurotrophin pathways appear to be extensively involved in the pathophysiology of neurodegenerative disorders (Parlato and Liss 2014) such as Alzheimer's (Hooper et al. 2008; Pei and Hugon 2008) and Parkinson's (Liu et al. 2013; Jiang et al. 2013; Li et al. 2014) disease, dissecting how lithium impacts these pathways may afford therapeutic targets for these diseases as well as for bipolar disorder (Chuang and Manji 2007).

3.8 Summary

The present chapter has reviewed the impacts of lithium on neurotransmission and cellular mechanism pathways underlying neuroprogression in bipolar disorder. Neuroprogression occurs through ongoing excitotoxicity leading to deficits in two key cellular transduction pathways: the excitotoxic damage-GSK-3β-mTOR and excitotoxic damage-GSK-3β-CREB-neurotrophin encoding pathways implicated in both neural degeneration and neuroprotection. Investigation into how the activity of these pathways leads to changes in the neurocircuitry involved in mood dysregulation and neurocognitive dysfunction will afford further insights into the aetiology of bipolar disorder. The therapeutic roles of lithium are to ameliorate compromised neurotransmission systems, cellular signal transduction pathways, neurocircuitry, mood regulation and neurocognitive function. In reviewing the impacts of lithium on the key neuroprogression pathways, it seems that both direct and indirect GSK-3β inhibition are key actions of lithium that lead to restoration of neurotransmission and correction of cellular dysfunction. These actions may be related to correcting the structure and function of neurocircuitry involved in mood regulation and neurocognition. Novel therapeutics targeting GSK-3β pathways may afford more tolerable and safer treatments with greater utility for currently untreated populations.

References

Adler CM, Levine AD, Delbello MP et al (2005) Changes in gray matter volume in patients with bipolar disorder. Biol Psychiatry 58:151–157

Ahluwalia P, Grewaal DS, Singhal RL (1981) Brain gabaergic and dopaminergic systems following lithium treatment and withdrawal. Prog Neuropsychopharmacol 5:527–530. doi:10.1016/0364-7722(81)90040-0

Alda M (2015) Lithium in the treatment of bipolar disorder: pharmacology and pharmacogenetics. Mol Psychiatry 20:661–670. doi:10.1038/mp.2015.4

Andreazza AC, Kapczinski F, Kauer-Sant'Anna M et al (2009) 3-Nitrotyrosine and glutathione antioxidant system in patients in the early and late stages of bipolar disorder. J Psychiatry Neurosci 34:263–271

Bearden CE, Thompson PM, Dalwani M et al (2007) Greater cortical gray matter density in lithium-treated patients with bipolar disorder. Biol Psychiatry 62:7–16. doi:10.1016/j.biopsych.2006.10.027

Beaulieu J-M, Sotnikova TD, Yao W-D et al (2004) Lithium antagonizes dopamine-dependent behaviors mediated by an AKT/glycogen synthase kinase 3 signaling cascade. Proc Natl Acad Sci U S A 101:5099–5104. doi:10.1073/pnas.0307921101

Bell EC, Willson MC, Wilman AH et al (2005a) Lithium and valproate attenuate dextroamphetamine-induced changes in brain activation. Hum Psychopharmacol 20:87–96. doi:10.1002/hup.665

Bell EC, Willson MC, Wilman AH et al (2005b) Differential effects of chronic lithium and valproate on brain activation in healthy volunteers. Hum Psychopharmacol 20:415–424. doi:10.1002/hup.710

Berk M (1999) Lamotrigine and the treatment of mania in bipolar disorder. Eur Neuropsychopharmacol 9(Supplement 4):S119–S123. doi:10.1016/s0924-977x(99)00025-5

Berk M (2009) Neuroprogression: pathways to progressive brain changes in bipolar disorder. Int J Neuropsychopharmacol 12:441–445. doi:10.1017/S1461145708009498

Berk M, Plein H, Belsham B (2000) The specificity of platelet glutamate receptor supersensitivity in psychotic disorders. Life Sci 66:2427–2432

Berk M, Dodd S, Kauer-Sant'Anna M et al (2007) Dopamine dysregulation syndrome: implications for a dopamine hypothesis of bipolar disorder. Acta Psychiatr Scand 116:41–49

Berk M, Brnabic A, Dodd S et al (2011a) Does stage of illness impact treatment response in bipolar disorder? Empirical treatment data and their implication for the staging model and early intervention. Bipolar Disord 13:87–98. doi:10.1111/j.1399-5618.2011.00889.x

Berk M, Kapczinski F, Andreazza AC et al (2011b) Pathways underlying neuroprogression in bipolar disorder: focus on inflammation, oxidative stress and neurotrophic factors. Neurosci Biobehav Rev 35:804–817. doi:10.1016/j.neubiorev.2010.10.001

Berman SB, Hastings TG (1999) Dopamine oxidation alters mitochondrial respiration and induces permeability transition in brain mitochondria: implications for Parkinson's disease. J Neurochem 73:1127–1137

Blond BN, Fredericks CA, Blumberg HP (2012) Functional neuroanatomy of bipolar disorder: structure, function, and connectivity in an amygdala-anterior paralimbic neural system. Bipolar Disord 14:340–355. doi:10.1111/j.1399-5618.2012.01015.x

Böer U, Cierny I, Krause D et al (2008) Chronic lithium salt treatment reduces CRE/CREB-directed gene transcription and reverses its upregulation by chronic psychosocial stress in transgenic reporter gene mice. Neuropsychopharmacology 33:2407–2415. doi:10.1038/sj.npp.1301640

Bonnín CM, Sanchez-Moreno J, Martínez-Arán A et al (2012) Subthreshold symptoms in bipolar disorder: impact on neurocognition, quality of life and disability. J Affect Disord 136:650–659. doi:10.1016/j.jad.2011.10.012

Bora E, Yücel M, Pantelis C (2009) Cognitive endophenotypes of bipolar disorder: a meta-analysis of neuropsychological deficits in euthymic patients and their first-degree relatives. J Affect Disord 113:1–20

Bora E, Fornito A, Yücel M, Pantelis C (2010) Voxelwise meta-analysis of gray matter abnormalities in bipolar disorder. Biol Psychiatry 67:1097–1105. doi:10.1016/j.biopsych.2010.01.020

Bown CD, Wang JF, Young LT (2003) Attenuation of N-methyl-D-aspartate-mediated cytoplasmic vacuolization in primary rat hippocampal neurons by mood stabilizers. Neuroscience 117:949–955

Brambilla P, Harenski K, Nicoletti M et al (2001) Differential effects of age on brain gray matter in bipolar patients and healthy individuals. Neuropsychobiology 43:242–247

Brambilla P, Perez J, Barale F et al (2003) GABAergic dysfunction in mood disorders. Mol Psychiatry 8:721–737. doi:10.1038/sj.mp.4001362

Brooks JO, Bonner JC, Rosen AC et al (2009) Dorsolateral and dorsomedial prefrontal gray matter density changes associated with bipolar depression. Psychiatry Res 172:200–204. doi:10.1016/j.pscychresns.2008.06.007

Brunello N, Tascedda F (2003) Cellular mechanisms and second messengers: relevance to the psychopharmacology of bipolar disorders. Int J Neuropsychopharmacol 6:181–189

Chalecka-Franaszek E, Chuang D-M (1999) Lithium activates the serine/threonine kinase Akt-1 and suppresses glutamate-induced inhibition of Akt-1 activity in neurons. Proc Natl Acad Sci U S A 96:8745–8750

Chang YC, Rapoport SI, Rao JS (2009) Chronic administration of mood stabilizers upregulates BDNF and bcl-2 expression levels in rat frontal cortex. Neurochem Res 34:536–541

Chen R-W, Chuang D-M (1999) Long term lithium treatment suppresses p53 and Bax expression but increases Bcl-2 expression. J Biol Chem 274:6039–6042. doi:10.1074/jbc.274.10.6039

Chen G, Rajkowska G, Du F et al (2000) Enhancement of hippocampal neurogenesis by lithium. J Neurochem 75:1729–1734. doi:10.1046/j.1471-4159.2000.0751729.x

Chen C-H, Suckling J, Lennox BR et al (2011) A quantitative meta-analysis of fMRI studies in bipolar disorder. Bipolar Disord 13:1–15. doi:10.1111/j.1399-5618.2011.00893.x

Chiu C-T, Chuang D-M (2010) Molecular actions and therapeutic potential of lithium in preclinical and clinical studies of CNS disorders. Pharmacol Ther 128:281–304. doi:10.1016/j.pharmthera.2010.07.006

Choi DW, Maulucci-Gedde M, Kriegstein AR (1987) Glutamate neurotoxicity in cortical cell culture. J Neurosci 7:357–368

Chuang D-M (2005) The antiapoptotic actions of mood stabilizers: molecular mechanisms and therapeutic potentials. Ann N Y Acad Sci 1053:195–204

Chuang D-M, Manji HK (2007) In search of the Holy Grail for the treatment of neurodegenerative disorders: has a simple cation been overlooked? Biol Psychiatry 62:4–6. doi:10.1016/j.biopsych.2007.04.008

Chuang D-M, Chen R-W, Chalecka-Franaszek E et al (2002) Neuroprotective effects of lithium in cultured cells and animal models of diseases. Bipolar Disord 4:129–136. doi:10.1034/j.1399-5618.2002.01179.x

Cousins DA, Butts K, Young AH (2010) The role of dopamine in bipolar disorder. Bipolar Disord 11:787–806

Cunha ABM, Frey BN, Andreazza AC et al (2006) Serum brain-derived neurotrophic factor is decreased in bipolar disorder during depressive and manic episodes. Neurosci Lett 398:215–219. doi:10.1016/j.neulet.2005.12.085

de Oliveira GS, Ceresér KM, Fernandes BS et al (2009) Decreased brain-derived neurotrophic factor in medicated and drug-free bipolar patients. J Psychiatr Res 43:1171–1174. doi:10.1016/j.jpsychires.2009.04.002

Diazgranados N, Ibrahim L, Brutsche NE et al (2010) A randomized add-on trial of an N-methyl-D-aspartate antagonist in treatment-resistant bipolar depression. Arch Gen Psychiatry 67:793–802. doi:10.1001/archgenpsychiatry.2010.90

Dixon JF, Hokin LE (1998) Lithium acutely inhibits and chronically up-regulates and stabilizes glutamate uptake by presynaptic nerve endings in mouse cerebral cortex. Proc Natl Acad Sci U S A 95:8363–8368

Einat H, Manji HK, Einat H, Manji HK (2006) Cellular plasticity cascades: genes-to-behavior pathways in animal models of bipolar disorder. Biol Psychiatry 59:1160–1171

Eldar-Finkelman H (2002) Glycogen synthase kinase 3: an emerging therapeutic target. Trends Mol Med 8:126–132

Emsell L, McDonald C (2009) The structural neuroimaging of bipolar disorder. Int Rev Psychiatry 21:297–313

Fernandes BS, Gama CS, Kauer-Sant'Anna M et al (2009) Serum brain-derived neurotrophic factor in bipolar and unipolar depression: a potential adjunctive tool for differential diagnosis. J Psychiatr Res 43:1200–1204

Ferrie L, Young AH, McQuade R (2005) Effect of chronic lithium and withdrawal from chronic lithium on presynaptic dopamine function in the rat. J Psychopharmacol 19:229–234. doi:10.1177/0269881105051525

Foland LC, Altshuler LL, Sugar CA et al (2008) Increased volume of the amygdala and hippocampus in bipolar patients treated with lithium. Neuroreport 19:221–224

Frank E, Nimgaonkar VL, Phillips ML, Kupfer DJ (2015) All the world's a (clinical) stage: rethinking bipolar disorder from a longitudinal perspective. Mol Psychiatry 20:23–31. doi:10.1038/mp.2014.71

Freland L, Beaulieu J-M (2012) Inhibition of GSK3 by lithium, from single molecules to signaling networks. Front Mol Neurosci 5:14. doi:10.3389/fnmol.2012.00014

Fries GR, Pfaffenseller B, Stertz L et al (2012) Staging and neuroprogression in bipolar disorder. Curr Psychiatry Rep 14:667–675. doi:10.1007/s11920-012-0319-2
Gambarana C, Ghiglieri O, Masi F et al (1999) The effects of long-term administration of rubidium or lithium on reactivity to stress and on dopamine output in the nucleus accumbens in rats. Brain Res 826:200–209. doi:10.1016/s0006-8993(99)01286-x
Ghasemi M, Dehpour AR (2011) The NMDA receptor/nitric oxide pathway: a target for the therapeutic and toxic effects of lithium. Trends Pharmacol Sci 32:420–434. doi:10.1016/j.tips.2011.03.006
Ghribi O, Herman MM, Spaulding NK, Savory J (2002) Lithium inhibits aluminum-induced apoptosis in rabbit hippocampus, by preventing cytochrome c translocation, Bcl-2 decrease, Bax elevation and caspase-3 activation. J Neurochem 82:137–145. doi:10.1046/j.1471-4159.2002.00957.x
Goodwin GM, Malhi GS (2007) What is a mood stabilizer? Psychol Med 37:609–614. doi:10.1017/S0033291706009305
Gould TD, Chen G, Manji HK (2002) Mood stabilizer psychopharmacology. Clin Neurosci Res 2:193–212. doi:10.1016/s1566-2772(02)00044-0
Gould TD, Picchini AM, Einat H, Manji HK (2006) Targeting glycogen synthase kinase-3 in the CNS: implication for the development of new treatments for mood disorders. CDT 7:1399–1409
Graham DG, Tiffany SM, Bell WR Jr, Gutknecht WF (1978) Autoxidation versus covalent binding of quinones as the mechanism of toxicity of dopamine, 6-hydroxydopamine, and related compounds toward C1300 neuroblastoma cells in vitro. Mol Pharmacol 14:644–653
Grima G, Benz B, Parpura V et al (2003) Dopamine-induced oxidative stress in neurons with glutathione deficit: implication for schizophrenia. Schizophr Res 62:213–224
Grimes CA, Jope RS (2001) CREB DNA binding activity is inhibited by glycogen synthase kinase-3 beta and facilitated by lithium. J Neurochem 78:1219–1232. doi:10.1046/j.1471-4159.2001.00495.x
Hajek T, Cullis J, Novak T et al (2012) Hippocampal volumes in bipolar disorders: opposing effects of illness burden and lithium treatment. Bipolar Disord 14:261–270
Hallahan B, Newell J, Soares JC et al (2011) Structural magnetic resonance imaging in bipolar disorder: an international collaborative mega-analysis of individual adult patient data. Biol Psychiatry 69:326–335. doi:10.1016/j.biopsych.2010.08.029
Hashimoto R, Takei N, Shimazu K et al (2002) Lithium induces brain-derived neurotrophic factor and activates TrkB in rodent cortical neurons: an essential step for neuroprotection against glutamate excitotoxicity. Neuropharmacology 43:1173–1179
Hokin LE, Dixon JF, Los GV (1996) A novel action of lithium: stimulation of glutamate release and inositol 1,4,5 trisphosphate accumulation via activation of the N-methyl D-aspartate receptor in monkey and mouse cerebral cortex slices. Adv Enzyme Regul 36:229–244. doi:10.1016/0065-2571(95)00021-6
Hooper C, Killick R, Lovestone S (2008) The GSK3 hypothesis of Alzheimer's disease. J Neurochem 104:1433–1439. doi:10.1111/j.1471-4159.2007.05194.x
Houenou J, Frommberger J, Carde S et al (2011) Neuroimaging-based markers of bipolar disorder: evidence from two meta-analyses. J Affect Disord 132:344–355. doi:10.1016/j.jad.2011.03.016
Ichikawa J, Dai J, Meltzer HY (2005) Lithium differs from anticonvulsant mood stabilizers in prefrontal cortical and accumbal dopamine release: role of 5-HT1A receptor antagonism. Brain Res 1049:182–190
Ikonomov OC, Manji HK (1999) Molecular mechanisms underlying mood stabilization in manic-depressive illness: the phenotype challenge. Am J Psychiatry 156:1506–1514
Jacobs D, Silverstone T (1986) Dextroamphetamine-induced arousal in human subjects as a model for mania. Psychol Med 16:323–329. doi:10.1017/S0033291700009132
Jiang J, Jiang J, Zuo Y, Gu Z (2013) Rapamycin protects the mitochondria against oxidative stress and apoptosis in a rat model of Parkinson's disease. Int J Mol Med 31:825–832. doi:10.3892/ijmm.2013.1280
Jope RS (1999) A bimodal model of the mechanism of action of lithium. Mol Psychiatry 4:21–25

Kapczinski F, Frey BN, Kauer-Sant'Anna M, Grassi-Oliveira R (2008) Brain-derived neurotrophic factor and neuroplasticity in bipolar disorder. Expert Rev Neurother 8:1101–1113
Kato T (2007) Mitochondrial dysfunction as the molecular basis of bipolar disorder: therapeutic implications. CNS Drugs 21:1–11
Kato T (2008) Molecular neurobiology of bipolar disorder: a disease of 'mood-stabilizing neurons'? Trends Neurosci 31:495–503
Kempton MJ, Geddes JR, Ettinger U et al (2008) Meta-analysis, database, and meta-regression of 98 structural imaging studies in bipolar disorder. Arch Gen Psychiatry 65:1017–1032
Kim HW, Rapoport SI, Rao JS (2010) Altered expression of apoptotic factors and synaptic markers in postmortem brain from bipolar disorder patients. Neurobiol Dis 37:596–603
Klein E, Melton DA (1996) A molecular mechanism for the effect of lithium on development. Proc Natl Acad Sci U S A 93:8455–8459
Lenox RH, McNamara RK, Papke RL, Manji HK (1998) Neurobiology of lithium: an update. J Clin Psychiatry 59(Suppl 6):37–47
Lewitzka U, Severus E, Bauer R et al (2015) The suicide prevention effect of lithium: more than 20 years of evidence-a narrative review. Int J Bipolar Disord 3:32. doi:10.1186/s40345-015-0032-2
Li D-W, Liu Z-Q, Chen W et al (2014) Association of glycogen synthase kinase-3beta with Parkinson's disease (review). Mol Med Rep 9:2043–2050. doi:10.3892/mmr.2014.2080
Lien R, Flaisher-Grinberg S, Cleary C et al (2008) Behavioral effects of Bcl-2 deficiency: implications for affective disorders. Pharmacol Rep 60:490–498
Liu K, Shi N, Sun Y et al (2013) Therapeutic effects of rapamycin on MPTP-induced Parkinsonism in mice. Neurochem Res 38:201–207. doi:10.1007/s11064-012-0909-8
Lyoo IK, Sung YH, Dager SR et al (2006) Regional cerebral cortical thinning in bipolar disorder. Bipolar Disord 8:65–74. doi:10.1111/j.1399-5618.2006.00284.x
Machado-Vieira R, Dietrich MO, Leke R et al (2007) Decreased plasma brain derived neurotrophic factor levels in unmedicated bipolar patients during manic episode. Biol Psychiatry 61:142–144
Machado-Vieira R, Manji HK, Zarate CAJ et al (2009) The role of lithium in the treatment of bipolar disorder: convergent evidence for neurotrophic effects as a unifying hypothesis. Bipolar Disord 11(Suppl 2):92–109
Maker HS, Weiss C, Silides DJ, Cohen G (1981) Coupling of dopamine oxidation (monoamine oxidase activity) to glutathione oxidation via the generation of hydrogen peroxide in rat brain homogenates. J Neurochem 36:589–593
Malhi GS, Chengappa KNR, Gershon S, Ghaemi SN (2011a) Atypical mood stabilizers: a new role for neuroleptics? Bipolar Disord 13:583–586. doi:10.1111/j.1399-5618.2011.00967.x
Malhi GS, Tanious M, Gershon S (2011b) The lithiumeter: a measured approach. Bipolar Disord 13:219–226. doi:10.1111/j.1399-5618.2011.00918.x
Malhi GS, Tanious M, Das P, Berk M (2012) The science and practice of lithium therapy. Aust N Z J Psychiatry 46:192–211. doi:10.1177/0004867412437346
Malhi GS, Tanious M, Das P et al (2013) Potential mechanisms of action of lithium in bipolar disorder. Current understanding. CNS Drugs 27:135–153. doi:10.1007/s40263-013-0039-0
Malhi GS, Byrow Y, Fritz K et al (2015) Mood disorders: neurocognitive models. Bipolar Disord 17(Suppl 2):3–20. doi:10.1111/bdi.12353
Malhi GS, Outhred T (2016a) Therapeutic Mechanisms of Lithium in Bipolar Diosrder; Recent Advances and Current Understanding. CNS Drugs. doi:10.1007/s40263-016-0380-1
Malhi GS, McAulay C, Gershon S, Geslier D, Fritz K, Das P, Outhred T (2016b) The Lithium Battery: assessing the neurocognitive profile of lithium in bipolar disorder. Bipolar Disord 18:102–115. doi:10.1111/bdi.12375
Manji HK, Chen G (2002) PKC, MAP kinases and the bcl-2 family of proteins as long-term targets for mood stabilizers. Mol Psychiatry 7(Suppl 1):S46–S56
Manji HK, Lenox RH (2000a) Signaling: cellular insights into the pathophysiology of bipolar disorder. Biol Psychiatry 48:518–530
Manji HK, Lenox RH (2000b) The nature of bipolar disorder. J Clin Psychiatry 61:42–57

Manji HK, Moore GJ, Chen G (2000) Lithium up-regulates the cytoprotective protein bcl-2 in the CNS in vivo: a role for neurotrophic and neuroprotective effects in manic depressive illness. J Clin Psychiatry 61:82–96

Marmol F (2008) Lithium: bipolar disorder and neurodegenerative diseases possible cellular mechanisms of the therapeutic effects of lithium. Prog Neuropsychopharmacol Biol Psychiatry 32:1761–1771

Mauer S, Vergne D, Ghaemi SN (2014) Standard and trace-dose lithium: a systematic review of dementia prevention and other behavioral benefits. Aust N Z J Psychiatry. doi:10.1177/0004867414536932

Merikangas KR, Akiskal HS, Angst J et al (2007) Lifetime and 12-month prevalence of bipolar spectrum disorder in the National Comorbidity Survey replication. Arch Gen Psychiatry 64:543–552. doi:10.1001/archpsyc.64.5.543

Michael N, Erfurth A, Ohrmann P et al (2003) Acute mania is accompanied by elevated glutamate/glutamine levels within the left dorsolateral prefrontal cortex. Psychopharmacology (Berl) 168:344–346. doi:10.1007/s00213-003-1440-z

Montezinho LP, Mørk A, Duarte CB et al (2007) Effects of mood stabilizers on the inhibition of adenylate cyclase via dopamine D2-like receptors. Bipolar Disord 9:290–297. doi:10.1111/j.1399-5618.2007.00354.x

Moore GJ, Cortese BM, Glitz DA et al (2009) A longitudinal study of the effects of lithium treatment on prefrontal and subgenual prefrontal gray matter volume in treatment-responsive bipolar disorder patients. J Clin Psychiatry 70:699–705

Ng WXD, Lau IY, Graham S, Sim K (2009) Neurobiological evidence for thalamic, hippocampal and related glutamatergic abnormalities in bipolar disorder: a review and synthesis. Neurosci Biobehav Rev 33:336–354

Nock MK, Hwang I, Sampson N et al (2009) Cross-national analysis of the associations among mental disorders and suicidal behavior: findings from the WHO World Mental Health Surveys. PLoS Med 6, e1000123. doi:10.1371/journal.pmed.1000123

Nonaka S, Hough CJ, Chuang D-M (1998) Chronic lithium treatment robustly protects neurons in the central nervous system against excitotoxicity by inhibiting N-methyl-d-aspartate receptor-mediated calciumâ€‰influx. Proc Natl Acad Sci U S A 95:2642–2647

Obata T (2002) Dopamine efflux by MPTP and hydroxyl radical generation. J Neural Transm 109:1159–1180

Ongür D, Drevets WC, Price JL (1998) Glial reduction in the subgenual prefrontal cortex in mood disorders. Proc Natl Acad Sci U S A 95:13290–13295. doi:10.1073/pnas.95.22.13290

Ongür D, Jensen JE, Prescot AP et al (2008) Abnormal glutamatergic neurotransmission and neuronal-glial interactions in acute mania. Biol Psychiatry 64:718–726. doi:10.1016/j.biopsych.2008.05.014

Orr CF, Rowe DB, Halliday GM (2002) An inflammatory review of Parkinson's disease. Prog Neurobiol 68:325–340

Outhred T, Kemp AH, Malhi GS (2014) Physiological correlates of bipolar spectrum disorders and their treatment. In: Kumari V, Boutros N, Bob P (eds) Current topics in behavioral neurosciences. Springer, Berlin, pp 1–56, 56

Parlato R, Liss B (2014) How Parkinson's disease meets nucleolar stress. Biochim Biophys Acta 1842:791–797. doi:10.1016/j.bbadis.2013.12.014

Pei J-J, Hugon J (2008) mTOR-dependent signalling in Alzheimer's disease. J Cell Mol Med 12:2525–2532. doi:10.1111/j.1582-4934.2008.00509.x

Phillips ML, Chase HW, Sheline YI et al (2015) Identifying predictors, moderators, and mediators of antidepressant response in major depressive disorder: neuroimaging approaches. Am J Psychiatry 172:124–138. doi:10.1176/appi.ajp.2014.14010076

Plein H, Berk M (2000) Changes in the platelet intracellular calcium response to serotonin in patients with major depression treated with electroconvulsive therapy: state or trait marker status. Int Clin Psychopharmacol 15:93–98

Post RM (2007) Kindling and sensitization as models for affective episode recurrence, cyclicity, and tolerance phenomena. Neurosci Biobehav Rev 31:858–873

Post RM, Jimerson DC, Bunney WE, Goodwin FK (1980) Dopamine and mania: behavioral and biochemical effects of the dopamine receptor blocker pimozide. Psychopharmacology (Berl) 67:297–305. doi:10.1007/bf00431272

Prickaerts J, Moechars D, Cryns K et al (2006) Transgenic mice overexpressing glycogen synthase kinase 3β: a putative model of hyperactivity and mania. J Neurosci 26:9022–9029. doi:10.1523/jneurosci.5216-05.2006

Rajkowska G (2002) Cell pathology in bipolar disorder. Bipolar Disord 4:105–116. doi:10.1034/j.1399-5618.2002.01149.x

Rajkowska G, Halaris A, Selemon LD (2001) Reductions in neuronal and glial density characterize the dorsolateral prefrontal cortex in bipolar disorder. Biol Psychiatry 49:741–752. doi:10.1016/s0006-3223(01)01080-0

Robinson SR (2000) Neuronal expression of glutamine synthetase in Alzheimer's disease indicates a profound impairment of metabolic interactions with astrocytes. Neurochem Int 36(4–5): 471–482

Rybakowski JK, Suwalska A (2010) Excellent lithium responders have normal cognitive functions and plasma BDNF levels. Int J Neuropsychopharmacol 13:617–622. doi:10.1017/S1461145710000404

Sarkar S, Krishna G, Imarisio S et al (2008) A rational mechanism for combination treatment of Huntington's disease using lithium and rapamycin. Hum Mol Genet 17:170–178

Sassi RB, Nicoletti M, Brambilla P et al (2002) Increased gray matter volume in lithium-treated bipolar disorder patients. Neurosci Lett 329:243–245

Schneider MR, Delbello MP, McNamara RK et al (2012) Neuroprogression in bipolar disorder. Bipolar Disord 14:356–374. doi:10.1111/j.1399-5618.2012.01024.x

Shiah I, Yatham LN (1998) GABA function in mood disorders: an update and critical review. Life Sci 63:1289–1303

Silverstone PH, McGrath BM, Kim H (2005) Bipolar disorder and myo-inositol: a review of the magnetic resonance spectroscopy findings. Bipolar Disord 7:1–10. doi:10.1111/j.1399-5618.2004.00174.x

Staunton DA, Magistretti PJ, Shoemaker WJ, Bloom FE (1982) Effects of chronic lithium treatment on dopamine receptors in the rat corpus striatum. I. Locomotor activity and behavioral supersensitivity. Brain Res 232:391–400. doi:10.1016/0006-8993(82)90282-7

Stokes AH, Hastings TG, Vrana KE (1999) Cytotoxic and genotoxic potential of dopamine. J Neurosci Res 55:659–665

Strakowski SM, Adler CM, Almeida J et al (2012) The functional neuroanatomy of bipolar disorder: a consensus model. Bipolar Disord 14:313–325. doi:10.1111/j.1399-5618.2012.01022.x

Tajes M, Yeste-Velasco M, Zhu X et al (2009) Activation of Akt by lithium: pro-survival pathways in aging. Mech Ageing Dev 130:253–261

Townsend J, Altshuler LL (2012) Emotion processing and regulation in bipolar disorder: a review. Bipolar Disord 14:326–339. doi:10.1111/j.1399-5618.2012.01021.x

Tramontina JF, Andreazza AC, Kauer-Sant'Anna M et al (2009) Brain-derived neurotrophic factor serum levels before and after treatment for acute mania. Neurosci Lett 452:111–113. doi:10.1016/j.neulet.2009.01.028

Tsaltas E, Kontis D, Boulougouris V, Papadimitriou GN (2009) Lithium and cognitive enhancement: leave it or take it? Psychopharmacology (Berl) 202:457–476

Tsapakis EM, Travis MJ (2002) Glutamate and psychiatric disorders. Adv Psychiatr Treat 8:189–197

Usher J, Menzel P, Schneider-Axmann T et al (2010) Increased right amygdala volume in lithium-treated patients with bipolar I disorder. Acta Psychiatr Scand 121:119–124

van Erp TGM, Thompson PM, Kieseppä T et al (2012) Hippocampal morphology in lithium and non-lithium-treated bipolar I disorder patients, non-bipolar co-twins, and control twins. Hum Brain Mapp 33:501–510. doi:10.1002/hbm.21239

Vargas C, Tannhauser M, Barros HMT (1998) Dissimilar effects of lithium and valproic acid on GABA and glutamine concentrations in rat cerebrospinal fluid. Gen Pharmacol Vasc Syst 30:601–604. doi:10.1016/s0306-3623(97)00328-5

Wise T, Radua J, Via E et al (2016) Common and distinct patterns of grey-matter volume alteration in major depression and bipolar disorder: evidence from voxel-based meta-analysis. Mol Psychiatry. doi:10.1038/mp.2016.72

Yksel C, Öngür D (2010) Magnetic resonance spectroscopy studies of glutamate-related abnormalities in mood disorders. Biol Psychiatry 68:785–794

Yucel K, Taylor VH, McKinnon MC et al (2008) Bilateral hippocampal volume increase in patients with bipolar disorder and short-term lithium treatment. Neuropsychopharmacology 33:361–367. doi:10.1038/sj.npp.1301405

Zipp F, Demisch L, Derouiche A, Fischer PA (1998) Glutamine synthetase activity in patients with Parkinson's disease. Acta Neurol Scand 97:300–302

Zoghbi HY, Orr HT (2000) Glutamine repeats and neurodegeneration. Annu Rev Neurosci 23:217–247. doi:10.1146/annurev.neuro.23.1.217

Zou J, Wang YX, Dou FF et al (2010) Glutamine synthetase down-regulation reduces astrocyte protection against glutamate excitotoxicity to neurons. Neurochem Int 56:577–584

Zuo DY, Wu YL, Yao WX et al (2007) Effect of MK-801 and ketamine on hydroxyl radical generation in the posterior cingulate and retrosplenial cortex of free-moving mice, as determined by in vivo microdialysis. Pharmacol Biochem Behav 86:1–7

The Effect of Lithium on Gene Expression Modulation

4

Alessio Squassina, Claudia Pisanu, and Martin Alda

Abstract

Lithium has been shown to influence the expression of hundreds of genes. Through this regulation, lithium interferes with a large number of cellular functions, including inositol metabolism, circadian rhythms, apoptosis and those neuroprotective pathways that are thought to play a role in the pathophysiology of bipolar disorder. The mechanisms through which lithium regulates gene expression are still not completely understood. However, converging data suggest that transcription factors and microRNAs, as well as epigenetic factors, may constitute key targets of lithium. In this chapter, we present and discuss the most compelling findings on lithium's effects on gene expression modulation, emphasizing those studies carried out in patient-derived cell lines, as these have the potential to highlight genes that may constitute biomarkers of the clinical efficacy of lithium.

Key Points

- The mood-stabilizing properties of lithium have largely been proved, but its mechanism of action is still not completely known.
- Lithium interferes with a large number of cellular processes and modulates the expression of hundreds of genes.

A. Squassina, PhD (✉)
Department of Biomedical Sciences, University of Cagliari, Cagliari, Italy

Department of Psychiatry, Dalhousie University, Halifax, NS, Canada
e-mail: squassina@unica.it

C. Pisanu, MD
Department of Biomedical Sciences, University of Cagliari, Cagliari, Italy
e-mail: malda@dal.ca

M. Alda, MD, FRCPC
Department of Psychiatry, Dalhousie University, Halifax, NS, Canada
e-mail: malda@dal.ca

G.S. Malhi et al. (eds.), *The Science and Practice of Lithium Therapy*,
DOI 10.1007/978-3-319-45923-3_4

- The inositol pathway is one of the best-studied targets of lithium, and in vivo and in vitro studies show that lithium directly inhibits inositol enzymes.
- It has been demonstrated that lithium is neuroprotective, a feature that partly depends on its effects on the expression of neurotrophic factors and pro- and anti-apoptotic genes.
- A growing body of evidence also shows that lithium modulates the expression of genes of the circadian system, the polyamine system and genes involved in neuronal plasticity.
- In their entirety, findings show that lithium modulates gene expression, and while the underlying mechanisms are still under investigation, interaction with epigenetics and non-coding RNAs seems to play a key role in lithium's mechanism of action.

4.1 Introduction

Lithium has been used in the management of bipolar disorder for over 50 years, being effective in the acute phases of illness (manic and depressive) and in the prevention of manic and depressive recurrences (Gitlin and Frye 2012; Malhi et al. 2012). Lithium is also effective in preventing suicidal behaviour, an effect of high importance in bipolar disorder in which suicide represents one of the leading causes of death (Baldessarini et al. 2006). We have yet to discover whether these clinical effects share the same pharmacological mechanisms.

Response to long-term lithium treatment is heritable, and a large body of evidence suggests that genetic determinants play a key role in modulating the degree of response. The first studies exploring the genetics of lithium response in bipolar disorder used the candidate gene approach. These studies explored specific biological hypotheses by genotyping polymorphisms located within or nearby genes potentially involved in the mechanism of action of lithium, or in the pathophysiology of bipolar disorder, using populations of patients with different responses to lithium and controls or families of affected individuals. The assumption of candidate gene association studies is that the polymorphisms selected for investigation could be directly or indirectly responsible for altered transcription activity or protein structure, therefore negatively affecting cell functioning. These alterations could ultimately be involved in the predisposition to the disorder or in the modulation of response to medications. Besides gene polymorphisms, candidate gene studies have also focused on the role of differences in gene expression levels in lithium response. Altered expression of a gene, or set of genes, could either be the cause or the consequence of dysregulations involved in modulating different responses to lithium treatment. Identification of genes differentially expressed between different groups of subjects acquires different biological meaning depending on the tissue used. For instance, studies of brain tissue provide information that is highly relevant for the understanding of the biological basis of a disorder, but these studies are usually not applicable for pharmacogenomics. On the other hand, peripheral tissues and patient-derived cell lines allow investigation of the effects of drugs in vivo or in vitro, and findings from these studies may lead to the identification of peripheral correlates of the phenotype, the so-called peripheral biomarkers.

Regardless of the tissue used, differential expression of genes between patients with different responses could be determined by different causes, such as the presence of polymorphisms in regulatory regions of the gene, or differences in epigenetic regulation of gene transcription. For the most part, the effects of lithium on gene expression have been investigated using animals, usually rats or mice, or human-derived cell lines. However, studies performed on patient-derived tissues represent the most valuable resource in this field, as they have the potential to identify markers of clinical response and drug targets. To date, these studies have provided intriguing findings, but a clear understanding of lithium's effects on gene expression, and how this correlates with clinical outcome, is still far from complete. A significant contribution to the field has been the advent of high-throughput technologies. These allow hypothesis-free approaches to explore the entire genome, generate new hypotheses and identify new genetic candidates for further investigation. Nowadays, DNA arrays allow the genotyping of several millions of single nucleotide polymorphisms (SNP) in one single array, providing a large set of genetic information with high coverage of the genome. On the other hand, RNA arrays (for messenger RNA or non-coding RNA) are designed to measure the expression of hundreds of thousands of genes in one single array and can be used with RNA extracted from any tissue. The number of studies applying these technologies on populations of patients characterized for lithium response is still scarce, but several international efforts are ongoing, and findings will soon be available.

In their entirety, data from candidate gene and hypothesis-generating studies in both animals and humans have provided evidence for the involvement of a number of genes and pathways in lithium's mechanism of action and in lithium response. It is well documented that lithium modulates the expression of hundreds of genes involved in key cellular processes, such as second messengers, transcription factors, elements involved in cell signalling, oncogenes, tumour-suppressor genes and many others. However, a proportion of these modifications are likely unspecific and not related to those mechanisms involved in the mood-stabilizing effects of lithium. It is therefore of crucial importance to identify which genes are key modulators of lithium's effects and how changes in their expression correlate with the clinical response. In this chapter, we outline the concepts of gene expression regulation and focus on genes and pathways for which decades of research have provided the most compelling findings. We give particular emphasis to findings from studies carried out on patient-derived tissues, as these studies have the potential to identify those genes and proteins involved in modulating the clinical efficacy of lithium.

4.2 Genes of the Inositol Phosphate Pathway

Early findings show that several enzymes of the inositol pathway are directly inhibited by lithium. As a consequence of this evidence, different strategies have been used over the years to explore and extend our understanding of the interaction between lithium and inositol and how this may correlate with clinical response to lithium treatment. To date, only a few hypothesis-based studies have investigated lithium's effects on the expression of inositol genes in patient populations, reporting for the most part negative findings. In this section we will discuss the inositol pathway and present findings from animal and human studies showing lithium's effects on inositol homeostasis, as elements of this pathway are also key up- or downstream effectors of other lithium targets, as discussed in later sections of this chapter.

Lithium directly interacts with enzymes of the inositol pathway. This knowledge led to the development of the 'inositol-depletion hypothesis' as a key component of the mechanism of action of lithium (Berridge 1989). The inositol phosphate pathway is a second messenger system with a crucial role in signal transduction. It comprises the lipid phosphatidylinositol 4,5-bisphosphate (PIP2) (a minor component of the cell membrane) and two second messengers derived from its hydrolysis: diacylglycerol (DAG) and inositol trisphosphate (IP3). Hydrolysis of PIP2 is catalyzed by the membrane enzyme phospholipase C (PLC) and is regulated by neurotransmitters such as serotonin and acetylcholine. DAG is located in the cell membrane, where it activates protein kinase C (PKC), while IP3 is a soluble molecule able to diffuse into the cell and increase intracellular calcium concentration. IP3 is metabolized to myo-inositol, which is recycled to synthesize new PIP2. Lithium at therapeutic concentrations directly inhibits two key enzymes of inositol metabolism: inositol monophosphatase (IMPase) and phosphatase inositol-1,4 bisphosphate 1-phosphatase (IPPase). IMPase inhibition causes a reduction of myo-inositol concentration (Allison and Stewart 1971) and, as a result, decreases inositol recycling.

According to the inositol-depletion hypothesis, lithium may exert its therapeutic effect in pathological states (leaving subjects with normal behaviour unaffected) via its uncompetitive inhibition of IMPase, which is more pronounced with high substrate concentration, as in the case of a pathologically overactivated system (Fauroux and Freeman 1999). Another hypothesis is that lithium-induced inositol depletion could be mediated by inhibition of glycogen synthase kinase 3 beta (GSK3B). Using the yeast *Saccharomyces cerevisiae* with a null mutation for GSK3B, Azab and coworkers showed that GSK3B is required for optimal inositol synthesis and loss of GSK3B activity leads to inositol depletion (Azab et al. 2007).

Interestingly, inhibition of the inositol pathway by lithium was also suggested by a gene expression study carried out in lymphoblasts from bipolar patients responsive to lithium and from healthy controls. Using a hypothesis-free approach, the study by Sun and coworkers (Sun et al. 2004) measured the expression of 2,400 cDNAs in lymphoblasts cultured with or without lithium chloride. Alpha1B-adrenoceptor (a1B-AR) was one of the genes most significantly downregulated by lithium exclusively in the bipolar sample. A1 adrenergic receptors are members of the G protein-coupled receptor superfamily. These receptors activate mitogenic responses and regulate growth and proliferation through the Gq/11 family of G proteins coupled with phospholipase C (PLC). Activation of PLC hydrolyses phosphatidylinositol 4,5-biphosphate (PIP2) to yield DAG and IP3, resulting in activation of PKC and release of intracellular calcium. The lithium-induced reduction of a1B-AR reported by Sun and colleagues may lead to a depletion of inositol levels, as it could be responsible for a reduction of inositol-generated second messengers. These findings suggest that lithium may regulate the inositol pathway not only via a direct inhibition of IMPase and IPPase but also by inhibition of a1B-AR expression.

In clinical studies, findings so far have been inconclusive: lithium has been reported to decrease brain myo-inositol levels in the right frontal lobe of bipolar patients (Moore et al. 1999), but other studies do not support this finding (Silverstone and McGrath 2009).

In summary, the interaction of lithium with the inositol pathway is one of the most corroborated hypotheses of lithium's mechanism of action, but to date there has been no compelling evidence that this system is implicated in modulating the clinical response to lithium treatment in bipolar patients.

4.3 Genes Involved in Apoptosis and Neuroprotection

The neuroprotective effects of lithium have been widely reported. These effects are mediated by the prevention of apoptotic-dependent cellular death through the regulation of neurotrophic factors and different signalling pathways. Data from animal and human studies show that modulation of the expression of genes implicated in neuronal growth is a key mechanism in lithium-induced neuroprotection. In this section we present converging evidence of the neuroprotective effect of lithium, focusing on studies showing lithium regulation of the expression of genes involved in neuroprotection, apoptosis and other neuronal functions.

It has been shown that lithium treatment interferes with glutamate activity in neurons.

Modulation of glutamate activity is an important mechanism in neuroprotection, as excessive release of glutamate induces excitotoxicity. Glutamate-induced excitotoxicity has been implicated in many neurodegenerative diseases. In this process, stimulation of N-methyl-D-aspartate (NMDA) receptors by glutamate causes an excessive influx of calcium into the cells, thus activating enzymes that damage cell structure. To test the hypothesis that lithium is protective against glutamate-induced excitotoxicity, Jakopec and colleagues explored the effect of lithium pretreatment on human glioblastoma cells exposed to cytotoxic concentrations of glutamate (Jakopec et al. 2008). Findings showed that lithium reverted glutamate effects through the modulation of the expression of genes with a critical role in cell survival and signalling, such as survivin and p21, two proteins with anti-apoptotic function.

It has also been suggested that lithium regulates mitochondrial function. In a study by Chen and coworkers, lithium doubled levels of B-cell lymphoma 2 (Bcl-2) in the frontal cortex of rats (Chen et al. 1999). Bcl-2 is a key member of the best characterized family of regulatory proteins involved in inducing or inhibiting apoptosis. Bcl-2 has a key role in promoting cell survival by preventing the mitochondrial release of cytochrome C. After being released into the cytoplasm in response to pro-apoptotic stimuli, cytochrome C triggers cellular death by activating caspases. Bcl-2 family members regulate this signalling pathway through the formation of complexes that stabilize the mitochondrial membrane (anti-apoptotic members) or increase its permeability (pro-apoptotic members). The exact mechanism by which lithium increases Bcl-2 expression is not completely understood. However, the study by Chen and colleagues suggested that the increase in Bcl-2 levels might be determined by an increased expression and DNA-binding activity of a Bcl-2 regulator, the transcription factor polyomavirus enhancer-binding protein (PEBP) 2B. Another study showed that lithium not only increases Bcl-2 levels but also prevents their reduction induced by 1-methyl-4-phenyl-1,2,3,6-tetrahydropyridine (MPTP) in mice (Youdim and Arraf 2004). Lithium treatment has also been shown to decrease the expression of the pro-apoptotic gene Bcl-2-associated X protein (Bax) (Sugawara et al. 2010) and protect against methamphetamine-induced toxicity, limiting the decrease of Bcl-2/Bax ratio in the prefrontal cortex of rats (Bachmann et al. 2009). Bax encodes for a member of the Bcl-2 protein family that promotes apoptosis by inducing cytochrome C release, caspase activation and DNA fragmentation through loss of integrity of the mitochondrial membrane. Bax forms a heterodimer with Bcl-2, inhibiting its anti-apoptotic activity (Yang and Korsmeyer 1996). Bax expression is induced by tumour protein p53 (p53), one of the most studied tumour-suppressor genes. p53 is a key regulator of cell survival, being involved in transcriptional activation of genes involved in arrest of the cell cycle, activation of DNA repair enzymes and, in the case of irreparable DNA damage, apoptosis. Interestingly, chronic treatment with lithium increases Bcl-2 expression and decreases mRNA and protein levels of Bax and p53 in cultured rat cerebellar granule cells (CGC), suggesting that the neuroprotective effect of lithium may be partially mediated by p53 inhibition (Chen and Chuang 1999).

Another element that could contribute to increased Bcl-2 levels after lithium treatment is the cAMP-responsive element-binding protein (CREB). The phosphorylated form of CREB (pCREB) acts in the nucleus as a transcription factor affecting the expression of a number of genes and has a critical role in promoting cell survival, as supported by the fact that genetically engineered mice with CREB deletion die immediately after birth (Rudolph et al. 1998). Effects of lithium on CREB expression were reported in a hypothesis-free study, where authors used microarrays to measure genome-wide expression levels in substantia nigra pars compacta of mice fed with an ordinary diet or a lithium diet (Farah et al. 2013). Mice were treated with saline or MPTP in a total of four groups. Results showed that levels of CREB were upregulated by lithium, downregulated by MPTP and maintained in mice treated with both MPTP and lithium.

As regards human studies, Grimes and Jope conducted several experiments in human neuroblastoma SH-SY5Y cells in order to understand the role of CREB in the mechanism of action of lithium (Grimes and Jope 2001). The results of these studies suggested that lithium could enhance CREB DNA-binding activity through inhibition of glycogen synthase kinase 3 beta (GSK3B), a serine/threonine kinase involved in neuronal cell development, apoptosis, Wnt signalling and regulation of transcription factors. To date, only one study has investigated the effect of lithium on CREB and pCREB in cell lines derived from bipolar patients (Alda et al. 2013).

This study showed that pCREB levels were increased in lymphoblasts from bipolar patient responders to lithium and in their relatives, as compared to controls. However, culturing lymphoblasts with lithium did not affect CREB and pCREB protein levels.

As mentioned earlier, lithium interaction with CREB is in part mediated by GSK3B. GSK3B is a key component of numerous signalling pathways and an important regulator of neuronal plasticity and cell survival. Its role in inducing apoptosis has been investigated in a large number of models of neuronal death (Hetman and Xia 2000; Li et al. 2000). GSK3B also represents one of the most extensively studied lithium targets. Lithium inhibits this enzyme directly, through competition with magnesium ions, and indirectly, increasing phosphorylation of a serine residue in the N-terminal domain. GSK3B is a critical activator of apoptosis through multiple mechanisms. In particular, GSK3B directly phosphorylates Bax (Linseman et al. 2004), thus activating caspases and apoptosis through the mechanism explained earlier. Moreover, GSK3B phosphorylates p53, therefore increasing its apoptotic action (Turenne and Price 2001). A recent study carried out in neuroblastoma SH-SY5Y cells showed that lithium and SB216763, an inhibitor of GSK3B, reduce mitochondrial translocation of p53 and block the mitochondrial apoptosis pathway induced by DNA-damaging agents doxorubicin, etoposide and camptothecin (Ngok-Ngam et al. 2013). These results suggest that inhibition of GSK3B could have a critical role in mediating lithium's neuroprotective effects. Lithium has also been shown to modulate the expression of other members of the Bcl-2 family. Chronic but not acute treatment with lithium or valproate increases Bcl-2-associated athanogene (BAG-1) expression in the rat hippocampus (Zhou et al. 2005). BAG-1 binds to Bcl-2, increasing its anti-apoptotic function and activating the extracellular signal-regulated kinase/mitogen-activated protein kinase (ERK/MAPK) cascade. Upregulation of BAG-1 is not induced by chronic treatment with amphetamine, haloperidol or fluvoxamine, suggesting a specific effect of the two mood stabilizers.

A significant regulation of the expression of pro- and anti-apoptotic genes has also been shown by a microarray expression study carried out in whole blood from 20 depressed bipolar patients treated with lithium in an 8-week open-label protocol (Lowthert et al. 2012). In this study, 127 genes showed differential expression in responders versus nonresponders, with significant upregulation of anti-apoptotic genes and downregulation of pro-apoptotic genes. Among the anti-apoptotic genes were Bcl-2 and insulin receptor substrate 2 (IRS2), while pro-apoptotic genes included Bcl-2-antagonist/killer 1 (BAK1) and Bcl-2-associated agonist of cell death (BAD).

Another element that has been suggested as being involved in mediating the neuroprotective effect of both lithium and valproate is SIX homeobox 1 (SIX1). The SIX1 gene codifies for a transcription factor involved in the regulation of cell proliferation, apoptosis and embryonic development, playing an important role in the development of organs. A recent study showed that lithium and valproate increase the expression of SIX1 in neuroblastoma SH-SY5Y cells and that pre-exposure to both drugs provides protection against staurosporine (STS)-induced apoptosis.

Moreover, removal of the SIX1 protein via siRNA antagonizes the ability of lithium and valproate to protect neuroblastoma cells from STS-induced apoptosis.

A large body of evidence supports the neuroprotective effects of lithium. Several potential effectors have been proposed and explored, but only a small number of these are supported by compelling findings. Bcl-2, CREB and GSK3B represent the most promising genes, as converging data from animal and human studies support an interaction between lithium and their expression and activity as well as their possible role in lithium-induced neuroprotection. Considering that these genes are members of a large number of pathways, through their regulation, lithium could interfere with many cellular functions involved in apoptosis and neuroprotection.

4.4 Genes Involved in Neuronal Plasticity

A large number of studies suggest that bipolar disorder is associated with impairments in neuronal plasticity, neurotrophic factors and glial and neuronal death in several brain areas, including those involved in mood regulation. Neurotrophic factors are critical regulators of plasticity and cell resilience in adult neurons and glia. Lithium has been shown to interfere with the expression of the key member of the nerve growth factor family, brain-derived neurotrophic factor (BDNF), a protein involved in promoting neuronal differentiation in the peripheral and central nervous systems. BDNF participates in axonal growth, in path finding and in the modulation of dendritic growth and morphology. BDNF activates tyrosine kinase receptor B (TrkB), by which it regulates intracellular signalling pathways involved in neuronal survival, synaptic plasticity and cognitive functions. A number of human studies have tested the hypothesis of the involvement of BDNF in the mechanism of action of lithium. Croce and coworkers showed that lithium abolished the excitotoxic effect of high concentrations of glutamate and that this effect was associated with increased BDNF protein and mRNA levels in human neuroblastoma cell lines (Croce et al. 2014). However, the most compelling findings supporting a lithium-BDNF interaction come from studies carried out in patient-derived cells. Tseng and colleagues (2008) explored the effect of lithium on BDNF levels in lymphoblasts derived from 12 lithium-responsive bipolar patients and 13 healthy controls. Bipolar subjects showed lower protein levels compared to controls. Interestingly, lithium treatment reduced these levels, which remained lower in the bipolar group.

In another study, Pandey and coworkers (2010) investigated differences in the BDNF gene and protein levels in blood cells from bipolar patients and controls and tested the effect of lithium. BDNF levels were lower in fresh lymphocytes and platelets from 26 paediatric bipolar patients compared to 21 healthy controls. Moreover, gene and protein levels were increased in lymphocytes sampled after 8 weeks of treatment with lithium, VPA or second-generation antipsychotics compared to lymphocytes sampled before lithium treatment. The studies by Tseng and Pandey show contrasting findings in respect to the effect of treatments on BDNF levels. However, the two studies present important differences. Tseng and colleagues explored the effect of lithium alone in lymphoblasts, whereas Pandey and colleagues investigated the effect of several mood stabilizers in fresh blood without testing treatment-specific effects. Another recent study reported higher BDNF plasma levels after 28 days of lithium monotherapy in bipolar patients (de Sousa et al. 2011). Even considering the discrepancies among studies, these findings converge towards a significant lithium-mediated modulation of BDNF expression. More data on the effect of lithium on BDNF are presented in Sect. 4.8, where we discuss lithium modulation of epigenetic regulation of BDNF transcription.

Other players in neuronal plasticity have been investigated in lithium response and the mechanism of action of lithium. Following an earlier linkage study suggesting the involvement of synapsin II (SYN2) in bipolar disorder and lithium response (Lopez et al. 2010), Cruceanu and colleagues (2012) explored the effects of in vitro lithium treatment on the SYN2 gene in lymphoblasts from bipolar patients, responders and nonresponders to lithium and healthy controls. SYN2 is a neuronal phosphoprotein involved in vesicle trafficking and synaptic plasticity. In the Cruceanu study, lithium treatment upregulated the expression of the two isoforms SYN2a and SYN2b in some lithium responders, while it downregulated it in others. Interestingly, lithium treatment in neuroblastoma cells increased SYN2a expression. These findings provide evidence for an involvement of SYN and regulation of neurotransmission and synaptogenesis in the mechanism of action of lithium.

Taken together, in vitro and in vivo studies suggest that genes involved in plasticity and neuronal growth, and in particular BDNF, could represent key players in lithium's mechanism of action.

4.5 Genes of the Circadian Rhythm System

Alteration of circadian rhythms, such as sleep disturbances and abnormalities of the sleep-wake cycle, has been reported in bipolar disorder (McCarthy and Welsh 2012). An extensive overview of the role of circadian rhythm dysregulation in bipolar disorder and lithium effects on this system is provided in Chapter 6 of this book. Here we focus on the most compelling findings supporting lithium effects on the expression of genes of the circadian system. Genes of this system have been studied to verify the hypothesis of a role for their altered expression level in the circadian rhythm disturbances observed in bipolar patients.

One of the most widely studied mechanisms regulating the expression of circadian clock genes is the feedback loop involving circadian locomotor output cycles kaput (CLOCK) protein, which plays a crucial role in promoting expression of downstream circadian genes (Takahashi et al. 2008). CLOCK encodes for a transcription factor that forms a heterodimer with the protein encoded by aryl hydrocarbon receptor nuclear translocator-like (Bmal1). During the day, the heterodimer CLOCK/Bmal1 promotes the expression of a number of genes such as period 1 (Per1), period 2 (Per2), period 3 (Per3), cryptochrome1 (Cry1) and cryptochrome2 (Cry2). PER and CRY proteins translocate into the nucleus where they inhibit their own transcription by interacting with the CLOCK/Bmal1 dimer. During the night, PER and CRY products are degraded, and the CLOCK/Bmal1 dimer is no longer inhibited. Moreover, Bmal1 expression is repressed by the nuclear hormone receptor Rev-Erb-α, which is involved in another feedback loop together with the RAR-related orphan receptor A (RORA) gene.

Evidence for an effect of lithium on genes of this system comes from both animal and human studies. In 2007, Roybal and colleagues (2007) showed that lithium reverted mania-like behaviours in mice with a CLOCK gene mutation. In another study, expression of Per2 and Cry1 was increased by lithium in murine fibroblasts, while Per3, Cry2, Bmal1 and Rev-Erb-α were reduced (Osland et al. 2011).

In regard to human studies, lithium was shown to reduce the expression of Bmal1 and Rev-Erb-α in fibroblasts of bipolar patients but not in healthy controls (Yang et al. 2009). One of the most accredited hypotheses is that lithium could interfere with circadian clock genes through the inhibition of GSK3B, although this hypothesis has yet to be confirmed (Mohawk et al. 2009; Kozikowski et al. 2011; Kinoshita et al. 2012; Li et al. 2012). In a recent study, lithium 1 mM increased Per2 rhythm amplitude in skin fibroblasts from healthy controls but not from bipolar patients, while lithium 10 mM increased period length in both bipolar cases and controls (McCarthy et al. 2013). PER3 and RORA genotypes predicted lithium period lengthening, whereas GSK3B genotype predicted rhythm effects of lithium, specifically among bipolar patients. Finally, lithium enhanced the resynchronization of damped rhythms, supporting the hypothesis that the effect on circadian period could mediate part of lithium's therapeutic effect in bipolar disorder.

To summarize, an altered circadian system has been widely reported in bipolar patients. Studies exploring the effects of lithium on the expression of genes of this system are scarce, but findings suggest that, through GSK3B and Rev-Erb-α, lithium could interfere with the circadian system.

4.6 Genes of the Polyamine System

Systems involved in modulating the response to stress stimuli have been widely studied in bipolar disorder. However, the effects of lithium on the genes of these systems have been scarcely investigated or reported. The polyamine system has been suggested to be dysfunctional in mood disorders and suicide, and altered levels of polyamines have been reported in these disorders (Fiori and Turecki 2008). Polyamines (PAs) are ubiquitous molecules responsible for the polyamine stress response (PSR), a physiological cascade of intracellular events activated by stressful stimuli involved in the restoring and maintenance of the organism's homeostatic equilibrium (Gilad and Gilad 1996). PAs are implicated in a variety of cell processes, including modulation of DNA, RNA and protein functions and ultimately neurotransmission (Igarashi and Kashiwagi 2010). A possible involvement of altered expression of genes of the PA system in suicide was first suggested by a microarray study exploring differences in genome-wide expression levels between suicide victims and controls. This study showed that the spermidine/spermine N1-acetyltransferase (SAT1) gene, encoding the rate-limiting enzyme of the catabolic and interconversion pathway of PAs, is downregulated in the cerebral cortex of suicide completers with and without major depression as compared to healthy controls, a finding that was replicated by independent investigations (Sequeira et al. 2006, 2007; Guipponi et al. 2009; Klempan et al. 2009). Interestingly, earlier studies showed that persistent PSR is blocked by chronic lithium treatment in the rat brain. This block results from an inhibition of the activity of the two PA enzymes, ornithine decarboxylase (ODC) and SAT1 (Gilad and Gilad 1996, 2003). However, the interaction between lithium and the PA system remains incompletely understood. A recent study (Squassina et al. 2013) showed that lithium treatment in vitro is able to increase SAT1 expression in lymphoblasts from bipolar patients with low and high genetic risk of suicide, but not in bipolar suicide completers. This study suggests that SAT1 expression is influenced by lithium and that its transcription regulation may be defective in suicide completers with bipolar disorder. Importantly, this study was carried out in peripheral cells derived from patients prior to a suicide attempt or completion. Taken together, the expression data published so far suggest that an altered expression of SAT1 may be a feature of the suicide brain and a potential target of lithium.

4.7 Possible Mechanisms Involved in Lithium Regulation of Gene Expression: microRNAs

In the previous sections of this chapter, we described the effects of lithium on the expression of genes coding for elements of pathways suggested to be involved in bipolar disorder or in the mechanism of action of lithium. These genes include transcription factors, through which lithium may regulate the expression of a vast number of downstream genes. However, transcription factors are not the only key players involved in the regulation of gene expression. Most of the transcribed portion of the human genome is not translated into proteins. This portion of non-coding RNA, which includes long and short RNAs, has fundamental importance in the modulation of gene expression. In this section we focus on short ncRNAs, called microRNAs (miRNAs). miRNAs are single-stranded molecules (~22 bases) of non-coding RNA present only in eukaryotes, where they regulate the expression of two-thirds of the transcriptome (Strachan and Read 2011). miRNAs mediate their regulatory effects by binding to the 3′ untranslated region (UTR) of target genes, leading to the degradation of messenger RNA (mRNA) and prevention of its translation into proteins. miRNA genes are often found in clusters. Their transcription is mediated by a RNA polymerase II, which generates a double-stranded precursor termed pri-miRNA. Pri-miRNAs are processed by the type III endonuclease Drosha, which produces several pre-miRNAs forming hairpin structures. Pre-miRNAs are capped and polyadenylated and then released in the cytoplasm, where they undergo processing by another type III endonuclease called Dicer. The cleavage by Dicer produces mature miRNAs that are incorporated into the RNA-induced silencing complex (RISC). RISC is a ribonucleoprotein complex able to bind target mRNAs by complementary base pairing.

The regulation activity mediated by miRNAs is sensitive to environmental changes and is independent from DNA sequence, a feature that has led to the annotation of miRNAs in the family of epigenetic processes. While the regulatory mechanisms mediated by miRNAs are still not completely understood, a growing body of evidence has shown that these short RNAs regulate key processes and functions of the central nervous system (Miller and Wahlestedt 2010). miRNAs are involved in neural development (De Pietri et al. 2008), synaptic plasticity (Schratt et al. 2006) and circadian rhythm regulation (Cheng et al. 2007).

Considering their role in processes potentially implicated in psychiatric disorders, miRNAs may constitute key effectors of psychotropic medications, through which these drugs could interfere with neuronal functions. As outlined in the preceding section, lithium induces neuroprotection through a number of mechanisms that are only partially known. To date, a number of animal and human studies have explored the role of miRNAs in lithium's mechanism of action and how this could affect neuroprotection.

In a recent study, Zhou and coworkers showed that both lithium and valproate influenced the expression of more than 30 miRNA genes in the rat hippocampus (Zhou et al. 2009). These miRNAs regulated pathways involved in neurogenesis, neurite outgrowth and brain and nervous system development. The most significant neuronal processes targeted by these miRNAs were axon guidance, beta-adrenergic transmission, Wnt/Beta-catenin and ERK and PTEN signalling, some of which had been previously reported to be influenced by lithium (Engel et al. 2009; Gould et al. 2007).

In the study by Zhou and colleagues, the authors combined predicted effectors of miRNAs significantly regulated by lithium and valproate with genome-wide scan data from the Wellcome Trust Case Control Consortium study (Wellcome Trust Case Control Consortium 2007). The combined analysis suggested that three miRNAs (miR-128a, miR-24 and miR-34a), downregulated by both treatments, negatively correlated with protein levels of their effectors dipeptidyl peptidase 10 (DPP10), thyroid hormone receptor beta (THRB) and metabotropic glutamate receptor 7 (GRM7). These three genes showed a trend for association with bipolar disorder in the WTCCC study. Lithium and valproate decreased the expression of miR-34a and increased the levels of GRM7 in primary cultures of hippocampal neurons. Interestingly, GRM7 was associated with bipolar disorder in previous studies (Saus et al. 2010; Wellcome Trust Case Control Consortium 2007), while other studies showed that lithium interferes with the glutamate system through AMPA and NMDA receptors (Du et al. 2004, 2008). Effects of lithium on miRNAs and their correlation with neuroprotection were also investigated in a study by Hunsberger and colleagues (2013). In this study, glutamate exposure was used to induce excitotoxicity in rat cerebellar granule cell cultures. These cells were treated with lithium and valproate. Combined treatment, which provided complete protection from glutamate excitotoxicity, altered the expression of 121 miRNAs, six of which were validated with real-time PCR, including miR-34a. The apoptotic effect of miR-34a was further tested in SH-SY5Y human neuroblastoma cells under basal conditions and following endoplasmic reticulum stress. Interestingly, miR-34a mimic enhanced the effect of endoplasmic reticulum stress in inducing apoptosis.

Taken together, findings from the Zhou and colleagues and Hunsberger and colleagues studies suggest that the neuroprotective effect of lithium might be at least partially mediated by its modulation of miRNA expression, which in turn regulate the expression of genes involved in neuronal activity. This assumption is further supported by findings from studies carried out in patient-derived tissues or cell lines. In a recent study, Chen and colleagues (2009) explored the effect of lithium treatment in vitro on the expression of a selected list of miRNAs in lymphoblasts from bipolar family members. Seven out of the 13 miRNAs measured were differentially expressed in lithium-treated versus non-treated lymphoblasts. This group of miRNAs included miR-221 and miR-34a, which were shown to be altered by lithium and valproate in the study by Zhou and colleagues (2009). Correlation analysis with mRNA levels showed inverse co-regulation of 29 and 10 predicted targets for miR-221 and miR-34a, respectively. Out of 39 targets, 5 were significantly enriched in gene ontology analyses and included adaptor-related protein complex 2 alpha 1 subunit (AP2A1), adaptor-related protein complex 2 sigma 1 subunit (AP2S1), CD2-associated protein (CD2AP), eukaryotic translation initiation factor 1 (EIF1) and vinculin (VCL). These genes belong to pathways involved in macromolecular complex assembly, protein complex assembly and cellular component assembly. The effect of treatments for bipolar mania, including lithium, was also tested in a recent study by Rong and colleagues (2011). This study measured plasma levels of miRNA-134 in 21 bipolar patients during mania and 21 matched controls and tested the correlation with treatments for bipolar disorder (lithium, valproate and chlorpromazine). miRNA-134 was previously reported to inhibit the expression of LIM domain kinase 1 (LimK1) gene, a protein kinase involved in dendritic spine development (Schratt et al. 2006). Levels of miRNA-134 were significantly lower in both drug-naive and drug-treated patients compared to controls. After 2 weeks of treatment, levels of miRNA-134 were higher (though still lower than in controls), suggesting that this miRNA may be associated with response to treatment in bipolar mania. While this study did not investigate specific effects of individual treatments, its findings suggest that regulation of miRNA-134 could represent a mechanism through which treatments for bipolar disorder influence neuroprotection and neurogenesis.

In their entirety, these findings suggest that regulation of miRNA expression could be one of the mechanisms through which lithium modulates neuronal functioning.

Despite the numerous facets that remain unclear, miRNAs represent a promising line of research that could lead to a better understanding of the mechanism of action of lithium. The number of miRNAs catalogued so far for *Homo sapiens* has exceeded 1,800 (http://www.mirbase.org), but it is reasonable to assume that new miRNAs will be discovered in the coming years. For the majority of known miRNAs, target genes are only predicted. Clear determination of miRNA/mRNA relationships would be essential to understand the role of miRNA-dependent regulation of gene expression in drug response and bipolar disorder. Achieving this goal is hampered by the complexity of the relationship between miRNAs and their effectors, as one single miRNA can interfere with the expression of many genes and one gene can be targeted by more than one miRNA. Nevertheless, the discovery of miRNAs and the evidence that lithium affects their expression are key steps in understanding the mechanism of action of lithium, as these short RNAs could represent the missing line in the complex picture of lithium-mediated regulation of gene expression.

4.8 Epigenetic Processes

As described in the preceding section, lithium modulates the expression of a large number of genes, including transcription factors and miRNAs. However, both transcription factors and miRNAs are products of gene transcription. While the mechanism by which lithium regulates the expression of these genes is still unknown, a growing body of research suggests that lithium interferes with epigenetic processes. Epigenetics is the dynamic regulation of gene expression without changing the DNA sequence. There is discussion about which processes should be included in the definition of epigenetics, but there is general agreement that chromatin modifications and DNA methylation constitute its core elements. Chromatin modifications are mediated by the two enzymes histone acetyltransferase (HAT) and histone deacetylase (HDAC), which are responsible for the acetylation and deacetylation of histones. Histone acetylation determines a decrease in the interaction of their N-termini with the negatively charged phosphate groups of DNA. The condensed chromatin is converted into a more relaxed structure, which is associated with greater levels of gene transcription. This relaxation can be reversed by HDAC activity. DNA methylation is probably the most widely investigated and understood epigenetic mechanism. This process consists of the transfer of a methyl group to the five termini of cytosines. Only cytosines with a downstream adjacent guanine (the so-called CpG sequences) are methylated. Methylation, mediated by DNA methyltransferases (DNMT), leads to the repression of gene expression.

Epigenetic processes are sensitive to environmental stimuli, thus representing the mechanism through which external events are translated into modifications inside the cells. There is evidence that epigenetic regulation specifically targeting neuroplasticity plays an important role in bipolar disorder (Laird 2003; Rutten and Mill 2009; Khare et al. 2011; Pidsley and Mill 2011; Fass et al. 2014). A growing body of evidence suggests that lithium may regulate gene expression by interacting with epigenetic processes. Based on previous data supporting a role for BDNF in lithium response, D'Addario and colleagues investigated the degree of methylation of the promoter region of the BDNF gene in peripheral blood mononuclear cells from bipolar patients and correlated these levels with pharmacological treatments (D'Addario et al. 2012). Levels of methylation were increased in bipolar II but not bipolar I patients, as compared to controls, and were negatively correlated with BDNF mRNA expression. Interestingly, a lower degree of methylation was observed in bipolar patients on treatment with either lithium or valproate as compared with patients on other pharmacological treatments. However, another study by Asai and colleagues (2013) reported no significant effect of lithium on the methylation of BDNF promoter. This study was carried out on SH-SY5Y human neuroblastoma cells treated with lithium, valproate or carbamazepine. Gene-specific analyses were carried out on BDNF and solute carrier family 6 member 4 (SLC6A4), both of which had been reported to show altered DNA methylation in bipolar disorder or suicide (Keller et al. 2010; Sugawara et al. 2011). Methylation levels of SLC6A4 were reduced in cells treated with all mood stabilizers, with the exception of valproate. Global differences in DNA methylation levels determined by pharmacological treatments were also tested. Compared to valproate and carbamazepine, lithium showed the largest effect on global DNA methylation, affecting the methylation of 483 genes, against 100 and 78 altered by valproate and carbamazepine, respectively. The three mood stabilizers tested induced more prominent over-methylation than hypo-methylation. Gene ontology analysis for genes exclusively regulated by lithium showed terms related to neuronal functions, including cell-cell signalling, synaptic transmission, locomotion and neurotransmitter activity.

To date, only one study has explored the effect of lithium on global methylation using tissues derived from bipolar patients characterized for their clinical response to lithium therapy. In this study, Huzayyin and colleagues (2014) explored the relationship between oxidative stress and DNA methylation in bipolar patients with excellent response to lithium and their unaffected and affected relatives, as well as healthy controls. The study was carried out in lymphoblasts cultured with or without lithium. To test the correlation between lithium treatment and methylation, levels of activity of 8-hydroxy-2-deoxyguanosine (8-OHdG), 5-methylcytosine (5-mc), mitochondrial complex I and glutathione peroxidase (GPx) were measured. These targets were selected based on their role in oxidative stress and DNA damage and on findings suggesting their potential involvement in the mechanism of action of lithium. 8-OHdG is one of the predominant forms of free radical-induced oxidative lesions, and oxidative damage has been reported in post-mortem brains from bipolar subjects (Buttner et al. 2007; Mustak et al. 2010; Che et al. 2010). DNA cytosine 5 methylation is a widely studied epigenetic pathway implicated in the control of gene expression and disease pathogenesis. Measurement of 5-mc activity is one of the most widely used methods to assess global methylation. Complex I is the first enzyme in the mitochondrial respiratory chain, and recent findings reported decreased activity of this enzyme in the prefrontal cortex of patients with bipolar disorder (Andreazza et al. 2010). GPx is among the most important mechanisms of defence for the mitochondria to avoid or repair oxidative modifications. Some studies have shown that lithium may exert an antioxidant effect through a stimulation of the glutathione antioxidant system (Shao et al. 2008).

In the study by Huzayyin and colleagues (2014), lower degrees of global methylation were observed in bipolar patients and their relatives as compared to controls, a difference which remained significant after lithium treatment. Lithium differentially affected methylation in bipolar patients and their relatives compared to controls, a result that could be related to lithium-induced biological adaptation in bipolar subjects after long-term treatment. Levels of 8-OHdG did not differ among groups, either before or after lithium treatment. On the contrary, activity of GPx was elevated in affected and unaffected relatives when compared to controls, while it was not different in bipolar patients. Taken together, findings from this study support the importance of differential DNA methylation in bipolar disorder and in lithium response.

Study of the epigenetics effects of lithium is a promising field in pharmacology. As we stated earlier in this chapter, lithium affects the expression of hundreds of genes, but little is known about the mechanisms involved. Epigenetic modifications could be involved in these effects, being responsible for the regulation of the expression of a large proportion of genes. While findings so far have been promising, more effort is needed to provide a broader and clearer picture of the complex interaction between lithium and epigenetic processes and how these could affect lithium response.

4.9 Summary

In summary, lithium affects the expression of multiple genes involved in a number of essential signalling and cellular processes, and most of these effects are likely to be intercorrelated. For a variety of reasons, it is at present difficult to identify those effects that are responsible for the clinical actions of lithium. Some of the difficulties result from an absence of well-validated animal models of bipolar disorder (and its particular features, such as recurrent clinical course or suicide behaviour). In addition, a number of in vitro studies of clinical samples have had to rely on peripheral tissues, not always differentiated based on the degree of clinical response. At least some of these difficulties may be overcome in the future by the development of stem cell-based methods, such as induced pluripotent stem cell-derived neurons. Characterization of lithium's mechanism of action will be an important step towards developing therapies that will be well tolerated while preserving the key effects of lithium.

References

Alda M, Shao L, Wang JF et al (2013) Alterations in phosphorylated cAMP response element-binding protein (pCREB) signaling: an endophenotype of lithium-responsive bipolar disorder? Bipolar Disord 15:824–831

Allison JH, Stewart MA (1971) Reduced brain inositol in lithium-treated rats. Nat New Biol 233:267–268

Andreazza AC, Shao L, Wang JF et al (2010) Mitochondrial complex I activity and oxidative damage to mitochondrial proteins in the prefrontal cortex of patients with bipolar disorder. Arch Gen Psychiatry 67:360–368

Asai T, Bundo M, Sugawara H et al (2013) Effect of mood stabilizers on DNA methylation in human neuroblastoma cells. Int J Neuropsychopharmacol 16:2285–2294

Azab AN, He Q, Ju S et al (2007) Glycogen synthase kinase-3 is required for optimal de novo synthesis of inositol. Mol Microbiol 63:1248–1258

Bachmann RF, Wang Y, Yuan P et al (2009) Common effects of lithium and valproate on mitochondrial functions: protection against methamphetamine-induced mitochondrial damage. Int J Neuropsychopharmacol 12(6):805–822

Baldessarini RJ, Tondo L, Davis P et al (2006) Decreased risk of suicides and attempts during long-term lithium treatment: a meta-analytic review. Bipolar Disord 8(5 Pt 2):625–639

Berridge MJ (1989) The Albert Lasker medical awards. Inositol trisphosphate, calcium, lithium, and cell signaling. JAMA 262:1834–1841

Buttner N, Bhattacharyya S, Walsh J et al (2007) DNA fragmentation is increased in non-GABAergic neurons in bipolar disorder but not in schizophrenia. Schizophr Res 93:33–41

Che Y, Wang JF, Shao L et al (2010) Oxidative damage to RNA but not DNA in the hippocampus of patients with major mental illness. J Psychiatr Neurosci 35:296–302

Chen RW, Chuang DM (1999) Long term lithium treatment suppresses p53 and Bax expression but increases Bcl-2 expression. A prominent role in neuroprotection against excitotoxicity. J Biol Chem 274:6039–6042

Chen G, Zeng WZ, Yuan PX et al (1999) The mood-stabilizing agents lithium and valproate robustly increase the levels of the neuroprotective protein bcl-2 in the CNS. J Neurochem 72:879–882

Chen H, Wang N, Burmeister M et al (2009) MicroRNA expression changes in lymphoblastoid cell lines in response to lithium treatment. Int J Neuropsychopharmacol 12:975–981

Cheng HY, Papp JW, Varlamova O et al (2007) MicroRNA modulation of circadian-clock period and entrainment. Neuron 54:813–829

Croce N, Mathe AA, Gelfo F et al (2014) Effects of lithium and valproic acid on BDNF protein and gene expression in an in vitro human neuron-like model of degeneration. J Psychopharmacol 28:964–972

Cruceanu C, Alda M, Grof P et al (2012) Synapsin II is involved in the molecular pathway of lithium treatment in bipolar disorder. PLoS One 7:e32680

D'Addario C, Dell'osso B, Palazzo MC et al (2012) Selective DNA methylation of BDNF promoter in bipolar disorder: differences among patients with BDI and BDII. Neuropsychopharmacology 37:1647–1655

De Pietri TD, Pulvers JN, Haffner C et al (2008) miRNAs are essential for survival and differentiation of newborn neurons but not for expansion of neural progenitors during early neurogenesis in the mouse embryonic neocortex. Development 135:3911–3921

de Sousa RT, van de Bilt MT, Diniz BS et al (2011) Lithium increases plasma brain-derived neurotrophic factor in acute bipolar mania: a preliminary 4-week study. Neurosci Lett 494:54–56

Du J, Gray NA, Falke CA et al (2004) Modulation of synaptic plasticity by antimanic agents: the role of AMPA glutamate receptor subunit 1 synaptic expression. J Neurosci 24:6578–6589

Du J, Creson TK, Wu LJ et al (2008) The role of hippocampal GluR1 and GluR2 receptors in manic-like behavior. J Neurosci 28:68–79

Engel SR, Creson TK, Hao Y et al (2009) The extracellular signal-regulated kinase pathway contributes to the control of behavioral excitement. Mol Psychiatry 14:448–461

Farah R, Khamisy-Farah R, Amit T et al (2013) Lithium's gene expression profile, relevance to neuroprotection A cDNA microarray study. Cell Mol Neurobiol 33:411–420

Fass DM, Schroeder FA, Perlis RH et al (2014) Epigenetic mechanisms in mood disorders: targeting neuroplasticity. Neuroscience 264:112–130

Fauroux CM, Freeman S (1999) Inhibitors of inositol monophosphatase. J Enzym Inhib 14(2):97–108

Fiori LM, Turecki G (2008) Implication of the polyamine system in mental disorders. J Psychiatry Neurosci 33:102–110

Gilad GM, Gilad VH (1996) Brain polyamine stress response: recurrence after repetitive stressor and inhibition by lithium. J Neurochem 67(5):1992–1996

Gilad GM, Gilad VH (2003) Overview of the brain polyamine-stress-response: regulation, development, and modulation by lithium and role in cell survival. Cell Mol Neurobiol 23(4-5): 637–649

Gitlin M, Frye MA (2012) Maintenance therapies in bipolar disorders. Bipolar Disord 14(Suppl 2):51–65

Gould TD, Einat H, O'Donnell KC et al (2007) Beta-catenin overexpression in the mouse brain phenocopies lithium-sensitive behaviors. Neuropsychopharmacology 32:2173–2183

Grimes CA, Jope RS (2001) CREB DNA binding activity is inhibited by glycogen synthase kinase-3 beta and facilitated by lithium. J Neurochem 78:1219–1232

Guipponi M, Deutsch S, Kohler K et al (2009) Genetic and epigenetic analysis of SSAT gene dysregulation in suicidal behavior. Am J Med Genet B Neuropsychiatr Genet 150B(6):799–807

Hetman M, Xia Z (2000) Signaling pathways mediating anti-apoptotic action of neurotrophins. Acta Neurobiol Exp (Wars) 60:531–545

Hunsberger JG, Fessler EB, Chibane FL et al (2013) Mood stabilizer-regulated miRNAs in neuropsychiatric and neurodegenerative diseases: identifying associations and functions. Am J Transl Res 5:450–464

Huzayyin AA, Andreazza AC, Turecki G et al (2014) Decreased global methylation in patients with bipolar disorder who respond to lithium. Int J Neuropsychopharmacol 17:561–569

Igarashi K, Kashiwagi K (2010) Modulation of cellular function by polyamines. Int J Biochem Cell Biol 42(1):39–51

Jakopec S, Karlovic D, Dubravcic K et al (2008) Lithium effect on glutamate induced damage in glioblastoma cells. Coll Antropol 32(Suppl 1):87–91

Keller S, Sarchiapone M, Zarrilli F et al (2010) Increased BDNF promoter methylation in the Wernicke area of suicide subjects. Arch Gen Psychiatry 67:258–267

Khare T, Pal M, Petronis A (2011) Understanding bipolar disorder: the epigenetic perspective. In: Behavioral neurobiology of bipolar disorder and its treatment. Springer Science + Business Media, New York pp 31–49

Kinoshita C, Miyazaki K, Ishida N (2012) Chronic stress affects PERIOD2 expression through glycogen synthase kinase-3beta phosphorylation in the central clock. Neuroreport 23:98–102

Klempan TA, Rujescu D, Merette C et al (2009) Profiling brain expression of the spermidine/spermine N1-acetyltransferase 1 (SAT1) gene in suicide. Am J Med Genet B Neuropsychiatr Genet 150B(7):934–943

Kozikowski AP, Gunosewoyo H, Guo S et al (2011) Identification of a glycogen synthase kinase-3beta inhibitor that attenuates hyperactivity in CLOCK mutant mice. ChemMedChem 6:1593–1602

Laird PW (2003) The power and the promise of DNA methylation markers. Nat Rev Cancer 3:253–266

Li M, Wang X, Meintzer MK et al (2000) Cyclic AMP promotes neuronal survival by phosphorylation of glycogen synthase kinase 3beta. Mol Cell Biol 20:9356–9363

Li J, Lu WQ, Beesley S et al (2012) Lithium impacts on the amplitude and period of the molecular circadian clockwork. PLoS One 7:e33292

Linseman DA, Butts BD, Precht TA et al (2004) Glycogen synthase kinase-3beta phosphorylates Bax and promotes its mitochondrial localization during neuronal apoptosis. J Neurosci 24:9993–10002

Lopez dL, Jaitovich-Groisman I, Cruceanu C et al (2010) Implication of synapse-related genes in bipolar disorder by linkage and gene expression analyses. Int J Neuropsychopharmacol 13:1397–1410

Lowthert L, Leffert J, Lin A et al (2012) Increased ratio of anti-apoptotic to pro-apoptotic Bcl2 gene-family members in lithium-responders one month after treatment initiation. Biol Mood Anxiety Disord 2:15

Malhi GS, Bargh DM, McIntyre R et al (2012) Balanced efficacy, safety, and tolerability recommendations for the clinical management of bipolar disorder. Bipolar Disord 14(Suppl 2):1–21

McCarthy MJ, Welsh DK (2012) Cellular circadian clocks in mood disorders. J Biol Rhythms 27:339–352

McCarthy MJ, Wei H, Marnoy Z et al (2013) Genetic and clinical factors predict lithium's effects on PER2 gene expression rhythms in cells from bipolar disorder patients. Transl Psychiatry 3:e318

Miller BH, Wahlestedt C (2010) MicroRNA dysregulation in psychiatric disease. Brain Res 1338:89–99

Mohawk JA, Miranda-Anaya M, Tataroglu O et al (2009) Lithium and genetic inhibition of GSK3beta enhance the effect of methamphetamine on circadian rhythms in the mouse. Behav Pharmacol 20:174–183

Moore GJ, Bebchuk JM, Parrish JK et al (1999) Temporal dissociation between lithium-induced changes in frontal lobe myo-inositol and clinical response in manic-depressive illness. Am J Psychiatry 156:1902–1908

Mustak MS, Hegde ML, Dinesh A et al (2010) Evidence of altered DNA integrity in the brain regions of suicidal victims of bipolar depression. Indian J Psychiatry 52:220–228

Ngok-Ngam P, Watcharasit P, Thiantanawat A et al (2013) Pharmacological inhibition of GSK3 attenuates DNA damage-induced apoptosis via reduction of p53 mitochondrial translocation and Bax oligomerization in neuroblastoma SH-SY5Y cells. Cell Mol Biol Lett 18:58–74

Osland TM, Ferno J, Havik B et al (2011) Lithium differentially affects clock gene expression in serum-shocked NIH-3T3 cells. J Psychopharmacol 25:924–933

Pandey GN, Dwivedi Y, Rizavi HS et al (2010) Brain-derived neurotrophic factor gene and protein expression in pediatric and adult depressed subjects. Prog Neuropsychopharmacol Biol Psychiatry 34:645–651

Pidsley R, Mill J (2011) Epigenetic studies of psychosis: current findings, methodological approaches, and implications for postmortem research. Biol Psychiatry 69:146–156

Rong H, Liu TB, Yang KJ et al (2011) MicroRNA-134 plasma levels before and after treatment for bipolar mania. J Psychiatr Res 45:92–95

Roybal K, Theobold D, Graham A et al (2007) Mania-like behavior induced by disruption of CLOCK. Proc Natl Acad Sci U S A 104(15):6406–6411

Rudolph D, Tafuri A, Gass P et al (1998) Impaired fetal T cell development and perinatal lethality in mice lacking the cAMP response element binding protein. Proc Natl Acad Sci U S A 95:4481–4486

Rutten BP, Mill J (2009) Epigenetic mediation of environmental influences in major psychotic disorders. Schizophr Bull 35:1045–1056

Saus E, Brunet A, Armengol L et al (2010) Comprehensive copy number variant (CNV) analysis of neuronal pathways genes in psychiatric disorders identifies rare variants within patients. J Psychiatr Res 44:971–978

Schratt GM, Tuebing F, Nigh EA et al (2006) A brain-specific microRNA regulates dendritic spine development. Nature 439:283–289

Sequeira A, Gwadry FG, Ffrench-Mullen JM et al (2006) Implication of SSAT by gene expression and genetic variation in suicide and major depression. Arch Gen Psychiatry 63(1):35–48

Sequeira A, Klempan T, Canetti L et al (2007) Patterns of gene expression in the limbic system of suicides with and without major depression. Mol Psychiatry 12(7):640–655

Shao L, Cui J, Young LT et al (2008) The effect of mood stabilizer lithium on expression and activity of glutathione s-transferase isoenzymes. Neuroscience 151:518–524

Silverstone PH, McGrath BM (2009) Lithium and valproate and their possible effects on the myo-inositol second messenger system in healthy volunteers and bipolar patients. Int Rev Psychiatry 21:414–423

Squassina A, Manchia M, Chillotti C et al (2013) Differential effect of lithium on spermidine/spermine N1-acetyltransferase expression in suicidal behaviour. Int J Neuropsychopharmacol 16:2209–2218

Strachan T, Read AP (2011) Human molecular genetics. Garland Science/Taylor & Francis Group, New York

Sugawara H, Iwamoto K, Bundo M et al (2010) Effect of mood stabilizers on gene expression in lymphoblastoid cells. J Neural Transm 117:155–164

Sugawara H, Iwamoto K, Bundo M et al (2011) Hypermethylation of serotonin transporter gene in bipolar disorder detected by epigenome analysis of discordant monozygotic twins. Transl Psychiatry 1:e24

Sun X, Young LT, Wang JF et al (2004) Identification of lithium-regulated genes in cultured lymphoblasts of lithium responsive subjects with bipolar disorder. Neuropsychopharmacology 29:799–804

Takahashi JS, Hong HK, Ko CH et al (2008) The genetics of mammalian circadian order and disorder: implications for physiology and disease. Nat Rev Genet 9(10):764–775

Tseng M, Alda M, Xu L et al (2008) BDNF protein levels are decreased in transformed lymphoblasts from lithium-responsive patients with bipolar disorder. J Psychiatry Neurosci 33:449–453

Turenne GA, Price BD (2001) Glycogen synthase kinase3 beta phosphorylates serine 33 of p53 and activates p53's transcriptional activity. BMC Cell Biol 2:12

Wellcome Trust Case Control Consortium (2007) Genome-wide association study of 14,000 cases of seven common diseases and 3,000 shared controls. Nature 447:661–678

Yang E, Korsmeyer SJ (1996) Molecular thanatopsis: a discourse on the BCL2 family and cell death. Blood 88:386–401

Yang S, Van Dongen HP, Wang K et al (2009) Assessment of circadian function in fibroblasts of patients with bipolar disorder. Mol Psychiatry 14:143–155

Youdim MB, Arraf Z (2004) Prevention of MPTP (N-methyl-4-phenyl-1,2,3,6-tetrahydropyridine) dopaminergic neurotoxicity in mice by chronic lithium: involvements of Bcl-2 and Bax. Neuropharmacology 46:1130–1140

Zhou R, Gray NA, Yuan P et al (2005) The anti-apoptotic, glucocorticoid receptor cochaperone protein BAG-1 is a long-term target for the actions of mood stabilizers. J Neurosci 25:4493–4502

Zhou R, Yuan P, Wang Y et al (2009) Evidence for selective microRNAs and their effectors as common long-term targets for the actions of mood stabilizers. Neuropsychopharmacology 34:1395–1405

Neuroimaging and Lithium

5

Josselin Houenou, Chloé Benizri, and Barbara Jakubowicz

Abstract

Neuroimaging holds the promise of being a means to improve our understanding of lithium's mechanism of action. Structural brain imaging studies of individuals exposed to lithium, mainly using magnetic resonance imaging (MRI), have repeatedly evidenced an increased volume of grey matter in cross-sectional, but also in longitudinal, studies. They suggest that lithium has a neurotrophic and neuroprotective effect, particularly in the hippocampus. This effect seems to be supported by lithium-treated rodent models. Findings in white matter are more heterogeneous and sparse. We still do not know, however, whether the neurotrophic effects associated with lithium are a direct consequence of the element or are secondary to symptomatic improvement, although the effect does seem lithium specific. Finally, in the near future, neuroimaging of lithium may assist the clinician through the identification of biomarkers of response to lithium and through the direct measurement of lithium brain concentrations using MRI spectroscopy.

J. Houenou (✉)
AP-HP, Hôpitaux Universitaires Mondor, DHU PePsy, Université Paris Est, Pôle de Psychiatrie, Créteil, France

INSERM U955, IMRB, Team 15 'Psychiatrie Translationnelle', 40 rue de Mesly, 94000 Créteil, France

Fondation Fondamental, Créteil, France

UNIACT, Neurospin, I2BM, CEA Saclay, Gif sur Yvette, France
e-mail: josselin.houenou@inserm.fr

C. Benizri
INSERM U955, IMRB, Team 15 'Psychiatrie Translationnelle', 40 rue de Mesly, 94000 Créteil, France

Fondation Fondamental, Créteil, France
e-mail: chloe.benizri@gmail.com

B. Jakubowicz
INSERM U955, IMRB, Team 15 'Psychiatrie Translationnelle', 40 rue de Mesly, 94000 Créteil, France

Fondation Fondamental, Créteil, France

UNIACT, Neurospin, I2BM, CEA Saclay, Gif sur Yvette, France
e-mail: jakubowicz.barbara@gmail.com

G.S. Malhi et al. (eds.), *The Science and Practice of Lithium Therapy*,
DOI 10.1007/978-3-319-45923-3_5

Key Points

- Neuroimaging studies suggest that lithium has a neurotrophic effect.
- This neurotrophic action of lithium mostly impacts grey matter.
- A key challenge is the discovery of biomarkers of response to lithium.
- MRI spectroscopy may provide a clearer understanding of lithium's mechanism of action.

5.1 Introduction

Lithium has been used for more than 60 years and is the standard treatment for bipolar disorder. However, its mechanism of action is still unknown. In this chapter, we describe neuroimaging studies exploring lithium's effects on brain structure and function. Most of these studies have been performed using a cross-sectional design and, along with a few longitudinal neuroimaging studies, bring evidence of a putative neuroprotective and neurotrophic effect of lithium. We will then discuss future developments of lithium imaging, such as magnetic resonance spectroscopy of lithium.

5.2 Lithium and Brain Volumes

Several cross-sectional studies have explored the associations between lithium medication and total brain, grey matter, white matter or regional volumes. Longitudinal neuroimaging studies of lithium's effects offer the advantage of better control of confounding factors and biases, such as selection effect (e.g. patients already under lithium may have different clinical and neuroimaging characteristics than other patients at baseline). Therefore longitudinal neuroimaging studies probably lead to a more valid understanding of lithium's mechanism of action than cross-sectional studies do.

Through both cross-sectional and longitudinal studies, we will focus successively on:

- Lithium neurotrophic properties
- Lithium specificity with respect to these properties

5.2.1 Neurotrophic Effects

The first study on lithium-induced increased volumes in the human brain was published in 2000 in *The Lancet*. This work used 3D magnetic resonance imaging (MRI) and brain segmentation to study pharmacologically induced increases in grey matter volume. It found a total increase of grey matter volume of about 3 % (24 cm^3)

after 4 weeks' treatment with lithium in ten patients with bipolar disorder (bipolar I patients suffering from depressive relapse). However, the authors found no modification in cerebral white matter. This suggested that lithium has neurotrophic effects (Moore et al. 2000).

Several cross-sectional studies have been performed using magnetic resonance spectroscopy of N-acetylaspartate (NAA) in patients receiving lithium. NAA, which is synthesised in mitochondria, is a putative marker of neuronal viability, helps to maintain myelin and is involved in neuron-to-glia signalling. In a first study, nine older adults with bipolar I disorder treated with lithium had in vivo measurements of serum and brain lithium levels taken using magnetic resonance spectroscopy of lithium, NAA and myo-inositol. The positive relationship identified between higher brain lithium levels and elevated NAA levels in older patients with bipolar disorder may support lithium neuroprotective, neurotrophic and mitochondrial function-enhancing effects (Forester et al. 2008). Another study evidenced that, after 4 weeks of lithium, this treatment was associated with an increase in both brain NAA levels in healthy controls and in patients with bipolar disorder (Moore et al. 2002). These findings provide indirect support for lithium having neuroprotective effects and for there being negative effects of the illness burden on prefrontal NAA levels in patients with bipolar disorder (Hajek et al. 2012).

5.2.1.1 Specificity of Effect

Two questions arise concerning the neurotrophic effect of lithium:

- Is it specific to lithium, or is this effect common across other medications for bipolar disorder?
- Does the effect have regional specificity?

Is the Neurotrophic Effect Specific to Lithium?

An animal study by Vernon and colleagues (Vernon et al. 2012) strongly suggests that lithium has specific neurotrophic brain effects. For 8 weeks, rodents received either lithium or haloperidol. Treatment was then interrupted for the following 8 weeks, while measurements were taken using longitudinal in vivo structural MRI. Finally, confirmation was obtained with postmortem findings using unbiased stereology. Chronic haloperidol treatment induced decreases in whole brain volume (−4 %) and cortical grey matter (−6 %), along with hypertrophy of the striatum (+14 %). In contrast, chronic lithium treatment induced increases in whole brain volume (+5 %) and cortical grey matter (+3 %), without a significant effect on striatal volume. Following 8 weeks of drug withdrawal, haloperidol-induced changes in brain volumes normalised, whereas the lithium-treated animals retained significantly greater total brain volumes, as confirmed postmortem. This animal model provides strong evidence for lithium having direct and specific neurotrophic effects.

In humans, a longitudinal study conducted at Washington University included 13 medication-free patients with bipolar disorder. An MRI exam was performed on all patients using a clinical 1.5 T whole-body scanner. They then received lithium medication for 3 months. Then a second MRI was performed. Results revealed lithium-induced increases in grey matter volume (2.56 %, 17.6 cm^3 increase). No white matter volume modification was observed. In contrast, valproate-treated patients with bipolar disorder did not show grey matter volume changes over time (Lyoo et al. 2010).

Similarly, according to an updated review of the effects of medication on neuroimaging findings in bipolar disorder, lithium use is associated with increased volumes—particularly in areas subserving emotion processing and mood regulation (the amygdala, hippocampus, anterior and subgenual cingulate cortex)—while the use of antipsychotic agents and anticonvulsants is generally not (Hafeman et al. 2012).

Is Lithium's Neurotrophic Effect Regionally Specific?

The current neurobiological models of emotion regulation describe two steps:

- An early pre-attentional step (tens of milliseconds after the stimulus). This automatic regulation step of emotions involves the classical deep limbic structures (the amygdala and hippocampus).
- A later step (several hundreds of milliseconds after the stimulus) of voluntary emotion regulation. It involves cognitive structures such as the prefrontal cortex and cingulate dorsal gyrus. These structures modulate the activity of the amygdala and hippocampus.

Compared to healthy controls, during emotional processing, patients with bipolar disorder exhibit an overactive ventral-limbic network (including regions such as the amygdala, the parahippocampal gyrus, the anterior and subgenual cingulate) (Malhi et al. 2015). This ventral-limbic network is poorly modulated by the ventral prefrontal cortices due to defective prefrontal-limbic connectivity in bipolar patients. This altered balance between limbic and prefrontal networks is thought to underlie the abnormal emotional reactivity present in bipolar disorder and also mood switches (Houenou et al. 2012; Phillips et al. 2008).

5.2.1.2 Effects on the Hippocampus

In two areas of the adult human brain (the olfactory bulb and the dentate gyrus of the hippocampus), new neurons are generated throughout life and form an integral part of the normal functional circuitry (Lledo et al. 2006). These areas are natural targets for the neurotrophic effect of lithium. Therefore, Baykara and colleagues focused on the hippocampus (located in the medial part of the temporal lobe, which is also a key component of the emotion regulation network) (Baykara et al. 2012). Seventeen adolescents aged between 13 and 19 with type I bipolar disorder were included in the study. Six of them took lithium for 1–14 months. After this, their right hippocampal volumes were found to be significantly larger than before treatment. The authors did not find any change in hippocampus size in groups taking valproate or second-generation antipsychotics (SGA). However, this finding must be considered with caution, because this was a secondary aim of the study. The main finding was that there was no significant difference between the right and left hippocampus volumes of patients with bipolar disorder and controls.

A longitudinal controlled study of bilateral hippocampal volume increase in patients with bipolar disorder who took long-term lithium treatment is more convincing. The 12 included patients were medication naive, having never received any psychopharmacological treatment for psychiatric illness before entry into the study. Lithium was introduced simultaneously at the time of the first MRI scan (one brain MRI exam at the beginning of the study, another one after 2 years and the last one after 4 years). Results showed that bilateral hippocampal volumes were larger in the lithium (+4–5 %) group than in the unmedicated group. This increase happened mostly during the first 2 years of MRI monitoring. There was no brain volume difference at the beginning of the study between bipolar disorder subjects and control subjects. This suggests that lithium did not restore a decreased hippocampal volume due to bipolar disorder but that this medication has neurotrophic properties (Yucel et al. 2007).

Another longitudinal study reached a similar conclusion (Hajek et al. 2012). Whereas the comparison of hippocampal volumes between the non-lithium patients and controls showed marked differences ($p<0.05$), there were no significant differences between young bipolar disorder patients and controls or between the lithium-treated bipolar disorder and control participants (0.50 and 0.30, respectively). These results are also in line with previous investigations, which have shown a lack of hippocampal volume differences amongst lithium-treated patients or patients at the beginning of illness and controls. The smaller hippocampal volumes in bipolar disorder patients selected because of minimal lifetime exposure to lithium may be secondary to the burden of illness (mean 25.6 ± 9.8 years of illness and 10.5 ± 5.1 episodes) (Hajek et al. 2012).

A recent study about familial contributions to hippocampal morphology in bipolar disorder found that lithium-treated bipolar I patients had significantly larger global hippocampal volume ($p=0.03$) compared to healthy controls (9 %), non-bipolar co-twins (12 %) and non-lithium-treated bipolar I patients (8 %). In contrast, hippocampal volumes in non-lithium-treated bipolar I patients did not differ from those of non-bipolar co-twins and control twins. This result supports the hypothesis of neurotrophic effects of lithium in the hippocampus (Lyoo et al. 2010).

5.2.1.3 Effects on Fronto-limbic Structure

The fronto-limbic networks play a critical role in emotion regulation and in bipolar disorder. A study comparing hippocampal and amygdala volumes in older bipolar patients with controls using high-resolution MRI scans reported smaller hippocampal volumes in bipolar patients ($p=0.001$) and also a smaller right amygdala ($p=0.01$). Total hippocampal volume was negatively associated with the duration of depressive ($p=0.035$) and manic ($p=0.027$) episodes but not lithium use. It suggests that the neuroprogressive course was related to the severity of the disorder. Amygdala volumes were not associated with the duration of mood episodes (Wijeratne et al. 2013).

To assess the association between therapeutic response to lithium and fronto-limbic volumes in bipolar I patients, another group recruited 24 bipolar I patients and 11 healthy controls who underwent MRI scans at baseline and 4 weeks later. The bipolar I patients received lithium during this 4-week period. Mood symptoms were rated with the Hamilton Depression and the Young Mania Rating Scales at baseline and after 4 weeks. Response was defined as a 50 % decrease on either scale. Over both time periods, nonresponders had a smaller right amygdala than healthy comparisons and euthymic bipolar I patients ($p=0.035$ and $p=0.003$, respectively). When baseline and after-treatment volumes were compared, there was a significant enlargement in the left dorsolateral prefrontal cortex in bipolar I patients who responded to treatment ($p=0.002$ and $p=0.006$, respectively). Left hippocampus and right anterior cingulate cortex volumes decreased in nonresponders ($p=0.02$ and $p=0.0001$, respectively) (Selek et al. 2013).

5.2.1.4 Effects on Grey Matter Total Volume

A few studies have observed that the neurotrophic effect of lithium might extend beyond the hippocampus. In the study by Moore and colleagues, 28 bipolar depressed subjects were investigated at baseline (medication-free) and after 4 weeks of lithium administration. Significant increases in total brain grey matter volume in bipolar subjects were observed after 4 weeks of lithium administration ($p=0.0043$) (Moore et al. 2009).

A more recent cross-sectional study, quoted previously, shows that there were no statistically significant differences in grey matter in bipolar disorder patients compared to controls. However, importantly, patients taking lithium had significantly more grey matter volume compared to patients not taking lithium ($p<0.002$) (Emsell et al. 2013). A recent review reported similar results with a positive correlation between the proportion of subjects taking lithium and grey matter volume (Hafeman et al. 2012).

5.2.1.5 Effects on White Matter

The data about the effects of lithium on white matter are more heterogeneous and depend on the MRI technique used. White matter contains axons that allow communication between the different brain areas. A recent cross-sectional study showed that patients with euthymic bipolar I disorder had diffuse global white matter deficits ($p=0.018$) as compared to healthy subjects. Increasing age was significantly correlated with decrease in white matter volume ($p=0.039$) in both populations, but this effect was more pronounced in bipolar disorder patients (Emsell et al. 2013).

Several studies have explored the effect of lithium on white matter volumes. Most of them describe no statistically significant association.

Recently, a team used a transversal design to compare patients with and without lithium using diffusion MRI, a technique that reveals microscopic details about white matter tissue architecture (Le Bihan and Johansen-Berg 2011). In this work about tract-specific white matter structural disruption in patients with bipolar disorder, lithium significantly influenced the results: lithium use was associated with normal diffusivity values in tracts connecting the amygdala with the prefrontal cortex, contrary to the no-lithium patients (Benedetti et al. 2011). In another study from the same group, in a group of 70 patients with bipolar disorder, lithium was associated with the functional integrity of the brain and restored white matter connections.

Moreover, this parameter was associated to the GSK3-β genotype. Lithium inhibits glycogen synthase kinase 3-beta (GSK3-β). The less active GSK3-β promoter gene variants have been associated with less detrimental clinical features of bipolar disorder. GSK3-β gene variants and lithium can influence brain grey matter structure in psychiatric conditions. A neuroimaging cross-sectional study suggests that GSK3-β inhibition and lithium could counteract the detrimental influences of bipolar disorder on white matter structure (Benedetti et al. 2013).

Taken as a whole, anatomical neuroimaging results suggest that lithium's long-term benefits may be due to a neurotrophic action, particularly in grey matter. Studies of white matter are more heterogeneous and sparse and have to be confirmed in larger and longitudinal studies.

5.3 Lithium and Brain Function

The networks impacted by bipolar disorder are mainly those related to emotional, mnesic, reward and psychomotor processes. It has been suggested that lithium could contribute to restoring the malfunctions that are observed in these processes.

5.3.1 Emotion Processing

Abnormalities in emotional processing are a key component of the symptomatology of bipolar disorders. A meta-analysis of MRI functional studies has shown that patients with bipolar disorder display decreased activation and diminution of grey matter volumes in a cortical-cognitive brain network associated with the regulation of emotions. By contrast, patients exhibited increased activation in ventral-limbic brain regions that mediate the experience of emotions and the generation of emotional responses. This meta-analysis provides evidence for functional and anatomical alterations in bipolar disorder in brain networks associated with the experience and regulation of emotions (Houenou et al. 2011).

Lithium, as the main mood stabiliser in bipolar disorder, could possibly have functional effects on these networks. We previously described the positive effect lithium has on cerebral volumes, particularly in the fronto-limbic structures, grey matter volumes and possibly white matter structures. Some studies have tried to draw parallels between these volume increases and improvement of emotional control and therefore response to treatment. In a previously described longitudinal study in depressed bipolar subjects taking lithium (Selek et al. 2013), the authors reported a positive correlation between clinical improvement of the patients and the increase in volume of their fronto-limbic structures.

Similarly, the increase in grey matter volume induced by lithium was associated with a positive clinical response as measured by the Hamilton Depression Scale in the Lyoo et al. study (Lyoo et al. 2010). However, no study has yet identified direct correlations between lithium, brain structure and function and better emotion regulation using specific tasks.

5.3.2 Mnesic Function

In parallel to the hippocampus MRI measures in the study by Yucel and colleagues discussed previously, the mnesic performances of patients were evaluated. During 4 years of follow-up, the hippocampal volume increased significantly ($p<0.01$), and was associated with an improvement in some verbal memory tasks, but was not correlated to the subject's mood state. This finding suggests that lithium, by inducing an increase in the hippocampal volume that is implicated in the emotional, cognitive and mnesic functions, may be involved in the maintenance of cognition in patients with bipolar disorder, whereas clinically a decline of cognitive performance is the more usual outcome (Yucel et al. 2007; Malhi et al. 2016).

5.4 Disputes over the Neurotrophic Effect of Lithium

Neuroimaging studies have reported that lithium can increase the volume of grey matter in the human brain, a finding that has been ascribed to the neurotrophic or neuroprotective effects of the drug. Lithium, however, might directly influence the intensity of the magnetic resonance signal, so that it may be possible that the volumetric findings are artefacts, being a consequence of altered image contrast.

To test this hypothesis, a Newcastle team conducted a longitudinal study on the neurotrophic effects of lithium by using two established analysis techniques: the voxel-based morphometry (the method usually used for this kind of protocol) and the normalisation/atrophy technique (a technique that operates by detecting changes in the position of the boundaries of the brain). Anatomical and quantitative MRI T1 scans were acquired in 31 healthy young men before and after taking either lithium or placebo for 11 days. Voxel-based morphometry revealed that grey matter volume was increased by lithium but not placebo ($p=0.001$), whereas the normalisation/atrophy technique showed no difference between lithium and placebo ($p=0.23$). Taking lithium reduced the T1 relaxation of the grey matter only ($p=0.008$). The authors concluded that atomic-level interactions combined with the classical biological effects of lithium contributed to the signal change. Thus the results of previous studies suggesting a neurotrophic effect for lithium could possibly be caused by poor interpretation linked to an artefact or variance in analysis method (Cousins et al. 2013).

These results are, however, subject to debate (Ferrier et al. 2013): studies of lithium's neurotrophic property, indirectly measured via N-acetylaspartate concentration (a marker of neuronal viability), suggest the opposite hypothesis (Moore et al. 2002). In addition, as previously reported by Vernon and colleagues (Vernon et al. 2012), there are some increases in whole brain volume and, more particularly, in grey matter areas in an animal model. These measurements were carried out with both an MRI device and postmortem examination in animals administered lithium for 8 weeks.

However, studies using MRI to examine the neurotrophic effect of lithium in humans have produced heterogeneous results: a study comparing 32 bipolar disorder lithium-naive patients and 32 healthy subjects found no structural or volumetric difference within the (left and right) hippocampus ($p=0.804$ and $p=0.375$, respectively) and (left and right) amygdala ($p=0.475$ and $p=0.489$, respectively) between the two groups. Future studies that further elucidate the impact of age, course of illness and medication on amygdala structure in bipolar disorder are warranted (Foland-Ross et al. 2013).

Further, the reported associations between hippocampal volumes and lithium medication may reflect the direct neurochemical effects of lithium or indirect effects secondary to symptomatic improvement, or prevention of relapse, of bipolar disorder. In the study by Hajek and colleagues examining the neuroprotective effect of lithium on hippocampal volumes in bipolar disorder, independent of long-term treatment response, the authors showed that the 19 patients with bipolar disorder and less than 3 months of lifetime lithium exposure (the non-lithium group) had smaller hippocampal volumes than the 37 bipolar disorder patients with at least 2 years of lithium treatment (the lithium group) ($p=0.009$). The association between lithium treatment and hippocampal volume appeared to be independent of the long-term treatment response and occurred even in subjects with episodes of bipolar disorder while on lithium. Thus, in this study, whatever its clinical efficacy, the hippocampal volume was bigger in patients taking lithium (Hajek et al. 2014).

5.5 The Future of Lithium Imaging

Except for familial aggregation of good response, few clinical markers allow the psychiatrist to predict response to lithium. Discovering biomarkers of a future good or bad response to lithium is therefore of critical value. Moreover, comparison between good and bad responders could allow us to better understand the conditions that are required for the efficient action of lithium. Finally, we do not know whether the correlation between cerebral volumes and lithium is an indirect effect, e.g. caused by the clinical response.

5.5.1 Neuroimaging and Response to Lithium

Interestingly, the longitudinal study by Moore and colleagues, described previously, found that only lithium responders had a significant increase in grey matter volume in the prefrontal cortex ($p=0.003$) and in the left subgenual prefrontal cortex volume ($p=0.0786$) after 4 weeks of lithium treatment (Moore et al. 2000). Similarly, we saw in the study by Lyoo and colleagues (Lyoo et al. 2010) that an improvement in depressive symptoms was associated with grey matter volume augmentation in bipolar disorder patients. This effect seemed to be specific to lithium, since the same protocol with valproate yielded different results. These two studies suggest potential biomarkers of a good response to lithium; however they are only assessable after the introduction of treatment (volume increase of grey matter).

According to the findings of another study (Selek et al. 2013), decreased left hippocampus and right anterior cingulate cortex volumes may be markers of non-response to lithium amongst patients. A smaller right amygdala may reflect symptomatic remission and, more interestingly, be a pretreatment marker of treatment non-response. Increases in the left prefrontal cortex and left dorsolateral prefrontal cortex as a result of lithium treatment may relate to lithium's neurotrophic effects. The authors suggest in their discussion that increases in brain volumes observed in patients responsive to lithium may be related to its neurotrophic properties. But this last point remains speculative. Indeed, one article, discussed above, found increases in hippocampal volume whether the patient was a good or bad responder to lithium (Hajek et al. 2014).

In summary, these studies suggest the existence of some neuroanatomical correlates for lithium response, but, to date, only one study has identified pretreatment markers of outcome. Further, these studies include very small samples, with short follow-up periods and heterogeneous samples, and only one study has a control group treated with another mood stabiliser.

5.5.2 Measures of the Intracerebral Lithium Concentration

The measurement of plasma lithium levels is currently the gold standard for monitoring treatment with lithium. Optimal plasma lithium levels should be in a range of therapeutic efficiency without overdosing (typically between 0.8 and 1.2 mmol/L for the long-release form) (see Appendix 3). Nevertheless, lithium crossing of the blood-brain barrier may vary within and between individuals. The intracerebral concentration of lithium is probably more relevant for lithium treatment monitoring.

A brain MRI technique—nuclear magnetic resonance spectroscopy imaging (MRSI)—may allow measurements of intracerebral concentrations of lithium in humans in vivo, by exploiting the magnetic properties of lithium atoms. This method does not require any injection of contrast but uses specific and expensive antennas, a long acquisition duration and complex post-processing; it is currently only used in a few research settings in the world.

The relevance of this technique has been demonstrated by a few studies. Forester and colleagues, in an early work, used lithium MRSI to show that intracerebral lithium concentration in the cingulate cortex was correlated with the rate of NAA (reflecting the number of live neurons) (Forester et al. 2008).

More recently, in 2012, a study demonstrated the first whole brain 'high spatial resolution' lithium MRSI in 15 patients with bipolar disorder treated with lithium. Brain lithium levels were strongly correlated with serum lithium concentration.

However, the lithium distribution was non-uniform throughout the brain, especially in subjects at a higher therapeutic serum lithium level. This finding may suggest that lithium targets specific brain sites that are associated with therapeutic effect. Only a small number of patients were included in this study, and more research should be conducted in bipolar disorder patients on lithium therapy using 3D Li MRSI (Lee et al. 2012).

These two studies show the possibility of studying intracerebral concentrations of lithium in vivo, through MRI spectroscopy, offering fresh perspectives on the understanding of its mechanism of action through its regional distribution and correlation with plasma measures. In subsequent work, prospective cohort studies of good and poor responders to lithium using spectroscopic technique could substantially increase our knowledge on this subject.

5.6 Conclusion

Neuroimaging studies strongly suggest that lithium has neurotrophic properties, particularly in the cerebral grey matter. We still have to identify the cell types involved and the link with its therapeutic effects. These studies also suggest that some anatomical parameters are correlated with response to lithium. Finally, MR spectroscopy may give clues to the concentration and distribution of intracerebral lithium and help us to better understand its mechanism of action. Brain imaging of lithium gives us a glimpse of possible future 'bedside' use to help us in predicting therapeutic response to lithium in patients with bipolar disorder. This could ultimately lead to a 'personalised' prescription of lithium.

5.7 Summary

Many studies have compared patients with bipolar disorder with and without lithium using structural MRI. These studies have had cross-sectional or longitudinal designs. Most of these studies have found increases in grey matter volumes in patients taking lithium, particularly in the hippocampus. Findings in white matter are far fewer and more disparate. A very recent technique, MR spectroscopy, may in the future allow a better understanding of lithium's mechanism of action, by describing its distribution in the brain and correlation of its intracerebral concentration with clinical plasma lithium levels. Further, a few studies are beginning to unravel biomarkers of response to lithium, paving the way to a personalised prescription of lithium.

References

Baykara B, Inal-Emiroglu N, Karabay N et al (2012) Increased hippocampal volumes in lithium treated adolescents with bipolar disorders: a structural MRI study. J Affect Disord 138:433–439

Benedetti F, Absinta M, Rocca MA et al (2011) Tract-specific white matter structural disruption in patients with bipolar disorder. Bipolar Disord 13:414–424

Benedetti F, Bollettini I, Barberi I et al (2013) Lithium and GSK3-beta promoter gene variants influence white matter microstructure in bipolar disorder. Neuropsychopharmacology 38:313–327

Cousins DA, Aribisala B, Nicol Ferrier I et al (2013) Lithium, gray matter, and magnetic resonance imaging signal. Biol Psychiatry 73:652–657

Emsell L, Leemans A, Langan C et al (2013) Limbic and callosal white matter changes in euthymic bipolar I disorder: an advanced diffusion magnetic resonance imaging tractography study. Biol Psychiatry 73:194–201

Ferrier NI, Blamire AM, Cousins DA (2013) Reply to: effects of lithium on magnetic resonance imaging signal might not preclude increases in brain volume after chronic lithium treatment. Biol Psychiatry 74:e41–e42

Foland-Ross LC, Thompson PM, Sugar CA et al (2013) Three-dimensional mapping of hippocampal and amygdalar structure in euthymic adults with bipolar disorder not treated with lithium. Psychiatry Res 211:195–201

Forester BP, Finn CT, Berlow YA et al (2008) Brain lithium, N-acetyl aspartate and myo-inositol levels in older adults with bipolar disorder treated with lithium: a lithium-7 and proton magnetic resonance spectroscopy study. Bipolar Disord 10:691–700

Hafeman DM, Chang KD, Garrett AS et al (2012) Effects of medication on neuroimaging findings in bipolar disorder: an updated review. Bipolar Disord 14:375–410

Hajek T, Cullis J, Novak T et al (2012) Hippocampal volumes in bipolar disorders: opposing effects of illness burden and lithium treatment. Bipolar Disord 14:261–270

Hajek T, Bauer M, Simhandl C et al (2014) Neuroprotective effect of lithium on hippocampal volumes in bipolar disorder independent of long-term treatment response. Psychol Med 44:507–517

Houenou J, Frommberger J, Carde S, et al (2011) Neuroimaging-based markers of bipolar disorder: evidence from two meta-analyses. J Affect Disord 132(3):344–355. doi:10.1016/j.jad.2011.03.016. Epub 2011 Apr 5

Houenou J, d'Albis MA, Vederine FE et al (2012) Neuroimaging biomarkers in bipolar disorder. Front Biosci (Elite Ed) 4:593–606

Le Bihan D, Johansen-Berg H (2011) Diffusion MRI at 25: exploring brain tissue structure and function. Neuroimage 61(2):324–341. doi:10.1016/j.neuroimage.2011.11.006. Epub 2011 Nov 20

Lee JH, Adler C, Norris M et al (2012) 4-T 7Li 3D MR spectroscopy imaging in the brains of bipolar disorder subjects. Magn Reson Med 68:363–368

Lledo PM, Alonso M, Grubb MS (2006) Adult neurogenesis and functional plasticity in neuronal circuits. Nat Rev Neurosci 7:179–193

Lyoo IK, Dager SR, Kim JE et al (2010) Lithium-induced gray matter volume increase as a neural correlate of treatment response in bipolar disorder: a longitudinal brain imaging study. Neuropsychopharmacology 35:1743–1750

Malhi GS, Byrow Y, Fritz K, Das P, Baune BT, Porter RJ, Outhred T (2015) Mood disorders: neurocognitive models. Bipolar Disord 17(suppl 2):3–20

Malhi GS, McAulay C, Gershon S, Gessler D, Fritz K, Das P, Outhred T (2016) The Lithium Battery: assessing the neurocognitive profile of lithium in bipolar disorder. Bipolar Disord 18:102–115

Moore GJ, Bebchuk JM, Wilds IB et al (2000) Lithium-induced increase in human brain grey matter. Lancet 356:1241–1242

Moore CM, Demopulos CM, Henry ME et al (2002) Brain-to-serum lithium ratio and age: an in vivo magnetic resonance spectroscopy study. Am J Psychiatry 159:1240–1242

Moore GJ, Cortese BM, Glitz DA et al (2009) A longitudinal study of the effects of lithium treatment on prefrontal and subgenual prefrontal gray matter volume in treatment-responsive bipolar disorder patients. J Clin Psychiatry 70:699–705

Phillips ML, Ladouceur CD, Drevets WC (2008) A neural model of voluntary and automatic emotion regulation: implications for understanding the pathophysiology and neurodevelopment of bipolar disorder. Mol Psychiatry 13(829):833–857

Selek S, Nicoletti M, Zunta-Soares GB et al (2013) A longitudinal study of fronto-limbic brain structures in patients with bipolar I disorder during lithium treatment. J Affect Disord 150:629–633

Vernon AC, Natesan S, Crum WR et al (2012) Contrasting effects of haloperidol and lithium on rodent brain structure: a magnetic resonance imaging study with postmortem confirmation. Biol Psychiatry 71:855–863

Wijeratne C, Sachdev S, Wen W et al (2013) Hippocampal and amygdala volumes in an older bipolar disorder sample. Int Psychogeriatr 25:54–60

Yucel K, McKinnon MC, Taylor VH et al (2007) Bilateral hippocampal volume increases after long-term lithium treatment in patients with bipolar disorder: a longitudinal MRI study. Psychopharmacology (Berl) 195:357–367

Lithium and Circadian Rhythms

6

Pierre Alexis Geoffroy and Bruno Etain

Abstract

Circadian rhythm abnormalities are associated with bipolar disorder pathogenesis, clinical manifestations and relapse. Lithium is the first-line treatment for bipolar disorder in preventing relapses and suicide and is a thoroughly investigated chronobiologic agent. Indeed, lithium acts at a physiological level on period, phase, amplitude and coupling of biological rhythms and at a molecular level on circadian genes and proteins. Further, lithium appears to interact with environmental light through the retinal–hypothalamic–pineal pathway to influence circadian rhythms. The therapeutic action of lithium in bipolar disorder patients would appear to be determined, at least partly, by its effects on biological rhythms.

P.A. Geoffroy (✉)
Inserm, U1144, Paris 75006, France

Université Paris Descartes, UMR-S 1144, Paris 75006, France

Université Paris Diderot, Sorbonne Paris Cité, UMR-S 1144, Paris 75013, France

Département de Psychiatrie et de Médecine Addictologique,
AP-HP, GH Saint-Louis – Lariboisière – F. Widal, Paris, Cedex 10 75475, France

Fondation FondaMental, Créteil 94000, France
e-mail: pierre.a.geoffroy@gmail.com

B. Etain
Fondation FondaMental, Créteil 94000, France

Université Paris Descartes, UMR-S 1144, Paris 75006, France

Université Paris Diderot, Sorbonne Paris Cité, UMR-S 1144, Paris 75013, France

Département de Psychiatrie et de Médecine Addictologique,
AP-HP, GH Saint-Louis – Lariboisière – F. Widal, Paris, Cedex 10 75475, France
e-mail: bruno.etain@inserm.fr

G.S. Malhi et al. (eds.), *The Science and Practice of Lithium Therapy*,
DOI 10.1007/978-3-319-45923-3_6

Key Points

- Circadian rhythm abnormalities are associated with bipolar disorder pathogenesis, clinical manifestations and relapses.
- Lithium is the first-line treatment for bipolar disorders for preventing mood relapses and suicide.
- Lithium mildly lengthens the circadian period of behavioural rhythms and delays the phase of behavioural and physiological circadian rhythms (such as sleep/wakefulness and body temperature rhythms) in mammals and humans.
- Pharmacogenetic studies have observed associations between response to lithium in bipolar disorders and several variants in circadian genes.
- Lithium affects the expression of numerous circadian genes, especially by inhibiting *GSK3β* expression and activating *CLOCK* transcription.

6.1 Introduction

Bipolar disorder is a severe psychiatric disorder that usually commences in early adulthood. It is characterized by recurrent manic and depressive episodes and affects 1–4 % of the general population worldwide (Merikangas et al. 2007). Among mental, neurological and substance-use disorders, bipolar disorder is the fourth most important contributor to the global disease burden (Collins et al. 2011). The severity and poor prognosis of bipolar disorder are linked to its high rate of recurrence, with a mean recurrence rate of 60 % within 2 years of an index episode, despite medication (Geddes and Miklowitz 2013). Bipolar disorder has a complex inheritance, implicating both genetic and environmental factors (Lichtenstein et al. 2009).

Cumulative evidence shows circadian rhythm abnormalities in association with bipolar disorder pathogenesis, clinical manifestations and relapse. Recent work shows abnormalities of circadian rhythms in bipolar disorder patients, both in the acute and remitted states of the disorder (McClung 2007, 2011; Murray and Harvey 2010; Etain et al. 2011). Instability in the circadian rhythm contributes to the underlying neurobiology and genetic susceptibility of bipolar disorder, making circadian rhythm instability a major avenue for research aiming to elucidate factors associated with bipolar disorder treatment response (Hasler et al. 2006).

In the 1880s, Carl Lange discovered that lithium salt could be used to treat bipolar disorder; this was rediscovered by John Cade in 1949, who validated its treatment efficacy. Since then, lithium has been the cornerstone of bipolar disorder treatment in all international therapeutic guidelines (Nivoli et al. 2010; Geoffroy et al. 2012; Malhi et al. 2015), being the first-line treatment for preventing bipolar disorder relapses of either polarity (Grunze et al. 2009, 2010; Yatham et al. 2009; Goodwin 2009). Furthermore, lithium is the only treatment shown to decrease suicide risk effectively (Yerevanian and Choi 2013). Although lithium's mechanism of action is not fully understood, its therapeutic action has been linked to its ability to modify circadian rhythms (Klemfuss 1992). This literature, in addition to contributing to a better knowledge of the mechanism of lithium's action, paves the way to identifying putative circadian biomarkers of response efficacy. This is of particular interest as, to date, few predictive markers of lithium's response efficacy exist, and none have been relevant enough to be implemented in clinical practice (Geoffroy et al. 2014a). Among these markers, a history of prophylactic response to lithium in first-degree relatives, course of episodes and complete remission between episodes has been proposed (Kleindienst et al. 2005).

In this context, circadian biomarkers could be promising candidate biomarkers for the study of an individual's response to lithium (Geoffroy et al. 2014b). Given the growing evidence emphasizing the clinical efficacy of lithium on acute episodes and relapse prevention, and its link to the stabilization of circadian rhythms, this chapter will specifically focus on lithium's action on circadian rhythms.

6.2 What Are Circadian Rhythms?

Circadian rhythms encompass all physiological (biological and behavioural) processes with approximately 24 h of periodicity. While large parts of the mammalian clock are yet to be unravelled, genetic, molecular, biochemical and behavioural observations in many mammalian species, including humans, have provided a widely accepted general clock model. Circadian timing in mammals is organized within a hierarchy of multiple circadian oscillators, powerfully co-ordinated by a highly specialized master pacemaker located in the suprachiasmatic nuclei (SCN) of the anterior hypothalamus (Reppert and Weaver 2002). This endogenous system led by the SCN can adapt to a daily synchronization entrained by environmental time signals such as daylight, environmental temperature, alimentation or social activities—also called social 'zeitgebers' (literally 'time givers') (Grandin et al. 2006). One of the most important synchronizers is light that comes from the retina to the SCN through the visual pathway, entraining circadian rhythms to the solar day. Beyond the SCN, peripheral tissue clocks also exist, which may either be synchronized with the SCN or independent from this master clock (Reppert and Weaver 2002) (Fig. 6.1).

Clock genes are widely expressed throughout peripheral organs and the brain, including forebrain regions that regulate mood, anxiety and cognition (McClung 2007; Murray and Harvey 2010). In both the SCN and peripheral clocks, the expression and rhythmic regulations of clock genes are generated via interconnected transcriptional–translational feedback loops (Reppert and Weaver 2002). Briefly, a first positive feedback loop is formed by two proteins, brain and muscle aryl hydrocarbon receptor nuclear translocator-like protein 1 (BMAL1) and circadian locomotor output cycle kaput (CLOCK), that heterodimerize and activate several clock genes such as period1 (PER1), period2 (PER2), cryptochrome1 (CRY1) and cryptochrome2 (CRY2) (Jin et al. 1999). A second negative feedback loop follows PER and CRY accumulation in the cytoplasm, which then translocate to the nucleus and inhibit their transcription (Jin et al. 1999). Numerous kinases are involved in

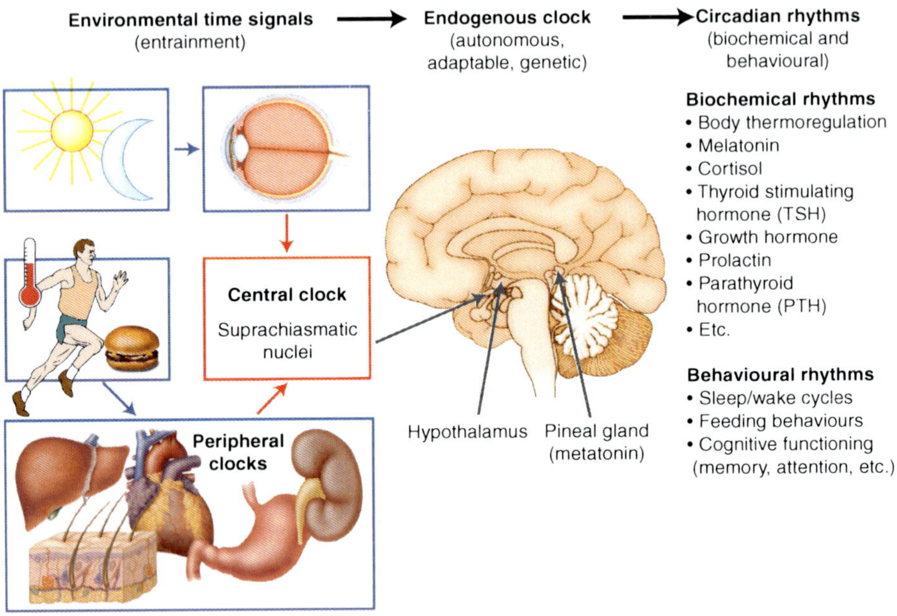

Fig. 6.1 Clock model of circadian rhythms

translational modifications of clock proteins regulating protein stability and translocation, including glycogen synthase kinase 3β (GSK3β), mitogen-activated protein kinase (MAPK), casein kinase (CK)I and CKII. Downstream of the core clock, thousands of clock-controlled genes (CCGs) oscillate rhythmically in most tissues (Reppert and Weaver 2002).

6.2.1 Lithium's Effect on Circadian Rhythms

Lithium's hypothetical mode of efficacy in bipolar disorder is, at least partly, via the stabilization of circadian rhythms. Indeed, lithium slightly lengthens the circadian period of behavioural rhythms and delays the phase of behavioural and physiological circadian rhythms (such as sleep/wakefulness and body temperature rhythms). These effects occur in many species, as well as humans, including those diagnosed with bipolar disorder (Mellerup et al. 1978; Kripke et al. 1978, 1979; Johnsson et al. 1979, 1980, 1983; Pflug and Engelmann 1987), including drosophila (Padiath et al. 2004; Dokucu et al. 2005), rats (Danguir et al. 1976; Kripke and Wyborney 1980; McEachron et al. 1981, 1982), hamsters (Klemfuss 1992; LeSauter and Silver 1993; Hafen and Wollnik 1994; Klemfuss and Kripke 1995), cats (Lanoir and Lardennois 1977), bats (Subbaraj 1981) and *Rhyparobia maderae* (or Madeira cockroach) (Hofmann et al. 1978). These studies consistently found that lithium lengthens the period of circadian rhythms in constant darkness, delays the phase of sleep/wakefulness rhythms, delays by 4–5 h the peak temperature elevation in the diurnal variation in body temperature and changes activity rhythm amplitude and duration. Chronic lithium treatment also contributes to circadian rhythmicity by improving the day-to-day stability of activity. Moreover, this increased stability appears to be dose related (Klemfuss and Kripke 1995). However, some inconsistencies between studies are observed, mostly regarding the direction of change in amplitude and the duration of activity. These discrepancies may be due to the type of animal studied (e.g. whether it is a nocturnal or diurnal mammal) and may also be evident when bipolar disorder patients are compared to healthy controls.

Using cellular models, Abe and colleagues demonstrated that lithium acts directly on the mammalian suprachiasmatic nuclei (SCN), considered to be the central pacemaker of circadian rhythms. They showed that lithium increased the circadian period of the firing rate rhythms of cultured SCN neurons, in a concentration-dependent manner (Abe et al. 2000). As such, this study indicates that lithium lengthens the period of behavioural circadian rhythms and the free-running period of neurons, by acting directly on the SCN.

Lithium acts on circadian rhythms not only at a behavioural but also at a biochemical level. Indeed, when lithium was administered to rats, compared to rats fed a control diet, McEachron and colleagues observed that plasma prolactin, PTH, corticosterone, aldosterone, serum calcium and magnesium and cerebellar calcium and magnesium showed larger delays in their circadian rhythms (McEachron et al. 1982). These findings reinforce the hypothesis that lithium delays circadian rhythms, including biochemical circadian rhythms. Further, it has been reported that lithium may shorten the duration of the nocturnal melatonin peak in rats (Werstiuk et al. 1984; Seggie et al. 1985). Subsequently, it was shown in lithium-treated rats that peak times of the circadian rhythms of glucose, calcium and potassium were delayed, respectively, by 3 h, 6 h and 6 h, whereas the circadian rhythms of malondialdehyde and lactic acid were advanced, respectively, by 9 h and 3 h (Subramanian et al. 1998).

In bipolar disorder patients, Kripke and colleagues investigated complete cycles of mania and depression, observing that patients whose circadian clocks appeared too slow were lithium nonresponders (NR), with lithium being effective in patients presenting with a 'circadian rhythm [that] free-ran fast', thanks to a slowing action on 'an overfast circadian clock to prevent desynchronization' (Kripke et al. 1978). In addition, in a comparison of cases with and without lithium, it was observed that internal desynchronization between the temperature rhythm and the sleep/wakefulness rhythm was only observed in patients without lithium (Johnsson et al. 1983). A desynchronization between two internal circadian rhythms may be a marker of individual vulnerability to mood symptoms.

Overall, the efficacy of lithium appears to lie in its ability to delay and resynchronize overly fast circadian rhythms. Bipolar disorder patients may present with an endogenous circadian oscillator (about a 24-h clock as for all mammals) that runs too fast or that has a phase that is too advanced.

6.2.2 Lithium's Influence on the Molecular Circadian Clock

Lithium also acts at a molecular level, impacting on the dynamics of clock genes and protein rhythms in the SCN and peripheral tissues (Li et al. 2012). Indeed, lithium is known to affect the expression of numerous key circadian genes, with a possible key effect mediated by inhibiting glycogen synthase kinase 3 beta (GSK3β) (Prickaerts et al. 2006; Jope and Roh 2006). GSK3β is a serine-threonine kinase involved in neuronal regulation by acting on a number of pathways that modulate cell survival, growth cone retraction, gene expression and microtubule formation (Harwood and Agam 2003). More recent findings demonstrate, in mammal cell lines, that the rhythmic activity of GSK3 isoforms is intrinsically and mutually connected to clock gene regulation (Besing et al. 2015).

By increasing GSK3β phosphorylation and inhibition, lithium thereby delays the phase of rhythmic clock gene expression, as demonstrated in studies using cultured cell lines (Iitaka et al. 2005) or animal models (Padiath et al. 2004). Lithium can act on enzymes that regulate the transcription of other circadian genes, either via direct inhibition, as with GSK3α and GSK3β, or indirectly by regulating other mechanisms such as the formation of a signalling complex comprised of beta-arrestin 2 (βArr2) and Akt (Freland and Beaulieu 2012). It has also been proposed that lithium influences circadian rhythms via its regulation of REV-ERBα, the protein product of the *NR1D1* gene. REV-ERBα is a phosphorylation target of GSK3. By inhibiting GSK3, lithium prevents the GSK3 phosphorylation of the REV-ERBα protein, which then degrades (Can et al. 2014).

Lithium also increases the oscillation amplitude of the PER2 protein rhythms in both central (SCN) and peripheral clocks (lung and lung fibroblasts) (Li et al. 2012). The authors hypothesized that this effect is mediated by lithium regulating *PER2*, transcriptionally or post-transcriptionally, given that lithium selectively increased PER2 mRNA levels (Osland et al. 2011). Osland and colleagues examined whether lithium affects the expression of genes regulating the circadian clock in cultured NIH-3 T3 cells synchronized by serum shock. They observed that lithium significantly increased the expression of *CRY1* and *PER2*, also prolonging the *PER2* period, and significantly decreased the expression of *PER3*, *CRY2*, *BMAL1*, *E4 binding protein 4* (*E4BP4*) and *REV-ERBα* (Osland et al. 2011). Lithium's effects on *PER2*, delaying the phase and lengthening the period of the PER2 rhythm, have been confirmed by several reports (Johansson et al. 2011). Moreover, McQuillin and colleagues have studied brain mRNA (39,000 genes) from ten mice after treatment with lithium compared with ten untreated controls (McQuillin et al. 2007). They observed, among the group of 121 genes with significant changes, that *PER2 and CRY1* demonstrated significant changes in gene expression that survived Bonferroni correction (McQuillin et al. 2007).

In addition, lithium improves diurnal activity rhythm and shows periodic activity alterations in transgenic mice with neuron-specific expression of mutant Polg (D181A)—an animal model with chronobiological abnormalities (Kato et al. 2007). Similarly, results in *Drosophila* show that lithium partially rescues the shortening of circadian period when the *GSK3β* gene is overexpressed in specific circadian pacemaker neurons (Dokucu et al. 2005). These authors also showed that lithium lengthens the period in GSK3β heterozygous mutants and double-time long mutants (Dokucu et al. 2005). Other circadian transgenic mice have been developed, including mice carrying a mutation in the *CLOCK* gene (Roybal et al. 2007). Interestingly, these transgenic mice display mania-like behaviours, including hyperactivity, decreased sleep, lowered depression-like behaviour and lower anxiety, coupled to increased reward value for cocaine, sucrose and medial forebrain bundle stimulation. These mania-like changes are rescued by chronic lithium administration (Roybal et al. 2007).

6.2.3 Results from Pharmacogenetics

Pharmacogenetics in bipolar disorder patients shows associations between lithium response and several variants in circadian genes. Among 88 patients with bipolar I, Benedetti and colleagues studied the association of the *GSK3β*(-50 T/C) polymorphism with the therapeutic response to lithium, demonstrating that the long-term response is influenced by the *GSK3β*-50 T/C polymorphism (Benedetti et al. 2005). However, later studies failed to replicate the influence of *GSK3β*-50 T/C on lithium efficacy (Szczepankiewicz et al. 2006), finding no association (Michelon et al. 2006). GSK3β is also known to phosphorylate and stabilize the orphan nuclear receptor REV-ERBα, one of the principal components of the circadian rhythm system that is involved in the cyclic regulation of brain and muscle Arnt-like protein-1 (BMAL1). Lithium induces the degradation of REV-ERBα, thereby regulating *BMAL1* gene expression, implicating REV-ERBα as a significant lithium target in its mechanism of action via circadian rhythm regulation (Yin et al. 2006). The association of the gene encoding for REV-ERBα (*NR1D1*) and the response to lithium in bipolar disorder patients have provided some controversial results. An initial study in 199 Sardinian bipolar disorder patients did not show any significant effect of any *NR1D1* polymorphisms with response to lithium (Manchia et al. 2009). However, Campos-de-Sousa and colleagues observed a significant association between the variant rs2314339 in *NR1D1* and lithium response (Campos-de-Sousa et al. 2010). Furthermore, the role for REV-ERBα in the therapeutic mechanism of lithium has recently been reinforced. McCarthy and colleagues conducted a candidate gene association study for 16 variants in seven circadian clock genes and lithium response in 282 Caucasian bipolar disorder patients (McCarthy et al. 2011). They found that a variant in the promoter of *NR1D1* (rs2071427) and a variant in *CRY1* (rs8192440) were nominally associated with the lithium response (McCarthy et al. 2011). Also, *GSK3β* and *NR1D1* genotypes, when considered together, predicted the response to lithium robustly and additively, with the lithium response being proportional to the number of lithium response-associated alleles (McCarthy et al. 2011). Finally, Rybakowski and colleagues, in 115 bipolar disorder patients, examined the association of the lithium response to four circadian genes: *CLOCK*, *ARNTL*, timeless circadian clock (*TIMELESS*) and *PER3*. They found an association with the response to lithium and polymorphisms in the *ARNTL* and *TIMELESS* genes, but not with *CLOCK* and *PER3* genes (Rybakowski et al. 2014).

These early results from pharmacogenetic studies of the role of the circadian system in lithium response are promising, but have to be seen as preliminary and requiring further replication.

Table 6.1 summarizes the circadian pharmacogenetic studies of response to lithium in bipolar disorder patients.

Table 6.1 A summary of the circadian pharmacogenetic studies of response to lithium in bipolar disorder patients

Gene	Sample	Association (yes/no)	Study design	Reference
NR1D1	199 BD lithium responders (57FR, 142PR+NR)	No	Retrospective	Manchia et al. (2009)
	170 BD	*Yes*[a]	Prospective	Campos-de-Sousa et al. (2010)
	282 BD (148R, 134NR)	*Yes*[b]	Retrospective	McCarthy et al. (2011)
GSK3β	88 BD I lithium responders	*Yes*[a]	Prospective	Benedetti et al. (2005)
	134 BD I	No	Retrospective	Michelon et al. (2006)
	89 BD (23 ER, 47 PR, 19 NR)	No	Retrospective	Szczepankiewicz et al. (2006)
	184 BD (92R, 92NR)	No	Retrospective	Bremer et al. (2007)
	282 BD (148R, 134NR)	*Yes*[b] in association with NR1D1 (trend alone)	Retrospective	McCarthy et al. (2011)
	101 BD (24ER, 51PR, 26NR)	No	Retrospective	Rybakowski et al. (2012)
CRY1	282 BD (148R, 134NR)	*Yes*[b]	Retrospective	McCarthy et al. (2011)
ARNTL	115 BD	*Yes*[b]	Retrospective	Rybakowski et al. (2014)
TIMELESS	115 BD	*Yes*[b]	Retrospective	Rybakowski et al. (2014)
PER3	115 BD	No	Retrospective	Rybakowski et al. (2014)
CLOCK	115 BD	No	Retrospective	Rybakowski et al. (2014)

BD bipolar disorder; *BD I* bipolar disorder I; *ER* excellent responders; *FR* full responders; *R* responders; *PR* partial or poor responders; *NR* nonresponders; *ARNTL* aryl hydrocarbon receptor nuclear translocator-like; *CLOCK* circadian locomotor output cycle kaput; *CRY1* cryptochrome1; *GSK3B* glycogen synthase kinase 3 beta; *NR1D1* nuclear receptor subfamily 1, group D, member 1; *PER* period circadian clock; *TIMELESS* timeless circadian clock

[a]Corrected for multiple testing (corrected association)

[b]Not corrected for multiple testing (nominal association)

6.3 Lithium and the Retinal–Hypothalamic–Pineal Pathway

We previously reported that lithium can shorten the duration of the nocturnal melatonin peak. Indeed, long-standing data show that lithium modifies the function of the retinal–hypothalamic–pineal pathway (Seggie et al. 1984).

Melatonin is a pineal gland secreted neurohormone that promotes sleep. The pineal gland is regulated by the SCN. Melatonin secretion follows a diurnal rhythm, with its synthesis determined by the activity of the serotonin N-acetyltransferase (AANAT) enzyme in a reaction that requires acetyl coenzyme A (AcCoA) (Hickman et al. 1999). Friedman and Yocca observed in 1981 that lithium causes an acute decrease in dark-induced activity of rat pineal AANAT. These authors also showed that chronic lithium treatment suppresses the amplitude and delays the peak of the diurnal cycle of AANAT activity (Friedman and Yocca 1981). Later these authors clarified that chronic lithium effects, which suppressed the peak activity and maximal concentrations of pineal AANAT and melatonin, were not related to changes in the precursor tryptophan indoles (5-hydroxytryptophan, serotonin and N-acetylserotonin), but to a lithium-induced desensitization of pineal beta adrenergic receptors (Yocca et al. 1983), the stimulation of which by norepinephrine is necessary for pineal melatonin synthesis. Later work confirmed that lithium changes the profile of the melatonin rhythm in the pineal gland, inducing a delay of the melatonin acrophase and producing another peak (Pablos et al. 1994).

Seggie and colleagues examined lithium's action on this pathway with albino rats that lack the enzyme for synthesis of eye pigment, which is important in the regulation of light cued rhythms (Seggie et al. 1982). They found that, compared to controls, rats treated with lithium had higher plasma corticosterone levels during the middle of the dark hours. The authors suggested that lithium broadens the 24-h rhythms leading to peaks that appear earlier and last longer (Seggie et al. 1982). Lithium also affects serum melatonin differently in albino compared to pigmented eye rats (Werstiuk et al. 1984). Further, lithium levels affect the circadian pattern of melatonin, notably during the dark period, including within the serum, retina and hypothalamus (Seggie et al. 1983). In addition, retinal melatonin levels, as with serum melatonin levels, were reduced by lithium.

Taken together, these findings in animals suggest that lithium modulates melatonin levels along the retinal–hypothalamic–pineal pathway, exerting its therapeutic action via alterations in the light sensitivity and light-induced norepinephrine receptor activity, thereby altering circadian rhythm synchronization (Seggie et al. 1987).

In healthy adults, data, in part, support such lithium effects on nocturnal melatonin secretion and light sensitivity. Hallam and colleagues showed that lithium has a significant effect on light sensitivity, but not on overall melatonin synthesis (Hallam et al. 2005).

Lithium may act by altering light sensitivity and hence entrainment of biological rhythms (Seggie 1988). These reported lithium effects on the retinal–hypothalamic–pineal pathway have been proposed to explain its therapeutic action in bipolar disorder patients that are supersensitive to light.

6.4 Summary

Lithium is a thoroughly investigated chronobiologic agent. The therapeutic action of lithium in bipolar disorder patients would appear to be determined, at least partly, by its effects on biological rhythms. Indeed, lithium acts at a physiological level on period, phase, amplitude and coupling of biological rhythms and at a molecular level on circadian genes and proteins. Further, lithium appears to interact with environmental light through the retinal–hypothalamic–pineal pathway to influence circadian rhythms.

Acknowledgements We would like to thank George Anderson at CRC Scotland & London for English editing services.

References

Abe M, Herzog ED, Block GD (2000) Lithium lengthens the circadian period of individual suprachiasmatic nucleus neurons. Neuroreport 11:3261–3264

Benedetti F, Serretti A, Pontiggia A et al (2005) Long-term response to lithium salts in bipolar illness is influenced by the glycogen synthase kinase 3-beta -50 T/C SNP. Neurosci Lett 376(1):51–55, Epub 2004 Dec 10

Besing RC, Paul JR, Hablitz LM et al (2015) Circadian rhythmicity of active GSK3 isoforms modulates molecular clock gene rhythms in the suprachiasmatic nucleus. J Biol Rhythm 30:155–160

Bremer T, Diamond C, McKinney R, Shehktman T, Barrett TB, Herold C, Kelsoe JR (2007) The pharmacogenetics of lithium response depends upon clinical co-morbidity. Mol Diagn Ther 11:161–170

Campos-de-Sousa S, Guindalini C, Tondo L et al (2010) Nuclear receptor rev-erb-{alpha} circadian gene variants and lithium carbonate prophylaxis in bipolar affective disorder. J Biol Rhythm 25:132–137. doi:10.1177/0748730410362713

Can A, Schulze TG, Gould TD (2014) Molecular actions and clinical pharmacogenetics of lithium therapy. Pharmacol Biochem Behav. doi:10.1016/j.pbb.2014.02.004

Collins PY, Patel V, Joestl SS et al (2011) Grand challenges in global mental health. Nature 475:27–30. doi:10.1038/475027a

Danguir J, Nicolaidis S, Perino-Martel MC (1976) Effects of lithium chloride on sleep patterns in the rat. Pharmacol Biochem Behav 5:547–550

Dokucu ME, Yu L, Taghert PH (2005) Lithium- and valproate-induced alterations in circadian locomotor behavior in Drosophila. Neuropsychopharm Off Publ Am Coll Neuropsychopharmacol 30:2216–2224. doi:10.1038/sj.npp.1300764

Etain B, Milhiet V, Bellivier F, Leboyer M (2011) Genetics of circadian rhythms and mood spectrum disorders. Eur Neuropsychopharmacol J Eur Coll Neuropsychopharmacol 21(Suppl 4):S676–S682. doi:10.1016/j.euroneuro.2011.07.007

Freland L, Beaulieu J-M (2012) Inhibition of GSK3 by lithium, from single molecules to signaling networks. Front Mol Neurosci 5:14. doi:10.3389/fnmol.2012.00014

Friedman E, Yocca FD (1981) The effect of chronic lithium treatment on rat pineal N-acetyltransferase rhythm. J Pharmacol Exp Ther 219:121–124

Geddes JR, Miklowitz DJ (2013) Treatment of bipolar disorder. Lancet 381:1672–1682. doi:10.1016/S0140-6736(13)60857-0

Geoffroy PA, Etain B, Henry C, Bellivier F (2012) Combination therapy for manic phases: a critical review of a common practice. CNS Neurosci Ther 18:957–964. doi:10.1111/cns.12017

Geoffroy P, Bellivier F, Leboyer M, Etain B (2014a) Can the response to mood stabilizers be predicted in bipolar disorder? Front Biosci (Elite Ed) 6:120–138. doi:10.2741/E696

Geoffroy PA, Etain B, Sportiche S, Bellivier F (2014b) Circadian biomarkers in patients with bipolar disorder: promising putative predictors of lithium response. Int J Bipolar Disord 2:5. doi:10.1186/2194-7511-2-5

Goodwin GM (2009) Evidence-based guidelines for treating bipolar disorder: revised second edition – recommendations from the British Association for psychopharmacology. J Psychopharmacol (Oxford, England) 23:346–388. doi:10.1177/0269881109102919

Grandin LD, Alloy LB, Abramson LY (2006) The social zeitgeber theory, circadian rhythms, and mood disorders: review and evaluation. Clin Psychol Rev 26:679–694. doi:10.1016/j.cpr.2006.07.001

Grunze H, Vieta E, Goodwin GM et al (2009) The World Federation of Societies of Biological Psychiatry (WFSBP) guidelines for the biological treatment of bipolar disorders: update 2009 on the treatment of acute mania. World J Biol Psychiatry Off J World Fed Soc Biol Psychiatry 10:85–116. doi:10.1080/15622970902823202

Grunze H, Vieta E, Goodwin GM et al (2010) The World Federation of Societies of Biological Psychiatry (WFSBP) guidelines for the biological treatment of bipolar disorders: update 2010 on the treatment of acute bipolar depression. World J Biol Psychiatry Off J World Fed Soc Biol Psychiatry 11:81–109. doi:10.3109/15622970903555881

Hafen T, Wollnik F (1994) Effect of lithium carbonate on activity level and circadian period in different strains of rats. Pharmacol Biochem Behav 49:975–983

Hallam KT, Olver JS, Horgan JE et al (2005) Low doses of lithium carbonate reduce melatonin light sensitivity in healthy volunteers. Int J Neuropsychopharmacol Off Sci J Coll Int Neuropsychopharmacol 8:255–259. doi:10.1017/S1461145704004894

Harwood AJ, Agam G (2003) Search for a common mechanism of mood stabilizers. Biochem Pharmacol 66:179–189

Hasler G, Drevets WC, Gould TD et al (2006) Toward constructing an endophenotype strategy for bipolar disorders. Biol Psychiatry 60:93–105. doi:10.1016/j.biopsych.2005.11.006

Hickman AB, Klein DC, Dyda F (1999) Melatonin biosynthesis: the structure of serotonin N-acetyltransferase at 2.5 A resolution suggests a catalytic mechanism. Mol Cell 3:23–32

Hofmann K, Günderoth-Palmowski M, Wiedenmann G, Engelmann W (1978) Further evidence for period lengthening effect of Li+ on circadian rhythms. Z Naturforsch C Biosci 33:231–234

Iitaka C, Miyazaki K, Akaike T, Ishida N (2005) A role for glycogen synthase kinase-3beta in the mammalian circadian clock. J Biol Chem 280:29397–29402. doi:10.1074/jbc.M503526200

Jin X, Shearman LP, Weaver DR et al (1999) A molecular mechanism regulating rhythmic output from the suprachiasmatic circadian clock. Cell 96:57–68

Johansson A-S, Brask J, Owe-Larsson B et al (2011) Valproic acid phase shifts the rhythmic expression of Period2::luciferase. J Biol Rhythms 26:541–551. doi:10.1177/0748730411419775

Johnsson A, Pflug B, Engelmann W, Klemke W (1979) Effect of lithium carbonate on circadian periodicity in humans. Pharmakopsychiatr Neuropsychopharmakol 12:423–425. doi:10.1055/s-0028-1094638

Johnsson A, Engelmann W, Pflug B, Klemke W (1980) Influence of lithium ions on human circadian rhythms. Z Naturforsch C Biosci 35:503–507

Johnsson A, Engelmann W, Pflug B, Klemke W (1983) Period lengthening of human circadian rhythms by lithium carbonate, a prophylactic for depressive disorders. Int J Chronobiol 8:129–147

Jope RS, Roh M-S (2006) Glycogen synthase kinase-3 (GSK3) in psychiatric diseases and therapeutic interventions. Curr Drug Targets 7:1421–1434

Kato T, Kubota M, Kasahara T (2007) Animal models of bipolar disorder. Neurosci Biobehav Rev 31:832–842. doi:10.1016/j.neubiorev.2007.03.003

Kleindienst N, Engel R, Greil W (2005) Which clinical factors predict response to prophylactic lithium? A systematic review for bipolar disorders. Bipolar Disord 7:404–417. doi:10.1111/j.1399-5618.2005.00244.x

Klemfuss H (1992) Rhythms and the pharmacology of lithium. Pharmacol Ther 56:53–78

Klemfuss H, Kripke DF (1995) Antimanic drugs stabilize hamster circadian rhythms. Psychiatry Res 57:215–222

Kripke DF, Wyborney VG (1980) Lithium slows rat circadian activity rhythms. Life Sci 26:1319–1321

Kripke DF, Mullaney DJ, Atkinson M, Wolf S (1978) Circadian rhythm disorders in manic-depressives. Biol Psychiatry 13:335–351

Kripke DF, Judd LL, Hubbard B et al (1979) The effect of lithium carbonate on the circadian rhythm of sleep in normal human subjects. Biol Psychiatry 14:545–548

Lanoir J, Lardennois D (1977) The action of lithium carbonate on the sleep-waking cycle in the cat. Electroencephalogr Clin Neurophysiol 42:676–690

LeSauter J, Silver R (1993) Lithium lengthens the period of circadian rhythms in lesioned hamsters bearing SCN grafts. Biol Psychiatry 34:75–83

Li J, Lu W-Q, Beesley S et al (2012) Lithium impacts on the amplitude and period of the molecular circadian clockwork. PLoS One 7:e33292. doi:10.1371/journal.pone.0033292

Lichtenstein P, Yip BH, Björk C et al (2009) Common genetic determinants of schizophrenia and bipolar disorder in Swedish families: a population-based study. Lancet 373:234–239. doi:10.1016/S0140-6736(09)60072-6

Malhi GS, Bassett D, Boyce P, Bryant R, Fitzgerald PB, Fritz K, Hopwood M, Lyndon B, Mulder R, Murray G, Porter R, Singh AB (2015) Royal Australian and New Zealand College of Psychiatrists clinical practice guidelines for mood disorders. Aust N Z J Psychiatry 49:1087–1206

Manchia M, Squassina A, Congiu D et al (2009) Interacting genes in lithium prophylaxis: preliminary results of an exploratory analysis on the role of DGKH and NR1D1 gene polymorphisms in 199 Sardinian bipolar patients. Neurosci Lett 467:67–71. doi:10.1016/j.neulet.2009.10.003

McCarthy MJ, Nievergelt CM, Shekhtman T et al (2011) Functional genetic variation in the Rev-Erbα pathway and lithium response in the treatment of bipolar disorder. Genes Brain Behav 10:852–861. doi:10.1111/j.1601-183X.2011.00725.x

McClung CA (2007) Circadian genes, rhythms and the biology of mood disorders. Pharmacol Ther 114:222–232. doi:10.1016/j.pharmthera.2007.02.003

McClung CA (2011) Circadian rhythms and mood regulation: insights from pre-clinical models. Eur Neuropsychopharmacol J Eur Coll Neuropsychopharmacol 21(Suppl 4):S683–S693. doi:10.1016/j.euroneuro.2011.07.008

McEachron DL, Kripke DF, Wyborney VG (1981) Lithium promotes entrainment of rats to long circadian light-dark cycles. Psychiatry Res 5:1–9

McEachron DL, Kripke DF, Hawkins R et al (1982) Lithium delays biochemical circadian rhythms in rats. Neuropsychobiology 8:12–29

McQuillin A, Rizig M, Gurling HMD (2007) A microarray gene expression study of the molecular pharmacology of lithium carbonate on mouse brain mRNA to understand the neurobiology of mood stabilization and treatment of bipolar affective disorder. Pharmacogenet Genomics 17:605–617. doi:10.1097/FPC.0b013e328011b5b2

Mellerup ET, Widding A, Wildschiødtz G, Rafaelsen OJ (1978) Lithium effect on temperature rhythm in psychiatric patients. Acta Pharmacol Toxicol (Copenh) 42:125–129

Merikangas KR, Akiskal HS, Angst J et al (2007) Lifetime and 12-month prevalence of bipolar spectrum disorder in the national comorbidity survey replication. Arch Gen Psychiatry 64:543–552. doi:10.1001/archpsyc.64.5.543

Michelon L, Meira-Lima I, Cordeiro Q et al (2006) Association study of the INPP1, 5HTT, BDNF, AP-2beta and GSK-3beta GENE variants and restrospectively scored response to lithium prophylaxis in bipolar disorder. Neurosci Lett 403:288–293. doi:10.1016/j.neulet.2006.05.001

Murray G, Harvey A (2010) Circadian rhythms and sleep in bipolar disorder. Bipolar Disord 12:459–472. doi:10.1111/j.1399-5618.2010.00843.x

Nivoli AMA, Murru A, Vieta E (2010) Lithium: still a cornerstone in the long-term treatment in bipolar disorder? Neuropsychobiology 62:27–35. doi:10.1159/000314307

Osland TM, Fernø J, Håvik B et al (2011) Lithium differentially affects clock gene expression in serum-shocked NIH-3T3 cells. J Psychopharmacol Oxf Engl 25:924–933. doi:10.1177/0269881110379508

Pablos MI, Santaolaya MJ, Agapito MT, Recio JM (1994) Influence of lithium salts on chick pineal gland melatonin secretion. Neurosci Lett 174:55–57

Padiath QS, Paranjpe D, Jain S, Sharma VK (2004) Glycogen synthase kinase 3beta as a likely target for the action of lithium on circadian clocks. Chronobiol Int 21:43–55

Pflug B, Engelmann W (1987) Chronobiology of lithium – studies on healthy, middle-aged women. Pharmacopsychiatry 20:181–188. doi:10.1055/s-2007-1017099

Prickaerts J, Moechars D, Cryns K et al (2006) Transgenic mice overexpressing glycogen synthase kinase 3beta: a putative model of hyperactivity and mania. J Neurosci Off J Soc Neurosci 26:9022–9029. doi:10.1523/JNEUROSCI.5216-05.2006

Reppert SM, Weaver DR (2002) Coordination of circadian timing in mammals. Nature 418:935–941. doi:10.1038/nature00965

Roybal K, Theobold D, Graham A et al (2007) Mania-like behavior induced by disruption of CLOCK. Proc Natl Acad Sci U S A 104:6406–6411. doi:10.1073/pnas.0609625104

Rybakowski JK, Czerski P, Dmitrzak-Weglarz M, Kliwicki S, Leszczynska-Rodziewicz A, Permoda-Osip A, Skibinska M, Suwalska A, Szczepankiewicz A, Hauser J (2012) Clinical and pathogenic aspects of candidate genes for lithium prophylactic efficacy. J Psychopharmacol (Oxford, England) 26:368–373

Rybakowski JK, Dmitrzak-Weglar M, Kliwicki S, Hauser J (2014) Polymorphism of circadian clock genes and prophylactic lithium response. Bipolar Disord 16:151–158. doi:10.1111/bdi.12136

Seggie J (1988) Lithium and the retina. Prog Neuropsychopharmacol Biol Psychiatry 12: 241–253

Seggie J, Werstiuk ES, Grota L (1982) Effect of chronic lithium treatment on twenty four hour variation in plasma and red blood cell lithium and sodium concentrations, drinking behavior, body weight, kidney weight, and corticosterone levels. Prog Neuropsychopharmacol Biol Psychiatry 6:455–458

Seggie J, Werstiuk E, Grota L, Brown GM (1983) Chronic lithium treatment and twenty-four hour rhythm of serum prolactin, growth hormone and melatonin in rats. Prog Neuropsychopharmacol Biol Psychiatry 7:827–830

Seggie J, Werstiuk ES, Grota L et al (1984) Lithium and melatonin in pigmented eye rats: effects of dose and time of day. Prog Neuropsychopharmacol Biol Psychiatry 8:733–736

Seggie J, Werstiuk ES, Joshi M (1985) Lithium and twenty-four hour rhythms of serum corticosterone, prolactin and growth hormone in pigmented eye rats. Prog Neuropsychopharmacol Biol Psychiatry 9:755–758

Seggie J, Werstiuk ES, Grota L (1987) Lithium and circadian patterns of melatonin in the retina, hypothalamus, pineal and serum. Prog Neuropsychopharmacol Biol Psychiatry 11:325–334

Subbaraj R (1981) Effect of lithium chloride on the circadian rhythm in the flight activity of the microchiropteran bat, Taphozous melanopogon. Z Für Naturforschung Sect C Biosci 36:1068–1071

Subramanian P, Menon VP, Arokiam FV et al (1998) Lithium modulates biochemical circadian rhythms in Wistar rats. Chronobiol Int 15:29–38

Szczepankiewicz A, Rybakowski JK, Suwalska A et al (2006) Association study of the glycogen synthase kinase-3beta gene polymorphism with prophylactic lithium response in bipolar patients. World J Biol Psychiatry Off J World Fed Soc Biol Psychiatry 7:158–161. doi:10.1080/15622970600554711

Werstiuk ES, Seggie J, Joshi M (1984) Lithium in pigmented eye rats: effects of dose and time of day on drinking, body weight, retinal and blood distribution. Prog Neuropsychopharmacol Biol Psychiatry 8:765–768

Yatham LN, Kennedy SH, Schaffer A et al (2009) Canadian Network for Mood and Anxiety Treatments (CANMAT) and International Society for Bipolar Disorders (ISBD) collaborative

update of CANMAT guidelines for the management of patients with bipolar disorder: update 2009. Bipolar Disord 11:225–255. doi:10.1111/j.1399-5618.2009.00672.x
Yerevanian BI, Choi YM (2013) Impact of psychotropic drugs on suicide and suicidal behaviors. Bipolar Disord 15:594–621. doi:10.1111/bdi.12098
Yin L, Wang J, Klein PS, Lazar MA (2006) Nuclear receptor Rev-erbalpha is a critical lithium-sensitive component of the circadian clock. Science 311:1002–1005. doi:10.1126/science.1121613
Yocca FD, de Paul Lynch V, Friedman E (1983) Effect of chronic lithium treatment on rat pineal rhythms: N-acetyltransferase, N-acetylserotonin and melatonin. J Pharmacol Exp Ther 226:733–737

Clinical Predictors Relevant to Lithium Response

7

Thomas Mauras, Sarah Sportiche, Sami Richa, and Marc Masson

Abstract

To adequately evaluate lithium response, one needs to consider the whole clinical profile of bipolar disorder; it is not sufficient just to consider symptoms. Clinical variables based on the anamnesis are still the best predictors of lithium response. A later age of disease onset, an episodic course characterized by a pattern of mania followed by depression, fewer hospitalizations preceding treatment and the absence of an episodic pattern of depression-mania interval and continuous cycling should give the physician some hope of making the right decisions. The possible future addition of biological or brain imaging signatures should provide valuable information that would help in several ways: by more powerfully predicting response in conjunction with genotypes, by serving as a biomarker of response in clinical trials and by revealing pathophysiological pathways from gene to clinical success.

T. Mauras, MD (✉)
Department of Psychiatry (Dr. Marcel), Centre Hospitalier Sainte Anne (Secteur 3, Dr Marcel), 1 rue Cabanis, 75014 Paris, France
e-mail: t.mauras@ch-sainte-anne.fr

S. Sportiche
Université Paris Descartes, UMR-S 1144, Paris 75006, France

Université Paris Diderot, Sorbonne Paris Cité, UMR-S 1144, Paris 75013, France

Département de Psychiatrie et de Médecine Addictologique, AP-HP, GH Saint-Louis – Lariboisière – F. Widal, Paris, Cedex 10 75475, France

S. Richa
Hôtel-Dieu de France, Service de psychiatrie, Université Saint-Joseph, Département de Psychiatrie, BP 166830 Beyrouth, Lebanon

M. Masson
Clinique du Château, 11bis rue de la Porte Jaune, 92380 Garches, France

Pôle de psychiatrie du Professeur Raphaël Gaillard, Service Hospitalo-Universitaire, Centre Hospitalier Sainte-Anne, 75014 Paris, France

G.S. Malhi et al. (eds.), *The Science and Practice of Lithium Therapy*,
DOI 10.1007/978-3-319-45923-3_7

Because of their relative homogeneity, lithium responders represent an important population for psychiatric research. This group of bipolar patients can thus be regarded as a good candidate population to open new fields of biological investigation, especially in the research of biomarkers of lithium response.

Key Points

- There is currently a lack of an agreed definition of 'lithium response'.
- Few clinical predictors are valid in the literature. For good response: episodic pattern of mania-depression-free interval and intermediate/high age of onset. For poorer response: high number of previous hospitalizations, episodic pattern of depression-mania-free interval and continuous cycling.
- Discrepancies in the literature may be explained by possible methodological bias of studies, the complex illness course of bipolar disorder and various therapeutic effects of lithium.
- It is important to identify a clinically homogeneous group of good lithium responders among bipolar patients to inform treatment and to identify future biomarkers of lithium response.

7.1 Introduction

Lithium occupies a special place among psychiatric treatments. It is one of the oldest medications used in the treatment of recurrent mood disorder and also is perhaps the most mysterious (Malhi 2010). Its effects were first recognized in patients with mania in a historical and seminal publication by Cade, which illustrated the successful use of the drug in the treatment of ten patients with mania (Cade 1949). In 1954, Schou and colleagues performed the first randomized controlled trial of lithium, during which they demonstrated that this treatment had a clear antimanic effect and was an alternative to electroconvulsive therapy (Schou et al. 1954).

However, it was not until the 1970s that Baastrup and Schou demonstrated that lithium confers protection against the recurrence of illness, first by using a non-blinded mirror design and then with a double-blind discontinuation study of the prophylactic qualities of lithium (Baastrup et al. 1970). In this study, among the 45 patients' lithium arm, none had recurrences of their mood illness, while 21 of 39 patients treated with placebo experienced one. The efficacy of lithium as a long-term prophylactic agent for the treatment of bipolar disorder was corroborated by a series of double-blind trials conducted in the 1970s and 1980s (Grof and Müller-Oerlinghausen 2009). Despite its well-established antimanic and prophylactic qualities, lithium has become the focus of controversies, and its effectiveness has been questioned with regard to other prophylactic treatments, which has led to an overall decrease in lithium prescriptions (Malhi and Gershon 2009). The reasons for this trend remain unclear; possible explanations include a lack of training of psychiatrists and/or 'aggressive marketing of alternative medication' (Young and Hammond 2007) and/or the potential for chronic side effects such as kidney damage.

Interestingly, despite this decreasing use of lithium, there is *increasing* evidence of its efficacy in the prophylactic treatment of bipolar disorder. Some recent trials and meta-analyses provide supportive data showing that lithium is an appropriate first-line agent for the prevention of manic and depressive episodes (Young and Hammond 2007; Geddes et al. 2010). Smith and colleagues used these studies in a meta-analysis evaluating the effectiveness of lithium as a maintenance treatment (Smith et al. 2007). This meta-analysis of 14 randomized controlled trials (eight included placebo-controlled trials) provides strong evidence for the prophylactic efficacy of lithium, which prevented relapse to any mood episode with a hazard ratio of 0.68. The overall prophylactic efficacy of lithium was largely explained by the reduction in manic relapses (hazard ratio 0.53, 95 % CI 0.35–0.79). Lithium-treated patients also had fewer depressive relapses, but this effect was smaller and not statistically significant.

So, after more than 60 years, lithium justifiably remains a first-line treatment for the long-term management of bipolar disorders (Malhi et al. 2015). Personalization of prescription incites us to identify the best therapeutic approach (monotherapy, association) for each patient. No fewer than 40 clinical predictors of lithium response have been advanced (Kleindienst et al. 2005a). For clinical practice, it is important to know the clinical profile of patients who should receive lithium prophylaxis as first-line treatment. But, first of all, two key questions have to be asked: How can we best define lithium response? And how can we get valid clinical predictors for lithium response?

7.2 How Can We Best Define Lithium Response?

Naturalistic analyses indicate that approximately only one third of bipolar patients achieve complete remission with lithium (Smith et al. 2007; Garnham et al. 2007), and lithium response is highly variable. How can these factors be better understood?

7.2.1 The Complexity of Bipolar Disorders

The first level of difficulty in evaluating lithium response is the complexity of bipolar disorder and its varied evolution: instability between bipolar I and bipolar II diagnosis, the unpredictability of the course of the illness (intensity of episodes, frequency of episodes), the induced mood swings, the presence of rapid cycling and psychiatric and physical comorbidities (Malhi et al. 2012).

7.2.2 Bipolar Spectrum Extension

A second level of difficulty is the extension of the bipolar spectrum to encompass a broad spectrum of disorders that are probably too far from the core to be considered as manic-depressive illness. Recent reports of reduced lithium efficacy might be attributable to the expansion of the diagnostic criteria of bipolar disorder in the Diagnostic and Statistical Manual-based systems (Baldessarini and Tondo 2000; Malhi and Porter 2014). The incorporation of too widely varied phenotypes of mood disorders may have skewed many trials on bipolar disorders, especially those evaluating therapeutic response.

Taking this into account, together with the first level of difficulty regarding complexity, does lithium resistance indeed exist? For some authors (with considerable clinical experience of bipolar disorder and practical use of lithium prescription), once the clinical profile of a good responder is established, there is a 'written guarantee' that lithium will provide good stabilization for years (Grof 2006). On the other hand, other authors describe secondary lithium resistance after years of prophylactic efficacy: more severe and/or more frequent breakthrough episodes appear progressively, a pattern consistent with the emergence of tolerance (Post 2012). Among lithium-refractory patients studied (Post 2010), 13.6% showed this phenomenon after being on lithium for an average of 6.6 years.

7.2.3 Various Therapeutic Effects

A third aspect for defining lithium response is the various therapeutic effects of lithium itself. Beyond the prophylactic effect, lithium has 'curative' properties for episodes, as well as clear clinical effects on suicide (Baldessarini et al. 2006) and impulsive behaviours. Each therapeutic property may be independent from one another. Lithium has proven useful in major depressive disorders, particularly with the treatment of resistant depression. To date, lithium's prophylactic and anti-suicidal effects are unique, and lithium differs from other mood stabilizers, as it reduces the risk of suicide not only through the prevention of mood episodes but also in lithium nonresponders, perhaps due to a specific anti-suicidal mechanism of action. This chapter focuses exclusively on the response to long-term treatment of patients with recurrent mood disorders.

7.2.4 From Monotherapy to Combination Therapy and Polypharmacy

In this respect, a fourth level of difficulty for defining lithium response becomes apparent. Prescription trends in the management of bipolar disorder have shifted away from monotherapy (even if this remains the patient's preference and the physician's aim) to combination therapy and polypharmacy. The trend is generally to use combinations of different putatitive mood stabilizers. The treatment of bipolar disorders for 4,700 patients in The Health Improvement Network (THIN) primary was analysed for over 15 years, between 1995 and 2009 (Hayes et al. 2011). In 1995 23 % were issued with two or more prescriptions for more than one psychotropic medication; by 2009 this had increased to 48 %. Seventeen percent of patients were prescribed lithium and an antipsychotic. Moreover, in 1995 5 % were prescribed an anticonvulsant (valproate, carbamazepine or lamotrigine) and an antipsychotic; by 2009 this had increased to 31 %. Lithium and an anticonvulsant were prescribed to 5 % of the population in 1995, and by 2009 this reached 12 %. This points to the possibility of a partial lithium response with lithium monotherapy and a full (or better) response due to the addition of another mood stabilizer with lithium. The role of psychoeducation and psychological-based therapies (cognitive behaviour therapy, cognitive remediation, interpersonal social rhythms therapy, mindfulness, etc.) might also be an important factor when evaluating the possibility of achieving a better, or a complete, lithium response.

7.2.5 Lithium Blood Levels

A fifth difficulty that should also be taken in account is the variability of lithium response due to its blood level. As lithium has a narrow therapeutic index, plasma lithium levels can be measured safely and accurately and are reasonable proxies for concentrations in the brain. Titration of plasma levels according to clinical need and symptomatic profile is a useful approach (Kleindienst et al. 2007), and this individualization of lithium dosage, on the basis of polarity, is likely to enhance efficacy and reduce side effects. Severus et al. (2009) has also shown that depression-prone bipolar disorder patients are likely to benefit from prophylactic lithium levels of 0.4–0.8 mmol/L, whereas those predisposed to mania tend to benefit from higher levels of 0.6–1.0 mmol/L. Thus, as mentioned by Malhi et al. (2011), lithium needs to be measured both literally in terms of its plasma levels and metaphorically, with respect to its clinical and functional effects. For that purpose, Malhi has designed a very useful tool, the 'lithiumeter', which provides a visual scale for gauging the optimal lithium plasma level, according to curative (on mania or depression) or prophylactic use.

7.2.6 Lithium Compliance

Finally, a sixth difficulty is probably treatment compliance itself, which plays a key role in the lithium response (Berk et al. 2010). Johnson and McFarland (1996) followed 1,500 patients treated with lithium and reported that the mean duration of continuous lithium adherence was 76 days. Thus, the potential benefits of lithium on recovery, preventing relapse and reducing mortality, are significantly undermined by poor adherence. Although clinicians commonly attribute their patients' medication non-adherence to side effects, a large survey of more than 3,000 patients suffering from mood disorders found that side effects are not a major determinant of adherence (Morselli and Elgie 2003). The results indicated that only 18.3 % of the patients stated that side effects were the main reason for treatment discontinuation. The major reasons cited for poor adherence were fear of becoming addicted to medication, poor medication information and fear of long-term side effects. This study found that about 35 % of the patients did not receive any information on the possible side effects of their medications and more than 50 % had not received guidance on side effect management. Lack of information may have contributed to the fear of side effects, and this fear may be a stronger predictor of non-adherence than the side effects themselves (Lingam and Scott 2002).

7.2.7 The Lack of an Agreed Definition for Lithium Response

Despite the six levels of difficulties just described, recently published studies define two categories of patients: 'responders' and 'nonresponders'. Given the complexity just discussed, it is not surprising that these definitions may vary from study to study. A classic clinical assessment can be undertaken using the Affective Morbidity Index (AMI) (Coppen et al. 1973) or with the Illness Severity Index (ISI) (Maj et al. 1984). The AMI takes into account the duration and the severity of an episode. Similarly, the ISI was defined as the frequency of affective episodes prior to starting lithium adjusted for age at the time lithium was started. Fifteen years ago, Canadian researchers introduced a scale allowing quantitative retrospective assessment of the quality of prophylactic lithium response (Manchia et al. 2013). This scale is referred to informally as 'the Alda scale'. Criterion A rates the degree of response (activity of the illness with an adequate lithium treatment) on a ten-point scale. Criterion B weighs clinical factors considered relevant in determining whether the observed clinical change is due to lithium treatment or not. Criterion B involves B1, the number of episodes off the treatment; B2, frequency of episodes off the treatment; B3, the duration of treatment; B4, compliance during period(s) of stability; and B5, the use of additional medications during periods of stability. The total score is obtained by subtracting B from A and is in the range of 0–10. It has been used widely, for example, in the Consortium of Lithium Genetics (ConLiGen) project, which sought to conduct a genome-wide association study (GWAS) in a large population of lithium-treated patients.

In clinical studies, the definition of lithium response has to specify the design and the aim of the trial: the diagnostic group studied (bipolar I or bipolar II), the long-term prophylactic response or the curative effectiveness of episodes, the number (and the name) of mood stabilizers associated with lithium, the association with psychoeducation and/or psychological and social therapies and the frequency of serum levels and their results. Remarkably, an agreed definition for lithium response(s) is yet to be written after more than 60 years of lithium prescription.

7.3 Obtaining Valid Clinical Response Predictors

The majority of potential predictors (such as the number of previous episodes) yield conflicting results, and some authors even disagree that a prediction can be established (Mander 1986). On the other hand, Grof (2006; Grof and Müller-Oerlinghausen 2009) proposed the term 'excellent lithium responders' for those patients whose lithium monotherapy dramatically changed the course of their illness (and consequently their lives) by the total prevention of further episodes. So can we predict which patient will respond and which will not? Two main issues should be addressed: the duration of lithium prescription and the methodology of the studies.

7.3.1 Duration of Lithium Prescription

First, time to response is a relevant factor that appears highly variable in clinical trials. The global consensus is that, in the first year of treatment, morbidity on lithium may still be high in patients who respond fully in the long term (Ahrens et al. 1993). The difference in time needed to obtain a significant response may reflect the heterogeneity of lithium's mechanisms of action as well as differences in metabolism between fast and slow responders. This difference could also be linked to compliance issues, time needed to achieve effective yet tolerable blood levels, psychological factors and interactions between the natural course of the illness and other environmental factors (addictions, antidepressant prescriptions, lack of insight, etc.). Consequently, to properly address these issues, prospective studies will be needed, with large samples of bipolar disorder patients treated with lithium.

Most recent clinical drug trials have been relatively short in duration and, therefore, do not accurately assess maintenance or prophylaxis. Moreover, other analyses that have tested lithium response have often relied on variables of convenience available in samples collected for other purposes. Fortunately, there are some exceptions, and we should acknowledge the work of some international groups, such as the International Group for the Study of Lithium-treated patients (IGSLI), which has collected data for decades (Grof 1994, 2006; Schultze et al. 2010).

7.3.2 Possible Methodological Biases

The methodology on lithium response must be proportionate with the complexity of the issue. To identify the predictors of lithium response, the patient's profile and treatment adequacy should be carefully assessed, since there are numerous confounding factors—and most variables are highly correlated. For example, Baethge et al. (2003) found that earlier lithium initiation was strongly associated with greater pretreatment illness intensity. This association leads to greater apparent reduction of morbidity during treatment versus before treatment and could theoretically indicate early treatment as a predictor of good response. However, greater morbidity probably encouraged earlier treatment; thus this inference could be misleading.

Trial methodology is crucial to the establishment of valid predictors. Some authors argue for establishing multivariate methods to take this complexity into account (Grof et al. 1993), while others highlight effect-size measure between studies in a meta-analytic approach (Kleindienst et al. 2005a, b). Each method has its own limitations. Multivariate analysis integrates multiple variables in the same model of analysis, and its hypothesis-free nature can potentially lead to identifying new variables associated with lithium response. The meta-analysis approach gives some strength to a candidate variable as a predictor; however it is subject to bias from aggregated data collected from different populations and bias from multiple comparisons.

As a unique definition for lithium response is impossible to find in the studies conducted to date, with variable durations of lithium prescription and with unavoidable methodological bias, one can easily understand why most of the clinical trials dedicated to lithium response may be skewed. The main results will be summarized in the following paragraphs; however one should keep in mind the potential reasons for these frequent contradictory findings.

7.4 Clinical Profile of Bipolar Disorders and Prediction of Lithium Response

From a broad literature review, we noticed some common clinical predictors of lithium response: the course of illness before lithium initiation, the family history (of bipolar disorder and of good lithium response), the first-episode polarity, the age of onset, the feature specificities during episodes, the potential for suicide, the course of the illness and the comorbidities (Tighe et al. 2011; Baldessarini and Tondo 2000).

7.4.1 Family History

Some studies have shown that patients who have relatives with bipolar disorder derive more benefit from lithium prophylaxis than do those whose family history is negative (Grof 1994). These early studies were retrospective, and it is important to note that other authors found contrary results (Coryell et al. 2000). Over the past two decades, there has been growing interest in longitudinal studies of the children of parents with bipolar disorders (Duffy et al. 2014). It has been advanced that a good response to lithium in a parent could be a predictor of response in the offspring. Even though this assumption is frequently supported by expert opinion, the available data remain tenuous (Duffy et al. 2007).

The response to long-term lithium may cluster in families (Grof 2006). Only lithium responders have been observed to have significant excesses of bipolar disorders in their family history. The first-degree relatives of bipolar patients responding to lamotrigine have an overabundance of anxiety disorders, panic attacks, substance abuse and alcohol addictions (Grof 2003).

7.4.2 Polarity of First Episode

Data concerning first-episode polarity associated with lithium response are sparse. Shapiro (1989) compared bipolar patients treated with lithium, imipramine or both on recurrence rate. He suggested that bipolar patients with a manic index have a better lithium response. However, the study by Yazici et al. (1999), including more than 300 bipolar patients, found a relationship between the index episode being manic and poor lithium response.

7.4.3 Age of Onset

Age of onset as a predictor of lithium response is one of the most replicated factors. Since Dunner and Fieve (1974), more than ten studies have highlighted a late illness onset as potentially protective against a recurrence under lithium. There is no threshold or absolute value, but a range could be extrapolated from the literature. Aagaard and Vestergaard (1990) found in a subgroup of 77 bipolar patients with good adherence that a younger age under 20 years old was associated with an eight-fold risk of poor response to lithium. Leboyer et al. (2005) supported the existence of three bipolar subgroups based on age of onset: early (17.4 ± 2.3 years), intermediate (25.1 ± 6.2 years) and late (40.4 ± 11.3 years) onset. Middle-/late-onset bipolar disorder also seems to be associated with a poor prognosis for lithium prophylaxis (Dell'Osso et al. 2009).

Thus, intermediate onset seems to be the range that benefits the most from lithium prescription. For example, the age of onset of the good response group was 26 ± 10 years compared to 23 ± 7 years for the poor response group (Yazici et al. 1999).

7.4.4 Specific Features During Episodes

Among the clinical features investigated, marked psychomotor retardation was found to be associated with better response (Ananth et al. 1979), and the presence of euphoric mania has been considered to predict a good prophylactic response to lithium (Calabrese et al. 1996). Thus, a classic presentation of the illness is historically associated with good lithium response.

On the other hand, patients with atypical features of the disorder, such as mood-incongruent psychotic symptoms, rapid cycling or residual symptoms after remission, seem to have a poorer lithium response.

The case of rapid cycling should be emphasized. Most lithium responders have experienced two or fewer episodes per year. Rapid or continuous cycling is less common in lithium responders, but a number of rapid-cycling patients respond well to lithium, particularly in the case of rapid cycling induced by antidepressants (Grof 2006). A recent study randomized rapid cyclers and found lithium and lamotrigine similarly effective (Calabrese et al. 2005).

The presence of atypical features (mainly mood-incongruent psychotic symptoms, interepisodic residual symptomatology, rapid cycling) influenced the time to first recurrence (Pfennig et al. 2010). The instantaneous risk of patients with at least two atypical features experiencing further episodes was 1.5-fold the risk of patients with a more classical clinical profile. However, data concerning atypical symptoms remain somewhat controversial. For example, a meta-analysis including more than 1,800 patients demonstrated that those with rapid-cycling bipolar disorder were less responsive to lithium but also to other mood stabilizers, compared to patients with classic cycling (Tondo et al. 2003).

7.4.5 Suicide

Lithium is the only psychotropic agent that has been shown to reduce the incidence of suicide and suicide attempts in patients suffering from mood disorders (Baldessarini et al. 2006). Overall, long-term lithium treatment resulted in a four- to fivefold lower risk of suicidal acts (Young and Hammond 2007). Interestingly, relatively few studies have examined the number of past suicidal behaviours as a predictor of lithium response. To our knowledge, only Yazici et al. (1999)) has addressed this question, and he found no difference between the good and poor lithium response groups.

7.4.6 Course of Illness and Pattern of Episodes

Since the seminal work of Angst (1978) to find the subtypes of bipolar disorders and the work of Kukopulos et al. (1980)) on the patterns of mood episodes, the major components on which description of the episode sequence is based are depression (D), mania (M) and the free interval (I) between any mood episode.

These components have been combined into a number of episode sequences such as MDI, which is characterized by a predominant pattern of mania-depression-free interval. Moreover, this sequence has been found to be strongly associated with lithium response (Grof et al. 1987). On the other hand, the opposite sequence DMI is associated with a poor lithium prophylaxis (Okuma 1993).

Moreover, several studies have also examined the number of previous hospitalizations for an affective episode as a potential predictor. They found a high number of hospitalizations to be closely related to poorer response to prophylactic lithium treatment (Maj 1990; Yazici et al. 1999).

7.4.7 Quality of Remission

The quality of the remission is another supposed predictor of prophylactic efficacy. For Grof (2006) it is the most important clinical predictor of good long-term lithium response. With others, he suggests evaluating the quality of the remission with the Minnesota Multiphasic Personality Inventory (MMPI) profile (Demidenko et al.

2004). A normal MMPI could be viewed as a quantitative confirmation that the patient is in remission. However, these proposals lead to two uncertainties. First, for the response prediction to be of clinical value, it has to be available before or shortly after initiation of the treatment response predictor. If conducted prematurely, the presence of residual symptoms on the assessment could indicate a preexisting active episode, rather bad news for prophylaxis. On the other hand, if conducted too late, the utility of the remission evaluation may be meaningless, since a patient who is in remission may have a better chance to remain in remission. Secondly, the MMPI is a psychological scale screening for a personality disorder, which is commonly recognized as a poor predictor of lithium prophylaxis and consequently could lead to interpretation bias.

7.4.8 Psychiatric Comorbidities

Indeed, poor responders show significantly greater personality disorders (Abou-Saleh 1993). Among the personality factors associated with the response, the most prominent ones are dominance versus stress and high emotivity and neuroticism (Kleindienst et al. 2005a, b). Concerning demographic factors, high social status, social support, compliance, being employed and a low number of life events were also associated with good response. Finally, substance abuse comorbidity is related to poor lithium response (O'Connell et al. 1991; Mander 1986).

7.4.9 Tentative Conclusions

To date, there has been only one meta-analysis, pooling around 2,000 potentially interesting studies and more than 40 potential variables associated with lithium response (Kleindienst et al. 2005a, b). This meta-analysis was commented upon and enriched by Tighe et al. (2011); however, since then no study has been conducted on the topic. In this valuable work, only five variables were identified as possible response predictors of prophylactic lithium: a good lithium response was associated with (1) an episodic pattern of mania-depression interval and (2) a high age of illness onset, whereas a poor lithium response was associated with (3) a high number of previous hospitalizations, (4) an episodic pattern of depression-mania interval and (5) continuous cycling.

7.5 Clinical Predictors: Crucial for Identifying Lithium Response Biomarkers?

Biological psychiatry expanded quickly after the 1970s. The resulting studies yielded a number of promising, but difficult to replicate, biological differences between lithium responders and nonresponders in terms of their performance on the dexamethasone suppression test and decrease in prolactin response to hypoglycaemia or folate deficiency (Yazici et al. 1999). Electroencephalography (EEG) was another tantalizing field of research. Reeves examined relative EEG differences in good responders compared to patients who responded to valproate for manic episode treatment (Calvert et al. 2006). EEG abnormalities isolated two distinctive groups. Seventy percent of the patients who responded to valproate were found to have EEG abnormalities, while only 30 % of those who responded to lithium had abnormal EEGs. Other neurophysiological variables such as event-related potentials (ERP) could be a promising predictor; however sufficient supporting data is still missing (Tighe et al. 2011; Frey et al. 2013).

The pharmacogenetics of lithium response has also been the subject of intense research. The potential number of candidate genes to study in relation to lithium response is somewhat disheartening, and many positive findings can appear plausible by a post hoc neurobiological explanation as relevant to lithium's effect (Alda 2015). In association studies, no positive findings have yet been replicated in at least two samples of 200 or more subjects. In genome-wide association analyses, only one study found a clear significant result: a polymorphism in the *GADL1* gene with an associated odds ratio of 82.2 to have a good response rate when you have a particular allele (Chen et al. 2014). This was indeed an interesting discovery. Unfortunately, the variant is absent in Caucasians and so a direct comparison with other genome-wide association studies is not possible. Disappointingly, other studies that tried to replicate it in East Asian populations have not been successful. However, the research of circadian biomarkers predictors to a good lithium response may be promising in the future (Geoffroy et al. 2014).

Neuroimaging is another promising tool for in vivo investigation of lithium response (Soares 2002). First, it allows the determination of structural correlations of lithium treatment. Lithium administration has been associated with increased grey matter volumes (Moore et al. 2000) and in key brain regions for mood disorder such as the hippocampus (Yucel et al. 2007), the amygdala (Foland et al. 2008) or the prefrontal cortex (Moore et al. 2009). In the latter study, they found that patients who responded to lithium had increased prefrontal grey matter compared with those who had a poor response. Although the role of a disconnected cortico-limbic circuitry in the development and maintenance of bipolar disorder has been proposed (Price and Drevets 2012; Adler et al. 2006), no convincing marker for lithium response has yet been proposed (Vai et al. 2014).

7.6 Summary

To estimate lithium response, one needs to consider the whole clinical profile of bipolar disorder; merely considering symptom presentation is not sufficient. Clinical variables based on the anamnesis are still the strongest predictors of lithium response. A later age of disease onset, an episodic course characterized by a pattern of mania followed by depression, fewer hospitalizations preceding treatment and the absence of an episodic pattern of depression-mania interval and continuous cycling should give some hope to the physician to make the right decisions. In future, the addition of biological or brain imaging signatures could provide valuable information that should help in several ways: by more powerfully predicting response in conjunction with genotypes, by serving as a biomarker of response in clinical trials and by revealing pathophysiological pathways from gene to clinical success (Grof 2006; Rohayem et al. 2008).

Lithium responders represent an important population for psychiatric research due to their potential homogeneity (Malhi and Geddes 2014). This group of bipolar patients can be regarded as a good candidate for the exploration of new fields of biological investigations, especially in the research of lithium response biomarkers.

References

Aagaard J, Vestergaard P (1990) Predictors of outcome in prophylactic lithium treatment: a 2-year prospective study. J Affect Disord 18(4):259–266

Abou-Saleh MT (1993) Who responds to prophylactic lithium therapy? Br J Psychiatry 163(Suppl 21):20–26

Adler CM, DelBello MP, Strakowski SM (2006) Brain network dysfunction in bipolar disorder. CNS Spectr 11(4):312–320

Ahrens B, Müller-Oerlinghausen B, Grof P (1993) Length of lithium treatment needed to eliminate the high mortality of affective disorders. Br J Psychiatry Suppl 21:27–29

Alda M (2015) Lithium in the treatment of bipolar disorder: pharmacology and pharmacogenetics. Mol Psychiatry (November 2014):1–10; 20(6):661–70.

Ananth J, Angelsmann F, Kiriakos R et al (1979) Prediction of lithium response. Acta Psychiatr Scand 60(3):279–286

Angst J (1978) The course of affective disorders. II. Typology of bipolar manic-depressive illness. Arch Psychiatr Nervenkr 226(1):65–73

Baastrup PC, Poulsen JC, Schou M et al (1970) Prophylactic lithium: double blind discontinuation in manic-depressive and recurrent-depressive disorders. Lancet 2(7668):326–330

Baethge C, Tondo L, Bratti IM et al (2003) Prophylaxis latency and outcome in bipolar disorders. Can J Psychiatry 48(7):449–457

Baldessarini RJ, Tondo L (2000) Does lithium treatment still work? Evidence of stable responses over three decades. Arch Gen Psychiatry 57(2):187–190

Baldessarini RJ, Tondo L, Davis P et al (2006) Decreased risk of suicides and attempts during long-terms lithium treatment: a meta-analytic review. Bipolar Disord 8(5 (p 2)):625–639

Berk L, Hallam KT, Colom F et al (2010) Enhancing medication adherence in patients with bipolar disorder. Hum Psychopharmacol 25(1):1–16

Cade JFJ (1949) Lithium salts in the treatment of psychotic excitement. Med J Aust 2:349–352

Calabrese JR, Fatemi SH, Kujawa M et al (1996) Predictors of response to mood stabilizers. J Clin Psychopharmacol 16(2 Suppl 1):24S–31S

Calabrese JR, Rapport DJ, Youngstrom EA et al (2005) New data on the use of lithium, divalproate and lamotrigine in rapid cycling bipolar disorder. Eur Psychiatry 20:92–95

Calvert NW et al (2006) The cost-effectiveness of lamotrigine in the maintenance treatment of adults with bipolar I disorder. JMCP 12(4):322–330

Chen CH, Lee CS, Lee MT et al (2014) Variant GADL1 and response to lithium therapy in bipolar I disorder. N Engl J Med 370(2):119–128

Coppen A, Peet M, Bailey J et al (1973) Double-blind and open prospective studies on lithium prophylaxis in affective disorders. Psychiatr Neurol Neurochir 76(6):501–510

Coryell W, Akiskal H, Leon AC et al (2000) Family history and symptom levels during treatment for bipolar I affective disorder. Biol Psychiatry 47(12):1034–1042

Dell'Osso B, Buoli M, Riundi R et al (2009) Clinical characteristics and long-term response to mood stabilizers in patients with bipolar disorder and different age at onset. Neuropsychiatr Dis Treat 5:399–404

Demidenko N, Grof P, Alda M et al (2004) MMPI as a measure of subthreshold and residual psychopathology among the offspring of lithium responsive and non-responsive bipolar parents. Bipolar Disord 6(4):323–328

Duffy A, Alda M, Milin R, Grof P (2007) A consecutive series of treated affected offspring of parents with bipolar disorder: is response associated with the clinical profile? Can J Psychiatry 52(6):369–376

Duffy A, Horrocks J, Doucette S et al (2014) The developmental trajectory of bipolar disorder. Br J Psychiatry 204(2):122–128

Dunner DL, Fieve RR (1974) Clinical factors in lithium carbonate prophylaxis failure. Arch Gen Psychiatry 30(2):229–233

Foland LC, Altshuler LL, Sugar CA et al (2008) Increased volume of the amygdala and hippocampus in bipolar patients treated with lithium. Neuroreport 19(2):221–224

Frey BN, Andreazzana AC, Houenou J et al (2013) Biomarkers in bipolar disorder: a positional paper from the International Society for Bipolar Disorders Biomarkers Task Force. Aust N Z J Psychiatry 47(4):321–332

Garnham J, Munro A, Slaney C et al (2007) Prophylactic treatment response in bipolar disorder: results of a naturalistic observation study. J Affect Disord 104(1-3):185–190

Geddes JR, Goodwin GM, Rendell J et al (2010) Lithium plus valproate combination therapy versus monotherapy for relapse prevention in bipolar I disorder (BALANCE): a randomised open-label trial. Lancet 375(9712):385–395

Geoffroy PA, Etain B, Sportiche S, Bellivier F (2014) Circadian biomarkers in patients with bipolar disorder: promising putative predictors of lithium response. Int J Bipolar Dis 2:5

Grof P (1994) Lithium response and genetics of affective disorders. J Affect Disord 32(2):85–95

Grof P (2003) Selecting effective treatments for bipolar disorders. J Clin Psychiatry 64(Supp 5):53–61

Grof P (2006) Responders to long-term lithium treatment. In: Bauer M, Grof P, Müller-Oerlinghausen B (eds) Lithium in neuropsychiatry. The comprehensive book. Informa Healthcare, Clochester, pp 157–178

Grof P, Müller-Oerlinghausen B (2009) A critical appraisal of lithium efficacy and effectiveness: the last 60 years. Bipolar Disord 11:10–19

Grof E, Haag M, Grof P et al (1987) Lithium response and the sequence of episode polarities: preliminary report on a Hamilton sample. Prog Neuropsychopharmacol Biol Psychiatry 11(2–3):199–203

Grof P, Alda M, Grof E et al (1993) The challenge of predicting response to stabilising lithium treatment. The importance of patient selection. Br J Psychiatry 21(Suppl):16–19

Hayes J, Prah P, Nazareth I et al (2011) Prescribing trends in bipolar disorder: cohort study in the United Kingdom THIN primary care database 1995–2009. PLoS One 6(12):e28725

Johnson RE, McFarland BH (1996) Lithium use and discontinuation in a health maintenance organization. Am J Psychiatry 153(8):993–1000

Kleindienst N, Engel R, Greil W (2005a) Which clinical factors predict response to prophylactic lithium? A systematic review for bipolar disorders. Bipolar Disord 7(5):404–417

Kleindienst N, Engel RR, Greil W (2005b) Psychosocial and demographic factors associated with response to prophylactic lithium: a systematic review for bipolar disorders. Psychol Med 35(12):1685–1694

Kleindienst N, Severus WE, Greil W (2007) Are serum lithium levels related to the polarity of recurrence in bipolar disorders? Evidence from a multicenter trial. Int Clin Psychopharmacol 22(3):125–131

Kukopulos A, Reginaldi D, Laddomada P et al (1980) Course of the manic-depressive cycle and changes caused by treatment. Pharmakopsychiatr Neuropsychopharmakol 13(4):156–167

Leboyer M, Henry C, Paillere-Martinot ML, Belliver F (2005) Age at onset in bipolar affective disorders: a review. Bipolar Disord 7(2):111–118

Lingam R, Scott J (2002) Treatment non-adherence in affective disorders. Acta Psychiatr Scand 105(3):164–172
Maj M (1990) Clinical prediction of response to lithium prophylaxis in bipolar patients: the importance of the previous pattern of course of the illness. Clin Neuropharmacol 13(Suppl 1):S66–S70
Maj M, Del Vecchio M, Starace F et al (1984) Prediction of affective psychoses response to lithium prophylaxis. The role of socio-demographic, clinical, psychological and biological variables. Acta Psychiatr Scand 69(1):37–44
Malhi GS (2010) History of the introduction of lithium into medicine and psychiatry. Acta Neuropsychiatr 22(3):161–161
Malhi GS, Geddes JR (2014) Carving bipolarity using a lithium sword. Br J Psychiatry 205:337–339. doi:10.1192/bjp.bp.114.148072
Malhi GS, Gershon S (2009) Ion men and their mettle. Aust N Z J Psychiatry 43:1091–1095
Malhi GS, Porter RJ (2014) Are 'buy-polar' forces and 'try-polar' thinking expanding bipolarity? Aust N Z J Psychiatry 48:697–700
Malhi GS, Tanious M, Gershon S (2011) The lithiumeter: a measured approach. Bipolar Disord 13(3):219–226
Malhi GS, Bargh DM, Cashman E, Frye MA, Gitlin M (2012) The clinical management of bipolar disorder complexity using a stratified model. Bipolar Disord 14(Suppl 2):66–89. doi:10.1111/j.1399-5618.2012.00993.x
Malhi GS, Bassett D, Boyce P, Bryant R, Fitzgerald PB, Fritz K, Hopwood M, Lyndon B, Mulder R, Murray G, Porter R, Singh AB (2015) Royal Australian and New Zealand College of Psychiatrists clinical practice guidelines for mood disorders. Aust N Z J Psychiatry 49:1087–1206
Manchia M, Adli M, Akula N et al (2013) Assessment of response to lithium maintenance treatment in bipolar disorder: a Consortium on Lithium Genetics (ConLiGen) report. PLoS One 8(6):e65636
Mander AJ (1986) Clinical prediction of outcome and lithium response in bipolar affective disorder. J Affect Disord 11(1):35–41
Moore GJ, Bebchuk JM, Wilds IB et al (2000) Lithium-induced increase in human brain grey matter. Lancet 356(9237):1241–1242
Moore GJ, Cortese BM, Giltz DA et al (2009) A longitudinal study of the effects of lithium treatment on prefrontal and subgenual prefrontal gray matter volume in treatment-responsive bipolar disorder patients. J Clin Psychiatry 70(5):699–705
Morselli PL, Elgie R (2003) GAMIAN-Europe/BEAM survey I-global analysis of a patient questionnaire circulated to 3450 members of 12 European advocacy groups operating in the field of mood disorders. Bipolar Disord 5(4):265–278
O'Connell RA, Mayo JA, Faltow L et al (1991) Outcome of bipolar disorder on long-term treatment with lithium. Br J Psychiatry 159:123–129
Okuma T (1993) Effects of carbamazepine and lithium on affective disorders. Neuropsychobiology 27(3):138–145
Pfennig A, Schlattmann P, Alda M et al (2010) Influence of atypical features on the quality of prophylactic effectiveness of long-term lithium treatment in bipolar disorders. Bipolar Disord 12(4):390–396
Post RM (2010) Letter to the editor regarding "A critical appraisal of lithium's efficacy and effectiveness: the last 60 years". Bipolar Disord 12(4):455–456; author reply 457–458
Post RM (2012) Acquired lithium resistance revisited: discontinuation-induced refractoriness versus tolerance. J Affect Disord 140(1):6–13
Price JL, Drevets WC (2012) Neural circuits underlying the pathophysiology of mood disorders. Trends Cogn Sci 16(1):61–71
Rohayem J, Baylé JF, Richa S (2008) Predictors of prophylactic response to lithium. Encéphale 34(4):394–399
Schou M, Juel-Nielsen N, Stromgren E, Voldby H (1954) The treatment of manic psychoses by the administration of lithium salts. J Neurol Neurosurg Psychiatry 17(4):250–260

Schultze TG, Alda M, Adli M et al (2010) The International Consortium on Lithium Genetics (ConLiGen): an initiative by the NIMH and IGSLI to study the genetic basis of response to lithium treatment. Neuropsychobiology 62(1):72–78

Severus WE et al (2009) Is the polarity of relapse/recurrence in bipolar-I disorder patients related to serum lithium levels? Results from an empirical study. J Affect Disord 115(3):466–470

Shapiro DR (1989) Response to maintenance therapy in bipolar illness. Arch Gen Psychiatry 46(5):401

Smith LA, Cornelius V, Warnock A et al (2007) Effectiveness of mood stabilizers and antipsychotics in the maintenance phase of bipolar disorder: a systematic review of randomized controlled trials. Bipolar Disord 9(4):394–412

Soares JC (2002) Can brain-imaging studies provide a "mood stabilizer signature?". Mol Psychiatry 7(Suppl 1):S64–S70

Tighe SK, Mahon PB, Potash JB (2011) Predictors of lithium response in bipolar disorder. Ther Adv Chron Dis 2(3):209–226

Tondo L, Hennen J, Baldessarini RJ (2003) Rapid-cycling bipolar disorder: effects of long-term treatments. Acta Psychiatr Scand 108(1):4–14

Vai B, Bollettini I, Benedetti F (2014) Corticolimbic connectivity as a possible biomarker for bipolar disorder. Expert Rev Neurother 14(6):631–650

Yazici O, Kora K, Uçok A et al (1999) Predictors of lithium prophylaxis in bipolar patients. J Affect Disord 55(2–3):133–142

Young AH, Hammond JM (2007) Lithium in mood disorders: increasing evidence base, declining use? Br J Psychiatry 191:474–476

Yucel K, McKinnon MC, Taylor VH et al (2007) Bilateral hippocampal volume increases after long-term lithium treatment in patients with bipolar disorder: a longitudinal MRI study. Psychopharmacology (Berl) 195(3):357–367

Lithium Response Variability (Pharmacogenomics Studies)

8

Alexandre Dayer, Nader Perroud, and Jean-Michel Aubry

Abstract

Lithium is considered to be the prototypical mood stabilizer. Naturalistic studies indicate that about one third of bipolar patients show a good or very good clinical response to lithium therapy. However, based on clinical evaluation, it remains difficult to reliably predict which patients are more likely to benefit from lithium therapy. In the era of personalized medicine, pharmacogenetics offers the hope that genetic markers could provide reliable predictors of individual responses to lithium therapy. As reviewed in this chapter, pharmacogenetic studies performed over the past two decades have unfortunately failed to produce predictive markers for lithium response that are meaningful for clinical practice. Lack of progress may be due to a variety of factors, including small sample sizes, clinical heterogeneity and variable definitions of lithium response versus non-response phenotypes. To circumvent these difficulties, large-scale international collaborative consortia are emerging, thus opening the possibility of performing genome-wide association studies on well-defined lithium-response phenotypes.

A. Dayer, MD
Mood Disorders Unit, Division of Psychiatric Specialties, Department of Mental Health and Psychiatry, University Hospitals of Geneva, Rue de Lausanne 20bis, CH-1201 Geneva, Switzerland
e-mail: alexandre.dayer@hcuge.ch

N. Perroud, MD
TRE Program, Division of Psychiatric Specialties, Department of Mental Health and Psychiatry, University Hospitals of Geneva, Rue de Lausanne 20bis, CH- 1201 Geneva, Switzerland
e-mail: nader.perroud@hcuge.ch

J.-M. Aubry, MD (✉)
Division of Psychiatric Specialties, Department of Mental Health and Psychiatry, University Hospitals of Geneva, and Faculty of Medicine, University of Geneva, Rue de Lusanne 20bis, Geneva 1201, Switzerland
e-mail: jean-michel.aubry@hcuge.ch

G.S. Malhi et al. (eds.), *The Science and Practice of Lithium Therapy*,
DOI 10.1007/978-3-319-45923-3_8

Key Points

- Reliable genetic predictors of individual responses to lithium therapy would be of great benefit to both patients and clinicians.
- Current findings based on small sample sizes and candidate-gene approaches lack robust replication.
- Large-scale international collaborative consortia are needed to replicate positive associations and to identify novel genetic targets.

8.1 Introduction

Lithium is considered to be the prototypical mood stabilizer. It has been shown to treat and prevent both manic and depressive phases of bipolar disorder (Fountoulakis et al. 2012; Fountoulakis and Vieta 2008; Suppes et al. 2005; Yatham et al. 2013a, b). Although lithium is a first-line treatment in bipolar disorder (Grandjean and Aubry 2009; Nivoli et al. 2011, 2012; Malhi et al. 2015), only a subset of patients shows a positive therapeutic response to it (Manchia et al. 2013). Naturalistic studies indicate that about one third of bipolar patients may achieve complete remission with lithium therapy (Garnham et al. 2007; Maj et al. 1998; Prien et al. 1974; Rybakowski et al. 2001; Solomon et al. 1995). However, the fraction of bipolar patients responding to lithium is likely to vary in a given clinical sample depending on whether bipolar II and subthreshold bipolar disorders are included. Furthermore, the definition of what constitutes a meaningful clinical therapeutic response varies in the literature (Geoffroy et al. 2014). At the clinical level, a number of variables have been associated with a poor lithium response (Grof et al. 1993). Generally, it has been shown that bipolar patients displaying an early onset of the disorder, an episodic clinical course, rapid cycling and mixed manic states are poor lithium responders (Grof et al. 1993; Strober et al. 1988). There is also some evidence that a positive therapeutic lithium response could run in families (Alda 2001). Bipolar patients responding to lithium appear to have a greater likelihood of having family relatives with a bipolar disorder responsive to lithium, thus suggesting that lithium responders may be genetically distinct (Aronoff and Epstein 1970; Duffy et al. 2007; Grof et al. 2002).

Determining the genetic factors that predict lithium responsiveness is of considerable interest, as it would facilitate the personalization of pharmacological treatment strategies for bipolar patients (Can et al. 2014; Geoffroy et al. 2014). Ideally, this would enable clinicians to maximize the benefits of lithium therapy in genetically prone responders and to minimize lithium side effects in genetically determined nonresponders by instead proposing to these patients alternative classes of mood stabilizers. The field of lithium pharmacogenomics is still in its infancy, and results are largely based on candidate-gene approaches performed on small sample sizes. Candidate genes have mainly been selected based on a priori hypotheses informed by preclinical data describing the molecular pathways modulated by lithium. In this chapter we review the available data on the pharmacogenetics of lithium response in bipolar patients and proceed by evaluating the evidence obtained in candidate genes involved in lithium-regulated signalling pathways.

8.2 The Inositol Pathway

Preclinical work over the past decades has led to the identification of a large number of molecular pathways modulated by therapeutic levels of lithium (0.6–1.2 mM) (Quiroz et al. 2004). Among these pathways, the impact of lithium on the regulation of inositol metabolism has been supported by a great deal of research (Manji et al. 1999; Quiroz et al. 2004; Williams et al. 2002). Activation of the inositol pathway classically occurs through G-protein-coupled receptors (GPCR) coupled to the enzyme phospholipase C (PLC). PLC activation leads to the hydrolysis of phosphatidylinositol 4,5-biphosphate (PIP2) into two important signalling molecules: inositol 1,4,5-triphosphate (IP3) and diacylglycerol (DAG). IP3 has multiple intracellular signalling properties, such as the induction of calcium release from the endoplasmic reticulum, whereas DAG activates downstream targets such as protein kinase C (PKC). IP3 is used to produce myo-inositol, which is recycled into PIP2 by phosphodiesterase enzymes, among which are inositol monophosphatase (IMPase) and inositol polyphosphate 1-phosphatase (IPPase).

Lithium is a positively charged ion that enters the cell and competes with the magnesium ion, which is used as a cofactor for enzymes such as IMPase and IPPase. By competing with magnesium, lithium is thought to inhibit IMPase and IPPase activity, thus decreasing the recycling of myo-inositol into PIP2 and leading to a state of relative inositol depletion. Evidence for this hypothesis is robust in rodent preclinical models following acute and chronic lithium administration (Allison et al. 1980; Allison and Stewart 1971; Lubrich et al. 1997; Sherman et al. 1981). In humans, decreased brain myo-inositol levels have been observed after chronic lithium therapy (Moore et al. 1999), whereas other studies have found no effect or even increased levels of myo-inositol after lithium therapy (Forester et al. 2008; Patel et al. 2006; Silverstone and McGrath 2009). Although there is a large amount of data indicating that lithium regulates the inositol pathway, there is only scant evidence in humans that differential clinical response to lithium therapy is determined by variations in genes belonging to the inositol pathway. In a small pilot sample of bipolar patients, a single nucleotide polymorphism (SNP) in the *INPP1* gene coding for the inositol polyphosphate 1-phosphatase enzyme (C973A) was found to be associated with lithium responsiveness (Steen et al. 1998). However, replication for this association failed in a second cohort tested in this initial report (Steen et al. 1998) and was not replicated in an independent study (Michelon et al. 2006). Apart from these initial findings, an interaction between a SNP in INPP1 and post-traumatic stress disorder in lithium responders with bipolar disorder has been reported (Bremer et al. 2007). In contrast to the *INPP1* gene, no association has been found between lithium response and genetic variation in other genes regulating the inositol pathway, such as the inositol monophosphatase *IMPA2* and *IMPA1* genes (Bremer et al. 2007; Michelon et al. 2006; Sjoholt et al. 2004), the phospholipase C-gamma 1 (*PLCG1*) gene (Ftouhi-Paquin et al. 2001) and the diacylglycerol kinase eta (*DGKH*) gene (Manchia et al. 2009b).

Taken together, available data indicate that genetic variation in genes involved in the inositol pathway does not appear to determine major differences in therapeutic responses to lithium. However, only a fraction of genetic variability in inositol-related genes has been tested, and large-scale studies are clearly needed to further explore additional targets in the inositol pathway and to potentially validate initial positive findings in the INPP1 gene.

8.3 Glycogen Synthase Kinase 3 and Circadian-Cycle-Related Genes

Glycogen synthase kinase 3 (GSK-3) is a constitutively active serine/threonine protein kinase that was initially found to deactivate glycogen synthase (Gould 2006). Lithium directly inhibits GSK-3 through competition with magnesium ions (Ryves and Harwood 2001) and indirectly through phosphorylation of N-terminal serines (Chalecka-Franaszek and Chuang 1999; Gould 2006; Gould and Manji 2005; Gurvich and Klein 2002). Given that GSK-3 regulates a variety of molecular pathways, including the Wnt and AKT signalling pathways (Chalecka-Franaszek and Chuang 1999; De Sarno et al. 2002), there has been a great deal of interest in understanding the molecular consequences of GSK-3 lithium-induced inhibition on neuronal function, signalling pathways and behavioural rodent phenotypes such as manic-like hyperlocomotion and depression-like behaviours (Beaulieu et al. 2004; Gould 2006; Gould et al. 2004; Polter et al. 2010; Prickaerts et al. 2006). Taken collectively, rodent studies support the role of GSK-3 in mediating lithium-induced behavioural responses (Gould et al. 2004), thus suggesting that genetic variants in *GSK-3* could be associated with differential lithium responsiveness in bipolar patients. An initial study reported that carriers of a rare functional variation in the promoter region of the beta isoform of *GSK-3* displayed a better respond to lithium therapy compared to carriers of a more common SNP allele (Benedetti et al. 2005). Interestingly, this finding could be subsequently replicated in an independent study (Adli et al. 2007). In addition to these two positive studies, an association between a SNP in *GSK-3beta* gene and lithium response in patients with bipolar disorder and post-traumatic stress disorder has been reported (Bremer et al. 2007). In contrast, no clear associations between polymorphisms in *GSK-3beta* were reported in other studies (McCarthy et al. 2011; Szczepankiewicz et al. 2006), thus casting doubts on the robustness of previous studies. Bipolar disorder is associated with dysregulation of circadian rhythms, and the *NR1D1* gene, encoding for the orphan nuclear receptor REV-ERBalpha, is a negative component of the circadian clock. Interestingly, REV-ERBalpha is a phosphorylation target of GSK-3beta, and its proteasomal degradation is regulated by lithium in cell cultures (Yin et al. 2006). Associations between genetic variation in *NR1D1* and lithium response have led to positive (Campos-de-Sousa et al. 2010; McCarthy et al. 2011) and negative results (Manchia et al. 2009b). Interestingly, genetic variations in *GSK-3beta* and *NR1D1* were found to act additively to predict response to lithium (McCarthy et al. 2011), suggesting that this pathway could be relevant for future studies. Taken collectively, published data are suggestive of a role for genetic variation in *GSK-3beta* in determining lithium response, but the robustness of these findings needs to be further confirmed in larger samples.

8.4 The Brain-Derived Neurotropic Factor (BDNF) Pathway

In both neuronal cultures in vitro and rodent models in vivo, lithium has been shown to be neuroprotective and to prevent the damaging effects of excitatory neurotoxicity and ischaemic insults (Chiu et al. 2013). Several molecular pathways could mediate these effects, including activation of pro-survival pathways such as the neurotrophin BDNF signalling pathway. For example, it has been shown that chronic lithium administration increases BDNF in brain regions such as the frontal cortex and the hippocampus (Angelucci et al. 2003; Fukumoto et al. 2001; Omata et al. 2008). Given the role of lithium in promoting BDNF signalling, genetic studies have focused on genetic variation in the *BDNF* gene and lithium responsiveness. The commonly studied *BDNF* val66met functional polymorphism has been investigated in several studies with mixed results. Initial studies reported an association between the *BDNF* val66met polymorphism and lithium responsiveness (Dmitrzak-Weglarz et al. 2008; Rybakowski et al. 2005b, 2007); however other studies failed to replicate these positive findings (Masui et al. 2006; Michelon et al. 2006). Some studies have focused on genetic variants in the *NTRK2* gene, which encodes for the TrkB protein, a key receptor binding to BDNF. Two independent studies have reported a significant association between *NTRK2* variants and lithium responsiveness (Bremer et al. 2007); however this result was not replicated in another sample (Dmitrzak-Weglarz et al. 2008; Wang et al. 2013). Taken collectively, there is some suggestive evidence in the field that genetic variation in BDNF signalling could account for differential clinical responsiveness to lithium, but these findings clearly need further validation in larger sample sizes.

8.5 The Monoamine Pathways

A large class of antidepressants exerts their action by selectively blocking the serotonin transporter, a key molecule regulating a variety of cellular and physiological processes in the brain and periphery (Murphy and Lesch 2008). Furthermore, manipulation of the serotonin system through tryptophan depletion induces depressive-related symptoms, thus suggesting a possible causal link between serotonin-related dysregulation and mood disorders (Smith et al. 1997; Cowen 2008). Finally, there is evidence of altered brain serotonin function in unmedicated depressed patients (Sharp and Cowen 2011; Cowen 2008). Given these arguments, mood disorders have traditionally been associated with disturbances in serotonin-related pathways (Manji et al. 2001). In the context of resistant depression, lithium has been used as an efficient augmentation therapy in association with antidepressants (Bauer et al. 2010; de Montigny et al. 1983). This has led to the hypothesis that lithium could exert these antidepressive properties by modulating the serotonin system. Studies in rodents indicate that lithium can increase serotonin levels in several brain regions (Baptista et al. 1990; Mork 1998). Furthermore, preclinical studies indicate that coadministration of lithium and antidepressants produces a synergetic effect on serotonergic transmission (Muraki et al. 2001; Okamoto et al. 1996; Wegener et al. 2003). Taken together, these preclinical data suggest that lithium could possibly exert some of its clinical efficacy through the modulation of the serotonin system, and this opens the possibility that genetic variation in serotonin-related genes could be involved in differential lithium responsiveness in bipolar patients.

In the field of mood disorders, the role of a functional genetic variant in the promoter region of the serotonin transporter (*5-HTTLPR*) has received great attention. A large body of research indicates that the short allele of *5-HTTLPR* (s-allele) interacts with early-life adversity to increase the risk of depression (Caspi et al. 2003; Karg et al. 2011; Uher et al. 2011). In addition, s-allele carriers display evidence of altered cortisol response to stress (Chen et al. 2014; Mueller et al. 2011; Way and Taylor 2010) or to tryptophan depletion (Cerit et al. 2013) as well as increased amygdala activation to negative emotional stimuli (Munafo et al. 2008). In addition, s-allele carriers from Caucasian, but not from Asian, origin appear to be less responsive to the antidepressant effects of selective serotonin reuptake inhibitors (Pollock et al. 2000; Porcelli et al. 2012; Smeraldi et al. 1998; Yoshida et al. 2002). In a similar manner, earlier studies have suggested that s-allele carriers could show decreased therapeutic responsiveness to lithium. Indeed homozygote carriers of the s-allele presenting either unipolar depression or bipolar disorder showed a decreased prophylactic response to lithium when prospectively assessed during more than 50 months (Serretti et al. 2001). An association between s-allele genotype and lithium non-response was replicated in an independent study (Rybakowski et al. 2005a). In contrast, opposite results were obtained in another study where s-allele carriers showed a better therapeutic response to lithium therapy compared to l-allele carriers (Serretti et al. 2004). These contradictory results have been further obscured by the fact that several studies revealed no association between lithium responsiveness and the *5-HTTLPR* genotype (Manchia et al. 2009a; Michelon et al. 2006). Thus, as described above with other genes investigated, the literature on *5-HTTLPR* genotype and lithium responsiveness is mixed and still inconclusive.

In addition to studies focused on the *5-HTTLPR*, lithium responsiveness in relationship to genetic variation in serotonin receptors has been investigated. Polymorphisms in several serotonin receptors, such as the *5-HT2A*, *5-HT2B*, and *5-HT2C*, were investigated in bipolar patients, and no variants were associated with lithium outcome in several independent studies (Dmitrzak-Weglarz et al. 2005; Manchia et al. 2009a; Serretti et al. 2000). Finally, it should be mentioned that lithium response in relation to genetic variation in other monoamine systems, such as the dopamine system, has been investigated. Similarly to the serotonin system, no clear association between genetic polymorphisms in different dopamine-related genes and lithium responsiveness has emerged. An initial report suggesting an association between a polymorphism in the *DRD1* gene coding for the dopamine receptor 1 and decreased responsiveness to lithium prophylaxis (Rybakowski et al. 2009) was not replicated in a subsequent study (Manchia et al. 2009a). No clear evidence for an association between therapeutic lithium response and genetic variation has been found in other monoamine-related genes, such as the tryptophan hydroxylase gene, the serotonin rate-limiting enzyme in serotonin synthesis (Serretti et al. 1999b); the *DRD2*, *DRD3* and *DRD4* genes coding for dopamine receptors (Manchia et al. 2009a; Serretti et al. 1998, 1999a); the dopamine transporter (*DAT*) gene (Manchia et al. 2009a); the catechol-O-methyltransferase (*COMT*) gene (Serretti et al. 2002); and the monoamine oxidase A (MAO-A) gene (Serretti et al. 2002; Turecki et al. 1999). Again, it should be mentioned that monoamine systems are highly complex and comprise many receptor subtypes with multiple splice variants. Much work thus remains to be done to map individual genetic variability in monoamine systems, thus leaving open the possibility that differential responses to the beneficial properties of lithium could still be partially determined by monoamine-related genetic differences.

8.6 Glutamate and GABA Pathways

A balance between inhibitory GABAergic and excitatory glutamatergic neurotransmissions is key for the proper function of neuronal circuits. The topic of antidepressant properties of drugs acting on the glutamatergic system has gained considerable interest recently with the observation that molecules targeting NMDA receptors, such as ketamine, can exert rapid antidepressant effects in unipolar and bipolar depression (Zarate et al. 2006, 2010, 2013). A rationale for a link between lithium and the glutamatergic system is provided by the observation that lithium may modulate NMDA/AMPA-mediated neurotransmission by modifying membrane expression of AMPA receptors in the hippocampus (Du et al. 2008; Gould et al. 2008). An association between increased risk for bipolar disorder and specific genetic polymorphisms in subunits of the NMDA receptors, such as the *GRIN2B* gene, has been reported, thus suggesting that this gene could be involved in lithium responsiveness (Martucci et al. 2006). However, no correlations between genetic variation in *GRIN2B* and lithium therapeutic responses were observed in a cohort of bipolar patients treated with lithium over 5 years (Szczepankiewicz et al. 2009b). Similarly, no evidence was found for an association between lithium response and polymorphisms in protein kinase FYN, which is involved in phosphorylating NMDA receptor subunits (Szczepankiewicz et al. 2009a). Suggestive evidence for an association between lithium response and genetic variation has been reported in *GRIA2*, a gene coding for AMPA receptor subunit (Perlis et al. 2009), and *CACNG2*, a gene coding for a voltage-dependent calcium channel involved in AMPA receptor trafficking (Silberberg et al. 2008). These studies clearly warrant further confirmation in independent samples. Regarding GABAergic pathways, candidate-gene approaches failed to demonstrate an association between lithium response and polymorphisms in

GABA-A receptor subunits including *GABRA3*, *GABRA5 and GABRB3* genes (Duffy et al. 2000). More recently, a study performed on bipolar patients of Han Chinese descent revealed a highly significant association ($p = 7.07 \times 10^{-50}$) between lithium response and two SNPs located in the glutamate decarboxylase-like protein 1 (*GADL1*) (Chen et al. 2014). The function of this protein is currently unknown, but it shares sequence homologies with glutamate decarboxylase isoforms, which normally catalyse the conversion of GABA to glutamate. However, a recent attempt to validate this finding in an Asian subsample of patients from the Consortium of Lithium Genetics (ConLiGen) failed (Hou et al. 2014). Future studies should thus be performed to further investigate the role of *GADL1* in lithium responsiveness and to provide a functional role for this protein in neuronal circuit function. Overall, convincing evidence is lacking for an association between lithium response and genes involved in glutamate or GABAergic signalling. However, given the complexity of these pathways, much work remains to be performed in these signalling systems.

8.7 Summary

Attempts in the preceding two decades to associate clinical lithium response with genetic variation in candidate genes involved in key signalling pathways modulated by lithium have failed to provide clinically meaningful results. As generally observed with genetic-association studies performed on small sample sizes, replication has been difficult to achieve, thus casting valid doubts on the robustness of positive published data. There is thus a critical need for larger sample sizes of bipolar patients with well-characterized lithium-response phenotypes in order to validate initial findings. Recently the field has moved forward towards achieving this goal with the emergence of large-scale international collaborative consortia. A genome-wide association study (GWAS) has been performed on more than 24,000 bipolar patients and controls (Muhleisen et al. 2014). Results have revealed a small number of hits reaching genome-wide statistical significance and, more generally, failed to validate candidate risk genes previously investigated before the era of GWAS studies. Interestingly, bipolar risk genes revealed in this recent study, such as *ANK3*, *CACNA1C* and *ODZ4* genes, were also identified in previous GWAS published in the field (Cross-Disorder Group of the Psychiatric Genomics Consortium and Genetic Risk Outcome of Psychosis (GROUP) Consortium 2013; Moskvina et al. 2009; Psychiatric GWAS Consortium Bipolar Disorder Working Group 2011; Schizophrenia Psychiatric Genome-Wide Association Study (GWAS) Consortium 2011).

With the aim of moving on to large-scale GWAS approaches, an international Consortium of Lithium Genetics (www.ConLiGen.org) emerged in 2009, enabling the collection of a sample size of more than 3,000 bipolar patients under lithium therapy and well characterized for their response to lithium (Manchia et al. 2013; Schulze et al. 2010). Given that variability in definitions of lithium response is an important confounding factor in pharmacogenetic studies, bipolar patients in the ConLiGen consortium are assessed using common quantitative scales, such as the Alda scale (Manchia et al. 2013).

In conclusion, there is hope that GWAS performed on larger sample sizes will help to determine whether some of the positive findings that have emerged using candidate-gene approaches can be replicated and whether novel targets can be identified. Overall, the hope is that a new generation of pharmacogenomic approaches will produce reliable and clinically meaningful genetic markers in order to predict which patients will benefit most from lithium prophylactic therapy.

References

Adli M, Hollinde DL, Stamm T, Wiethoff K, Tsahuridu M, Kirchheiner J et al (2007) Response to lithium augmentation in depression is associated with the glycogen synthase kinase 3-beta -50T/C single nucleotide polymorphism. Biol Psychiatry 62(11):1295–1302. doi:10.1016/j.biopsych.2007.03.023

Alda M (2001) Genetic factors and treatment of mood disorders. Bipolar Disord 3(6):318–324. doi:10.1034/j.1399-5618.2001.30606.x

Allison JH, Stewart MA (1971) Reduced brain inositol in lithium-treated rats. Nat New Biol 233(43):267–268

Allison JH, Boshans RL, Hallcher LM, Packman PM, Sherman WR (1980) The effects of lithium on myo-inositol levels in layers of frontal cerebral cortex, in cerebellum, and in corpus callosum of the rat. J Neurochem 34(2):456–458

Angelucci F, Aloe L, Jimenez-Vasquez P, Mathe AA (2003) Lithium treatment alters brain concentrations of nerve growth factor, brain-derived neurotrophic factor and glial cell line-derived neurotrophic factor in a rat model of depression. Int J Neuropsychopharmacol 6(3):225–231. doi:10.1017/S1461145703003468

Aronoff MS, Epstein RS (1970) Factors associated with poor response to lithium carbonate: a clinical study. Am J Psychiatry 127(4):472–480

Baptista TJ, Hernandez L, Burguera JL, Burguera M, Hoebel BG (1990) Chronic lithium administration enhances serotonin release in the lateral hypothalamus but not in the hippocampus in rats. A microdialysis study. J Neural Transm Gen Sect 82(1):31–41

Bauer M, Adli M, Bschor T, Pilhatsch M, Pfennig A, Sasse J et al (2010) Lithium's emerging role in the treatment of refractory major depressive episodes: augmentation of antidepressants. Neuropsychobiology 62(1):36–42. doi:10.1159/000314308

Beaulieu JM, Sotnikova TD, Yao WD, Kockeritz L, Woodgett JR, Gainetdinov RR et al (2004) Lithium antagonizes dopamine-dependent behaviors mediated by an AKT/glycogen synthase kinase 3 signaling cascade. Proc Natl Acad Sci U S A 101(14):5099–5104. doi:10.1073/pnas.0307921101

Benedetti F, Serretti A, Pontiggia A, Bernasconi A, Lorenzi C, Colombo C et al (2005) Long-term response to lithium salts in bipolar illness is influenced by the glycogen synthase kinase 3-beta -50 T/C SNP. Neurosci Lett 376(1):51–55. doi:10.1016/j.neulet.2004.11.022

Bremer T, Diamond C, McKinney R, Shehktman T, Barrett TB, Herold C et al (2007) The pharmacogenetics of lithium response depends upon clinical co-morbidity. Mol Diagn Ther 11(3):161–170. doi:10.1007/BF03256238

Campos-de-Sousa S, Guindalini C, Tondo L, Munro J, Osborne S, Floris G et al (2010) Nuclear receptor rev-erb-{alpha} circadian gene variants and lithium carbonate prophylaxis in bipolar affective disorder. J Biol Rhythm 25(2):132–137. doi:10.1177/0748730410362713

Can A, Schulze TG, Gould TD (2014) Molecular actions and clinical pharmacogenetics of lithium therapy. Pharmacol Biochem Behav. doi:10.1016/j.pbb.2014.02.004

Caspi A, Sugden K, Moffitt TE, Taylor A, Craig IW, Harrington H et al (2003) Influence of life stress on depression: moderation by a polymorphism in the 5-HTT gene. Science 301(5631):386–389. doi:10.1126/science.1083968

Cerit H, Jans LA, Van der Does W (2013) The effect of tryptophan on the cortisol response to social stress is modulated by the 5-HTTLPR genotype. Psychoneuroendocrinology 38(2):201–208. doi:10.1016/j.psyneuen.2012.05.016
Chalecka-Franaszek E, Chuang DM (1999) Lithium activates the serine/threonine kinase Akt-1 and suppresses glutamate-induced inhibition of Akt-1 activity in neurons. Proc Natl Acad Sci U S A 96(15):8745–8750
Chen CH, Lee CS, Lee MT, Ouyang WC, Chen CC, Chong MY et al (2014) Variant GADL1 and response to lithium therapy in bipolar I disorder. N Engl J Med 370(2):119–128. doi:10.1056/NEJMoa1212444
Chiu CT, Wang Z, Hunsberger JG, Chuang DM (2013) Therapeutic potential of mood stabilizers lithium and valproic acid: beyond bipolar disorder. Pharmacol Rev 65(1):105–142. doi:10.1124/pr.111.005512
Cowen PJ (2008) Serotonin and depression: pathophysiological mechanism or marketing myth? Trends Pharmacol Sci 29(9):433–436. doi:10.1016/j.tips.2008.05.004
Cross-Disorder Group of the Psychiatric Genomics Consortium, Genetic Risk Outcome of Psychosis (GROUP) Consortium (2013) Identification of risk loci with shared effects on five major psychiatric disorders: a genome-wide analysis. Lancet 381(9875):1371–1379. doi:10.1016/S0140-6736(12)62129-1
de Montigny C, Cournoyer G, Morissette R, Langlois R, Caille G (1983) Lithium carbonate addition in tricyclic antidepressant-resistant unipolar depression. Correlations with the neurobiologic actions of tricyclic antidepressant drugs and lithium ion on the serotonin system. Arch Gen Psychiatry 40(12):1327–1334
De Sarno P, Li X, Jope RS (2002) Regulation of Akt and glycogen synthase kinase-3 beta phosphorylation by sodium valproate and lithium. Neuropharmacology 43(7):1158–1164. doi:10.1016/S0028-3908(02)00215-0
Dmitrzak-Weglarz M, Rybakowski JK, Suwalska A, Slopien A, Czerski PM, Leszczynska-Rodziewicz A et al (2005) Association studies of 5-HT2A and 5-HT2C serotonin receptor gene polymorphisms with prophylactic lithium response in bipolar patients. Pharmacol Rep 57(6):761–765
Dmitrzak-Weglarz M, Rybakowski JK, Suwalska A, Skibinska M, Leszczynska-Rodziewicz A, Szczepankiewicz A et al (2008) Association studies of the BDNF and the NTRK2 gene polymorphisms with prophylactic lithium response in bipolar patients. Pharmacogenomics 9(11):1595–1603. doi:10.2217/14622416.9.11.1595
Du J, Creson TK, Wu LJ, Ren M, Gray NA, Falke C et al (2008) The role of hippocampal GluR1 and GluR2 receptors in manic-like behavior. J Neurosci 28(1):68–79. doi:10.1523/JNEUROSCI.3080-07.2008
Duffy A, Turecki G, Grof P, Cavazzoni P, Grof E, Joober R et al (2000) Association and linkage studies of candidate genes involved in GABAergic neurotransmission in lithium-responsive bipolar disorder. J Psychiatry Neurosci 25(4):353–358
Duffy A, Alda M, Milin R, Grof P (2007) A consecutive series of treated affected offspring of parents with bipolar disorder: is response associated with the clinical profile? Can J Psychiatry 52(6):369–376
Forester BP, Finn CT, Berlow YA, Wardrop M, Renshaw PF, Moore CM (2008) Brain lithium, N-acetyl aspartate and myo-inositol levels in older adults with bipolar disorder treated with lithium: a lithium-7 and proton magnetic resonance spectroscopy study. Bipolar Disord 10(6):691–700. doi:10.1111/j.1399-5618.2008.00627.x
Fountoulakis KN, Vieta E (2008) Treatment of bipolar disorder: a systematic review of available data and clinical perspectives. Int J Neuropsychopharmacol 11(7):999–1029. doi:10.1017/S1461145708009231
Fountoulakis KN, Kasper S, Andreassen O, Blier P, Okasha A, Severus E et al (2012) Efficacy of pharmacotherapy in bipolar disorder: a report by the WPA section on pharmacopsychiatry. Eur Arch Psychiatry Clin Neurosci 262(Suppl 1):1–48. doi:10.1007/s00406-012-0323-x

Ftouhi-Paquin N, Alda M, Grof P, Chretien N, Rouleau G, Turecki G (2001) Identification of three polymorphisms in the translated region of PLC-gamma1 and their investigation in lithium responsive bipolar disorder. Am J Med Genet 105(3):301–305. doi:10.1002/ajmg.1326

Fukumoto T, Morinobu S, Okamoto Y, Kagaya A, Yamawaki S (2001) Chronic lithium treatment increases the expression of brain-derived neurotrophic factor in the rat brain. Psychopharmacology (Berl) 158(1):100–106. doi:10.1007/s002130100871

Garnham J, Munro A, Slaney C, Macdougall M, Passmore M, Duffy A et al (2007) Prophylactic treatment response in bipolar disorder: results of a naturalistic observation study. J Affect Disord 104(1–3):185–190. doi:10.1016/j.jad.2007.03.003

Geoffroy PA, Bellivier F, Leboyer M, Etain B (2014) Can the response to mood stabilizers be predicted in bipolar disorder? Front Biosci (Elite Ed) 6:120–138

Gould TD (2006) Targeting glycogen synthase kinase-3 as an approach to develop novel mood-stabilising medications. Expert Opin Ther Targets 10(3):377–392. doi:10.1517/14728222.10.3.377

Gould TD, Manji HK (2005) Glycogen synthase kinase-3: a putative molecular target for lithium mimetic drugs. Neuropsychopharmacology 30(7):1223–1237. doi:10.1038/sj.npp.1300731

Gould TD, Zarate CA, Manji HK (2004) Glycogen synthase kinase-3: a target for novel bipolar disorder treatments. J Clin Psychiatry 65(1):10–21

Gould TD, O'Donnell KC, Dow ER, Du J, Chen G, Manji HK (2008) Involvement of AMPA receptors in the antidepressant-like effects of lithium in the mouse tail suspension test and forced swim test. Neuropharmacology 54(3):577–587. doi:10.1016/j.neuropharm.2007.11.002

Grandjean EM, Aubry JM (2009) Lithium: updated human knowledge using an evidence-based approach: part I: clinical efficacy in bipolar disorder. CNS Drugs 23(3):225–240. doi:10.2165/00023210-200923030-00004

Grof P, Alda M, Grof E, Fox D, Cameron P (1993) The challenge of predicting response to stabilising lithium treatment. The importance of patient selection. Br J Psychiatry Suppl 21:16–19

Grof P, Duffy A, Cavazzoni P, Grof E, Garnham J, MacDougall M et al (2002) Is response to prophylactic lithium a familial trait? J Clin Psychiatry 63(10):942–947

Gurvich N, Klein PS (2002) Lithium and valproic acid: parallels and contrasts in diverse signaling contexts. Pharmacol Ther 96(1):45–66. doi:10.1016/S0163-7258(02)00299-1

Hou L, Heilbronner U, Rietschel M, Kato T, Kuo PH, McMahon FJ et al (2014) Variant GADL1 and response to lithium in bipolar I disorder. N Engl J Med 370(19):1857–1859. doi:10.1056/NEJMc1401817#SA4

Karg K, Burmeister M, Shedden K, Sen S (2011) The serotonin transporter promoter variant (5-HTTLPR), stress, and depression meta-analysis revisited: evidence of genetic moderation. Arch Gen Psychiatry 68(5):444–454. doi:10.1001/archgenpsychiatry.2010.189

Lubrich B, Patishi Y, Kofman O, Agam G, Berger M, Belmaker RH et al (1997) Lithium-induced inositol depletion in rat brain after chronic treatment is restricted to the hypothalamus. Mol Psychiatry 2(5):407–412

Maj M, Pirozzi R, Magliano L, Bartoli L (1998) Long-term outcome of lithium prophylaxis in bipolar disorder: a 5-year prospective study of 402 patients at a lithium clinic. Am J Psychiatry 155(1):30–35

Malhi GS, Bassett D, Boyce P, Bryant R, Fitzgerald PB, Fritz K, Hopwood M, Lyndon B, Mulder R, Murray G, Porter R, Singh AB (2015) Royal Australian and New Zealand College of Psychiatrists clinical practice guidelines for mood disorders. Aust N Z J Psychiatry 49:1087–1206

Manchia M, Congiu D, Squassina A, Lampus S, Ardau R, Chillotti C et al (2009a) No association between lithium full responders and the DRD1, DRD2, DRD3, DAT1, 5-HTTLPR and HTR2A genes in a Sardinian sample. Psychiatry Res 169(2):164–166. doi:10.1016/j.psychres.2008.12.004

Manchia M, Squassina A, Congiu D, Chillotti C, Ardau R, Severino G et al (2009b) Interacting genes in lithium prophylaxis: preliminary results of an exploratory analysis on the role of DGKH and NR1D1 gene polymorphisms in 199 Sardinian bipolar patients. Neurosci Lett 467(2):67–71. doi:10.1016/j.neulet.2009.10.003

Manchia M, Adli M, Akula N, Ardau R, Aubry JM, Backlund L et al (2013) Assessment of response to lithium maintenance treatment in bipolar disorder: a Consortium on Lithium Genetics (ConLiGen) report. PLoS One 8(6), e65636. doi:10.1371/journal.pone.0065636

Manji HK, McNamara R, Chen G, Lenox RH (1999) Signalling pathways in the brain: cellular transduction of mood stabilisation in the treatment of manic-depressive illness. Aust N Z J Psychiatry 33(Suppl):S65–S83

Manji HK, Drevets WC, Charney DS (2001) The cellular neurobiology of depression. Nat Med 7(5):541–547. doi:10.1038/87865

Martucci L, Wong AH, De Luca V, Likhodi O, Wong GW, King N et al (2006) N-methyl-D-aspartate receptor NR2B subunit gene GRIN2B in schizophrenia and bipolar disorder: polymorphisms and mRNA levels. Schizophr Res 84(2–3):214–221. doi:10.1016/j.schres.2006.02.001

Masui T, Hashimoto R, Kusumi I, Suzuki K, Tanaka T, Nakagawa S et al (2006) Lithium response and Val66Met polymorphism of the brain-derived neurotrophic factor gene in Japanese patients with bipolar disorder. Psychiatr Genet 16(2):49–50. doi:10.1097/01.ypg.0000180680.72922.57

McCarthy MJ, Nievergelt CM, Shekhtman T, Kripke DF, Welsh DK, Kelsoe JR (2011) Functional genetic variation in the Rev-Erbalpha pathway and lithium response in the treatment of bipolar disorder. Genes Brain Behav 10(8):852–861. doi:10.1111/j.1601-183X.2011.00725.x

Michelon L, Meira-Lima I, Cordeiro Q, Miguita K, Breen G, Collier D et al (2006) Association study of the INPP1, 5HTT, BDNF, AP-2beta and GSK-3beta GENE variants and retrospectively scored response to lithium prophylaxis in bipolar disorder. Neurosci Lett 403(3):288–293. doi:10.1016/j.neulet.2006.05.001

Moore GJ, Bebchuk JM, Parrish JK, Faulk MW, Arfken CL, Strahl-Bevacqua J et al (1999) Temporal dissociation between lithium-induced changes in frontal lobe myo-inositol and clinical response in manic-depressive illness. Am J Psychiatry 156(12):1902–1908

Mork A (1998) Effects of lithium treatment on extracellular serotonin levels in the dorsal hippocampus and wet-dog shakes in the rat. Eur Neuropsychopharmacol 8(4):267–272. doi:10.1016/S0924-977X(97)00085-0

Moskvina V, Craddock N, Holmans P, Nikolov I, Pahwa JS, Green E et al (2009) Gene-wide analyses of genome-wide association data sets: evidence for multiple common risk alleles for schizophrenia and bipolar disorder and for overlap in genetic risk. Mol Psychiatry 14(3):252–260. doi:10.1038/mp.2008.133

Mueller A, Armbruster D, Moser DA, Canli T, Lesch KP, Brocke B et al (2011) Interaction of serotonin transporter gene-linked polymorphic region and stressful life events predicts cortisol stress response. Neuropsychopharmacology 36(7):1332–1339. doi:10.1038/npp.2011.11

Muhleisen TW, Leber M, Schulze TG, Strohmaier J, Degenhardt F, Treutlein J et al (2014) Genome-wide association study reveals two new risk loci for bipolar disorder. Nat Commun 5:3339. doi:10.1038/ncomms4339

Munafo MR, Brown SM, Hariri AR (2008) Serotonin transporter (5-HTTLPR) genotype and amygdala activation: a meta-analysis. Biol Psychiatry 63(9):852–857. doi:10.1016/j.biopsych.2007.08.016

Muraki I, Inoue T, Hashimoto S, Izumi T, Ito K, Koyama T (2001) Effect of subchronic lithium treatment on citalopram-induced increases in extracellular concentrations of serotonin in the medial prefrontal cortex. J Neurochem 76(2):490–497

Murphy DL, Lesch KP (2008) Targeting the murine serotonin transporter: insights into human neurobiology. Nat Rev Neurosci 9(2):85–96. doi:10.1038/nrn2284

Nivoli AM, Colom F, Murru A, Pacchiarotti I, Castro-Loli P, Gonzalez-Pinto A et al (2011) New treatment guidelines for acute bipolar depression: a systematic review. J Affect Disord 129(1–3):14–26. doi:10.1016/j.jad.2010.05.018

Nivoli AM, Murru A, Goikolea JM, Crespo JM, Montes JM, Gonzalez-Pinto A et al (2012) New treatment guidelines for acute bipolar mania: a critical review. J Affect Disord 140(2):125–141. doi:10.1016/j.jad.2011.10.015

Okamoto Y, Motohasi N, Hayakawa H, Muraoka M, Yamawaki S (1996) Addition of lithium to chronic antidepressant treatment potentiates presynaptic serotonergic function without changes in serotonergic receptors in the rat cerebral cortex. Neuropsychobiology 33(1):17–20

Omata N, Murata T, Takamatsu S, Maruoka N, Mitsuya H, Yonekura Y et al (2008) Neuroprotective effect of chronic lithium treatment against hypoxia in specific brain regions with upregulation of cAMP response element binding protein and brain-derived neurotrophic factor but not nerve growth factor: comparison with acute lithium treatment. Bipolar Disord 10(3):360–368. doi:10.1111/j.1399-5618.2007.00521.x

Patel NC, DelBello MP, Cecil KM, Adler CM, Bryan HS, Stanford KE et al (2006) Lithium treatment effects on Myo-inositol in adolescents with bipolar depression. Biol Psychiatry 60(9):998–1004. doi:10.1016/j.biopsych.2006.07.029

Perlis RH, Smoller JW, Ferreira MA, McQuillin A, Bass N, Lawrence J et al (2009) A genome-wide association study of response to lithium for prevention of recurrence in bipolar disorder. Am J Psychiatry 166(6):718–725. doi:10.1176/appi.ajp.2009.08111633

Pollock BG, Ferrell RE, Mulsant BH, Mazumdar S, Miller M, Sweet RA et al (2000) Allelic variation in the serotonin transporter promoter affects onset of paroxetine treatment response in late-life depression. Neuropsychopharmacology 23(5):587–590. doi:10.1016/S0893-133X(00)00132-9

Polter A, Beurel E, Yang S, Garner R, Song L, Miller CA et al (2010) Deficiency in the inhibitory serine-phosphorylation of glycogen synthase kinase-3 increases sensitivity to mood disturbances. Neuropsychopharmacology 35(8):1761–1774. doi:10.1038/npp.2010.43

Porcelli S, Fabbri C, Serretti A (2012) Meta-analysis of serotonin transporter gene promoter polymorphism (5-HTTLPR) association with antidepressant efficacy. Eur Neuropsychopharmacol 22(4):239–258. doi:10.1016/j.euroneuro.2011.10.003

Prickaerts J, Moechars D, Cryns K, Lenaerts I, van Craenendonck H, Goris I et al (2006) Transgenic mice overexpressing glycogen synthase kinase 3beta: a putative model of hyperactivity and mania. J Neurosci 26(35):9022–9029. doi:10.1523/JNEUROSCI.5216-05.2006

Prien RF, Caffey EM Jr, Klett CJ (1974) Factors associated with treatment success in lithium carbonate prophylaxis. Report of the Veterans Administration and National Institute of Mental Health collaborative study group. Arch Gen Psychiatry 31(2):189–192

Psychiatric GWAS Consortium Bipolar Disorder Working Group (2011) Large-scale genome-wide association analysis of bipolar disorder identifies a new susceptibility locus near ODZ4. Nat Genet 43(10):977–983. doi:10.1038/ng.943

Quiroz JA, Gould TD, Manji HK (2004) Molecular effects of lithium. Mol Interv 4(5):259–272. doi:10.1124/mi.4.5.6

Rybakowski JK, Chlopocka-Wozniak M, Suwalska A (2001) The prophylactic effect of long-term lithium administration in bipolar patients entering treatment in the 1970s and 1980s. Bipolar Disord 3(2):63–67

Rybakowski JK, Suwalska A, Czerski PM, Dmitrzak-Weglarz M, Leszczynska-Rodziewicz A, Hauser J (2005a) Prophylactic effect of lithium in bipolar affective illness may be related to serotonin transporter genotype. Pharmacol Rep 57(1):124–127

Rybakowski JK, Suwalska A, Skibinska M, Szczepankiewicz A, Leszczynska-Rodziewicz A, Permoda A et al (2005b) Prophylactic lithium response and polymorphism of the brain-derived neurotrophic factor gene. Pharmacopsychiatry 38(4):166–170. doi:10.1055/s-2005-871239

Rybakowski JK, Suwalska A, Skibinska M, Dmitrzak-Weglarz M, Leszczynska-Rodziewicz A, Hauser J (2007) Response to lithium prophylaxis: interaction between serotonin transporter and BDNF genes. Am J Med Genet B Neuropsychiatr Genet 144B(6):820–823. doi:10.1002/ajmg.b.30420

Rybakowski JK, Dmitrzak-Weglarz M, Suwalska A, Leszczynska-Rodziewicz A, Hauser J (2009) Dopamine D1 receptor gene polymorphism is associated with prophylactic lithium response in bipolar disorder. Pharmacopsychiatry 42(1):20–22. doi:10.1055/s-0028-1085441

Ryves WJ, Harwood AJ (2001) Lithium inhibits glycogen synthase kinase-3 by competition for magnesium. Biochem Biophys Res Commun 280(3):720–725. doi:10.1006/bbrc.2000.4169

Schizophrenia Psychiatric Genome-Wide Association Study (GWAS) Consortium (2011) Genome-wide association study identifies five new schizophrenia loci. Nat Genet 43(10):969–976. doi:10.1038/ng.940

Schulze TG, Alda M, Adli M, Akula N, Ardau R, Bui ET et al (2010) The International Consortium on Lithium Genetics (ConLiGen): an initiative by the NIMH and IGSLI to study the genetic basis of response to lithium treatment. Neuropsychobiology 62(1):72–78. doi:10.1159/000314708

Serretti A, Lilli R, Lorenzi C, Franchini L, Smeraldi E (1998) Dopamine receptor D3 gene and response to lithium prophylaxis in mood disorders. Int J Neuropsychopharmacol 1(2):125–129. doi:10.1017/S1461145798001230

Serretti A, Lilli R, Lorenzi C, Franchini L, Di Bella D, Catalano M et al (1999a) Dopamine receptor D2 and D4 genes, GABA(A) alpha-1 subunit genes and response to lithium prophylaxis in mood disorders. Psychiatry Res 87(1):7–19. doi:10.1016/S0165-1781(99)00056-6

Serretti A, Lilli R, Lorenzi C, Gasperini M, Smeraldi E (1999b) Tryptophan hydroxylase gene and response to lithium prophylaxis in mood disorders. J Psychiatr Res 33(5):371–377. doi:10.1016/S0022-3956(99)00013-8

Serretti A, Lorenzi C, Lilli R, Smeraldi E (2000) Serotonin receptor 2A, 2C, 1A genes and response to lithium prophylaxis in mood disorders. J Psychiatr Res 34(2):89–98. doi:10.1016/S0022-3956(00)00004-2

Serretti A, Lilli R, Mandelli L, Lorenzi C, Smeraldi E (2001) Serotonin transporter gene associated with lithium prophylaxis in mood disorders. Pharmacogenomics J 1(1):71–77

Serretti A, Lorenzi C, Lilli R, Mandelli L, Pirovano A, Smeraldi E (2002) Pharmacogenetics of lithium prophylaxis in mood disorders: analysis of COMT, MAO-A, and Gbeta3 variants. Am J Med Genet 114(4):370–379. doi:10.1002/ajmg.10357

Serretti A, Malitas PN, Mandelli L, Lorenzi C, Ploia C, Alevizos B et al (2004) Further evidence for a possible association between serotonin transporter gene and lithium prophylaxis in mood disorders. Pharmacogenomics J 4(4):267–273. doi:10.1038/sj.tpj.6500252

Sharp T, Cowen PJ (2011) 5-HT and depression: is the glass half-full? Curr Opin Pharmacol 11(1):45–51. doi:10.1016/j.coph.2011.02.003

Sherman WR, Leavitt AL, Honchar MP, Hallcher LM, Phillips BE (1981) Evidence that lithium alters phosphoinositide metabolism: chronic administration elevates primarily D-myo-inositol-1-phosphate in cerebral cortex of the rat. J Neurochem 36(6):1947–1951

Silberberg G, Levit A, Collier D, St Clair D, Munro J, Kerwin RW et al (2008) Stargazin involvement with bipolar disorder and response to lithium treatment. Pharmacogenet Genomics 18(5):403–412. doi:10.1097/FPC.0b013e3282f974ca

Silverstone PH, McGrath BM (2009) Lithium and valproate and their possible effects on the myo-inositol second messenger system in healthy volunteers and bipolar patients. Int Rev Psychiatry 21(4):414–423. doi:10.1080/09540260902962214

Sjoholt G, Ebstein RP, Lie RT, Berle JO, Mallet J, Deleuze JF et al (2004) Examination of IMPA1 and IMPA2 genes in manic-depressive patients: association between IMPA2 promoter polymorphisms and bipolar disorder. Mol Psychiatry 9(6):621–629. doi:10.1038/sj.mp.4001460

Smeraldi E, Zanardi R, Benedetti F, Di Bella D, Perez J, Catalano M (1998) Polymorphism within the promoter of the serotonin transporter gene and antidepressant efficacy of fluvoxamine. Mol Psychiatry 3(6):508–511

Smith KA, Fairburn CG, Cowen PJ (1997) Relapse of depression after rapid depletion of tryptophan. Lancet 349(9056):915–919. doi:10.1016/S0140-6736(96)07044-4

Solomon DA, Keitner GI, Miller IW, Shea MT, Keller MB (1995) Course of illness and maintenance treatments for patients with bipolar disorder. J Clin Psychiatry 56(1):5–13

Steen VM, Lovlie R, Osher Y, Belmaker RH, Berle JO, Gulbrandsen AK (1998) The polymorphic inositol polyphosphate 1-phosphatase gene as a candidate for pharmacogenetic prediction of lithium-responsive manic-depressive illness. Pharmacogenetics 8(3):259–268

Strober M, Morrell W, Burroughs J, Lampert C, Danforth H, Freeman R (1988) A family study of bipolar I disorder in adolescence. Early onset of symptoms linked to increased familial loading and lithium resistance. J Affect Disord 15(3):255–268

Suppes T, Dennehy EB, Hirschfeld RM, Altshuler LL, Bowden CL, Calabrese JR et al (2005) The Texas implementation of medication algorithms: update to the algorithms for treatment of bipolar I disorder. J Clin Psychiatry 66(7):870–886

Szczepankiewicz A, Rybakowski JK, Suwalska A, Skibinska M, Leszczynska-Rodziewicz A, Dmitrzak-Weglarz M et al (2006) Association study of the glycogen synthase kinase-3beta gene polymorphism with prophylactic lithium response in bipolar patients. World J Biol Psychiatry 7(3):158–161. doi:10.1080/15622970600554711

Szczepankiewicz A, Skibinska M, Suwalska A, Hauser J, Rybakowski JK (2009a) The association study of three FYN polymorphisms with prophylactic lithium response in bipolar patients. Hum Psychopharmacol 24(4):287–291. doi:10.1002/hup.1018

Szczepankiewicz A, Skibinska M, Suwalska A, Hauser J, Rybakowski JK (2009b) No association of three GRIN2B polymorphisms with lithium response in bipolar patients. Pharmacol Rep 61(3):448–452

Turecki G, Grof P, Cavazzoni P, Duffy A, Grof E, Ahrens B et al (1999) MAOA: association and linkage studies with lithium responsive bipolar disorder. Psychiatr Genet 9(1):13–16

Uher R, Caspi A, Houts R, Sugden K, Williams B, Poulton R et al (2011) Serotonin transporter gene moderates childhood maltreatment's effects on persistent but not single-episode depression: replications and implications for resolving inconsistent results. J Affect Disord 135(1–3):56–65. doi:10.1016/j.jad.2011.03.010

Wang Z, Fan J, Gao K, Li Z, Yi Z, Wang L et al (2013) Neurotrophic tyrosine kinase receptor type 2 (NTRK2) gene associated with treatment response to mood stabilizers in patients with bipolar I disorder. J Mol Neurosci 50(2):305–310. doi:10.1007/s12031-013-9956-0

Way BM, Taylor SE (2010) The serotonin transporter promoter polymorphism is associated with cortisol response to psychosocial stress. Biol Psychiatry 67(5):487–492. doi:10.1016/j.biopsych.2009.10.021

Wegener G, Bandpey Z, Heiberg IL, Mork A, Rosenberg R (2003) Increased extracellular serotonin level in rat hippocampus induced by chronic citalopram is augmented by subchronic lithium: neurochemical and behavioural studies in the rat. Psychopharmacology (Berl) 166(2):188–194. doi:10.1007/s00213-002-1341-6

Williams RS, Cheng L, Mudge AW, Harwood AJ (2002) A common mechanism of action for three mood-stabilizing drugs. Nature 417(6886):292–295. doi:10.1038/417292a

Yatham LN, Kennedy SH, Parikh SV, Schaffer A, Beaulieu S, Alda M et al (2013a) The evolution of CANMAT bipolar disorder guidelines: past, present, and future. Bipolar Disord 15(1):58–60. doi:10.1111/bdi.12038

Yatham LN, Kennedy SH, Parikh SV, Schaffer A, Beaulieu S, Alda M et al (2013b) Canadian Network for Mood and Anxiety Treatments (CANMAT) and International Society for Bipolar Disorders (ISBD) collaborative update of CANMAT guidelines for the management of patients with bipolar disorder: update 2013. Bipolar Disord 15(1):1–44. doi:10.1111/bdi.12025

Yin L, Wang J, Klein PS, Lazar MA (2006) Nuclear receptor Rev-erbalpha is a critical lithium-sensitive component of the circadian clock. Science 311(5763):1002–1005. doi:10.1126/science.1121613

Yoshida K, Ito K, Sato K, Takahashi H, Kamata M, Higuchi H et al (2002) Influence of the serotonin transporter gene-linked polymorphic region on the antidepressant response to fluvoxamine in Japanese depressed patients. Prog Neuropsychopharmacol Biol Psychiatry 26(2):383–386. doi:10.1016/S0278-5846(01)00287-1

Zarate CA Jr, Singh JB, Carlson PJ, Brutsche NE, Ameli R, Luckenbaugh DA et al (2006) A randomized trial of an N-methyl-D-aspartate antagonist in treatment-resistant major depression. Arch Gen Psychiatry 63(8):856–864. doi:10.1001/archpsyc.63.8.856

Zarate C Jr, Machado-Vieira R, Henter I, Ibrahim L, Diazgranados N, Salvadore G (2010) Glutamatergic modulators: the future of treating mood disorders? Harv Rev Psychiatry 18(5):293–303. doi:10.3109/10673229.2010.511059

Zarate C, Duman RS, Liu G, Sartori S, Quiroz J, Murck H (2013) New paradigms for treatment-resistant depression. Ann N Y Acad Sci 1292:21–31. doi:10.1111/nyas.12223

Lithium Response Variability: New Avenues and Hypotheses

9

Frank Bellivier and Cynthia Marie-Claire

Abstract

Lithium is the leading treatment for relapse prevention in bipolar disorders, but response prediction remains an important challenge for clinicians. Clinical predictors have poor sensitivity and specificity, and biomarkers predicting lithium response are lacking. This chapter reviews future research directions that may improve the prediction of lithium response. One important area for research is investigation of the blood and brain pharmacokinetics of lithium in humans, in particular the identification of the brain regions where lithium exerts its therapeutic action. Consistent data suggest the existence of variations in the lithium concentration between brain regions, but the association between brain lithium distribution and therapeutic response remains to be studied. It is also reasonable to think that active mechanisms at brain barriers underlie the brain distribution of lithium. Finally, a growing body of scientific data is available on the molecular signature of the lithium response, as well as lithium-targeted signalling pathways. All these areas of research are likely to open new avenues for identifying

F. Bellivier, MD, PhD (✉)
INSERM UMR-S1144, Paris Descartes University & Paris Diderot University,
PRES Sorbonne Paris Cité, Paris, France

UFR de Médecine, Denis Diderot University, Paris, France

FondaMental Foundation, Creteil, France

Psychiatry Department, AP-HP GH Saint-Louis – Lariboisiere – F. Widal & INSERM UMR-S 1144, 200, rue du Fg Saint-Denis, Paris 75475, Cedex 10, France
e-mail: frank.bellivier@inserm.fr

C. Marie-Claire, PhD
INSERM UMR-S1144, Paris Descartes University & Paris Diderot University,
PRES Sorbonne Paris Cité, Paris, France
e-mail: cynthia.marie-claire@inserm.fr

G.S. Malhi et al. (eds.), *The Science and Practice of Lithium Therapy*,
DOI 10.1007/978-3-319-45923-3_9

predictive biomarkers of the prophylactic response of lithium, thus facilitating the development of personalized therapeutic strategies.

Key Points

- The therapeutic response and side effects of lithium are highly variable and unpredictable.
- The blood and brain pharmacokinetics of extended-release lithium remain understudied.
- Active mechanisms at brain blood barriers regulate lithium access to the brain.
- How lithium is distributed in the brain and where it exerts its actions remain largely unknown.

9.1 Introduction

Lithium is the leading treatment for relapse prevention in bipolar disorders, with many patients remaining asymptomatic for several years or even decades. However, a significant subgroup of patients experience high relapse rates despite lithium treatment, and clinicians still struggle to accurately predict which patients will respond without prescribing a lengthy lithium trial. In addition, biomarkers predicting lithium response are lacking. This chapter briefly reviews data on (i) the variability of the lithium therapeutic response and its side effects, (ii) clinical predictors of lithium response, and (iii) biological predictors of lithium response. It then discusses the blood and brain pharmacokinetics of lithium in humans, brain regions where lithium exerts its therapeutic action, variations in the lithium concentration between brain regions, association between brain lithium distribution and therapeutic response, possible active mechanisms at the brain barriers underlying the brain distribution of lithium, and finally the molecular signature of the lithium response, as well as lithium-targeted signalling pathways. All these areas of research are likely to open new avenues to identifying biomarkers of the lithium prophylactic response. Improved prediction of lithium's prophylactic efficacy remains a major requirement to prevent patients from participating in inefficient, lengthy lithium trials and to develop personalized strategies.

9.2 Lithium Response Variability

Although lithium is the first-line agent for the management of bipolar disorder, the drug does not work for all patients. Geddes and colleagues (2004) published a meta-analysis of five randomized controlled trials comparing prophylactic lithium therapy with placebo in bipolar disorder and found that lithium is more effective than placebo in preventing recurrence of illness, with 60% in the lithium group remaining well over 1–2 years. More recently, another meta-analysis of six studies of lithium in the treatment of acute mania found that 47 % of patients responded, compared with 32 % of the controls (Yildiz et al. 2011). Naturalistic studies of lithium-treated patients indicated that complete remission was obtained in only one-third of the patients (Baldessarini and Tondo 2000; Garnham et al. 2007; Maj et al. 1995; Solomon et al. 1995). These studies also indicate that the illness course remains unmodified in up to 40 % of lithium-treated patients, despite good treatment adherence. A similar level of variability is also observed for the side-effect profile of lithium (Dols et al. 2013).

The evaluation of lithium's prophylactic efficacy requires at least 2 years of treatment. Therefore, the identification of predictive biomarkers is a major requirement to reduce inefficient and lengthy therapeutic trials for patients. As relapse prevention is the leading determinant of good outcomes in bipolar disorders, the prediction of efficacy and drug tolerance is a major issue for prognosis. This field of research benefits from recent developments in brain imaging technology, pharmacokinetics modelling, and genomics and post-genomics that may shed light on undiscovered areas. In this chapter, we review those areas of research that are in need of further study.

9.2.1 Predictors of Therapeutic Response

The course of illness prior to the initiation of lithium treatment has been well studied, with findings indicating that an episodic course of bipolar disorder is correlated with response (Garnham et al. 2007; Passmore et al. 2003). The major episode sequence 'mania-depression-free interval' has also been found to correlate with a better prophylactic response rate in comparison with the 'depression-mania-free interval' sequence (Kleindienst et al. 2005). Other indicators that correlate positively with prophylactic response include the following: positive family history of bipolar disorder (Grof et al. 1994, 2009; Passmore et al. 2003), absence of comorbidity, rapid cycling or mixed features (Backlund et al. 2009; Grof et al. 2009), and family history of lithium response (McKnew et al. 1981; Grof et al. 2009; Duffy et al. 2014). Conflicting results have been obtained with regard to the association between a late age at onset and good lithium response. There are studies that are both positive (Tondo et al. 2001; Kleindienst et al. 2005; Coryell et al. 2000; Yazici et al. 1999) and negative (Dunner and Fieve 1974; Kato et al. 2000; Maj et al. 1986; Okuma 1993; Rybakowski et al. 2007; Sarantidis and Waters 1981; Yang 1985). However, the picture remains complex because age at onset is associated with other characteristics that may influence the level of lithium response, such as age at first lithium prescription, duration of the disorder prior to lithium introduction, or a family history of bipolar disorder (correlated with a lower age at onset). Studies on the predominant polarity in relation to lithium prophylactic responses have also produced conflicting results (Kleindienst et al. 2005; Yazici et al. 1999).

In summary, differential responses to prophylactic lithium identify a subtype of bipolar disorder. The search for clinical biomarkers of lithium response has thus far remained inconclusive. However, recent familial studies have identified new avenues of clinical research. The traits of adult lithium responders are also observed in their affected children. In prospective studies, the children of bipolar parents developed symptoms earlier, with marked differences between the offspring of lithium responders and nonresponders (Duffy et al. 2014). The illness evolves progressively, from nonspecific sleep and anxiety disorders to mood symptoms and then, often starting in adolescence, to major depressive and later activated episodes of mania or psychosis.

9.2.2 Side Effects

Side effects are among the factors most frequently associated with poor adherence to treatment, which in turn influences treatment efficacy. Lithium's tolerance profile is characterized by neurological, gastrointestinal, metabolic, thyroid, dermatological, kidney, cognitive, sexual, haematological, liver, and teratogenic side effects. Studies on the prevalence and management of these various side effects remain heterogeneous and difficult to translate into clinical practice. Moreover, there is high interindividual variability in vulnerability to these side effects; responses are poorly correlated to lithium plasma levels (Sproule 2002) and are generally unpredictable (Dols et al. 2013). (For further information on side effects of lithium, see Chap. 15.)

9.3 Blood/Brain Pharmacokinetics

The management of lithium treatment is complicated by several uncertainties related to pharmacokinetics (PK) and pharmacodynamics (PD) issues: (i) animal studies and preliminary data in humans (Komoroski et al. 1997; Lee et al. 2012; Ramaprasad et al. 1992) indicate variations in the lithium concentration in different brain regions. (ii) The brain regions where lithium exerts its therapeutic action, as well as its mechanism of action, remain largely unknown. (iii) The brain pharmacokinetics of lithium do not correlate with its plasma pharmacokinetics (time to reach maximal concentration and elimination half-life), suggesting active mechanisms underlying the brain distribution of lithium through the brain barriers. To date, the role of the blood-brain barrier (BBB) and the blood-cerebrospinal fluid barrier (BCSFB) in brain lithium PD/PK relationships has never been studied. Moreover, these potential transport mechanisms through the brain barriers, due to their region-specific expression and/or activity, may be responsible for the heterogeneity in lithium brain distribution. (iv) Extended-release lithium (XR lithium), the most commonly used form of lithium salts, has not been well studied in terms of its PK parameters, and monitoring parameters have been estimated from old PK studies of immediate-release lithium (IR lithium). Further investigations using modern tools to identify the PK model parameters of XR lithium, as well as PK/PD relationships, may help to optimize the long-term management of XR lithium-treated patients (see Chap. 2).

9.4 Blood Lithium Pharmacokinetics

For a chronic treatment regime with twice-daily lithium dosing, classic PK studies recommend the use of the minimum serum lithium concentration (C_{min}, termed 'residual') measured 12 h after the last dose to estimate the exposure to lithium (Amdisen 1977). However, these studies are old, and none of them has identified other PK parameters that can be used in clinical practice (Sproule 2002). Population PK studies of immediate release (IR) lithium performed in adults observed that the clearance of lithium is between 1 and 2 l/h and that the half-life is between 17 and 27 h (Chen et al. 2000; Gaillot et al. 1979; Jermain et al. 1991; Lehmann et al. 1988; Turck et al. 2000; Yang et al. 1991; Hunter 1988). The main covariates that explain the maximum variance of the clearance of lithium are age, body weight, and serum creatinine concentration (or its clearance) (Jermain et al. 1991; Yukawa et al. 1993).

Using the erythrocyte lithium concentration (supposed to reflect intracellular lithium concentration) as a measure of lithium exposure, instead of plasma concentration, was first proposed in 1973 by Mendels and Frazer. Indeed, they reported a positive correlation between lithium preventive efficacy and a high erythrocyte/plasma lithium concentration ratio (Mendels and Frazer 1973). However, the accuracy of this parameter has not been validated by other studies (Rohayem et al. 2008), and no consensus currently exists on the use of this ratio as a reliable indicator of desirable prophylactic responses to lithium.

Subsequently, several authors have attempted to correlate lithium erythrocyte concentration and side effects/toxicity. Indeed, several authors have reported that lithium neurotoxicity (Dysken et al. 1979; Elizur et al. 1982; Olie et al. 1982) or changes in electroencephalography (EEG) results (Dasso et al. 1977; Zakowska-Dabrowska and Rybakowski 1973) correlate better with erythrocyte concentrations than they do with plasma lithium concentrations. In particular, the erythrocyte lithium concentration (but not plasma concentration) is higher in patients with side effects (polyuria, polydipsia, and tremors), compared to subjects free from such symptoms (Hewick and Murray 1976). However, this observation was not replicated in subsequent studies (Johnston et al. 1979; Ong 1983). The poor correlation between erythrocyte and brain lithium concentrations is a possible explanation for these conflicting results (Kato et al. 1993). The same group reported consistent data indicating a poor correlation between brain concentrations measured by Li-7 magnetic resonance spectroscopy (Li-7 MRS) and tremors (Kato et al. 1996).

PK parameters from the five available studies on XR lithium (lithium LP carbonate, 400 mg per 24 h, once daily) are summarized in Table 9.1. The PK data for XR lithium is limited and difficult to introduce into clinical practice, because all the reported evaluations were performed 12 h after lithium administration (daily intake), and little is known about how these results reflect the global exposure to lithium. In addition, the dosage required to achieve therapeutic levels is highly variable and unpredictable (Price and Heninger 1994).

Overall, the literature does not contain any comprehensive PK studies of XR lithium (once daily). The plasma lithium therapeutic range used in clinical practice

Table 9.1 Available studies on XR lithium (once daily)

Study population	No of subjects	Type of XR Li	Reference	Observed pharmacokinetics parameters
Normal control subjects	6	Lithium carbonate LP 400 mg (after 12 h)	Hunter (1988)	$C_{max} = 1.49 \pm 0.31$ mmol/L $T_{max} = 2.1 \pm 0.5$ h $t_{1/2} = 27.6 \pm 9.1$ h $V = 50.9$ L CL/F = 1.25 ± 0.4 L/h ASC (0–24) = 22.8 ± 4.9 mmol/L/h
Normal control subjects	11	Lithium carbonate LP 450 mg (after 12 h)	Kristoff et al. (1986)	15 % increase in concentration with ibuprofen 400 mg co-prescription
Normal control subjects	16	Lithium acetate 536 mg (after 12 h)	Turck et al. (2000)	21 % increase in concentration and AUC with meloxicam (20 mg) co-prescription Lithium PK parameters (XR lithium monotherapy group): C_{pre}, ss = 0.54 mmol/L C_{max}, ss = 0.97 mmol/L ASC ss = 7.75 mmol/L/h T_{max}, ss = 1.5 h, $C_{av} = 0.66$ mmol/L C_{min}, ss = 0.35 mmol/L CL/F = 30.2 mL/min, PTF = 96 % Ae = 14.0 mmol (100 % of dose) CLr = 30.2
Bipolar patients (including 31 subjects with lithium toxicity)	50	Lithium carbonate LP 400 mg or 200 mg (after 12 h)	ElDesoky et al. (2008)	CL = 0.51 L/h with 12.7 % BSV V1 (fixed) = 15.2 L Q (fixed) = 7.44 L/h V2 (fixed) = 6.7 L
Bipolar patients	42	Lithium LP citrate (after 24 h)	Hoegberg et al. (2012)	Comparison of lithium carbonate IR and lithium citrate Fr = 0.33 vs 0.33 V1 (L/kg) = 0.74 vs 0.74 ka (/h) = 2.79 vs 0.75 $f = 1.00$ vs 0.85 DS (mM) = 0.0036 vs 0.006

[a]Mean ± standard deviation; *LP* extended release

and measured at 12 h is based on approximate estimations of lithium plasma targets from old studies of IR lithium formulations (residual concentration).

The PK-PD relationship for XR lithium is another understudied area. Most of the published data have studied the PK-PD relationships for the IR formulation of lithium, and the results are sometimes contradictory (Higuchi et al. 1988; Terao et al. 1999; Wright and Crismon 2000; Chiu et al. 2007; Abou-Auda et al. 2008; Hoegberg et al. 2012; Huang et al. 2008; Sproule 2002). Indeed, the relationship between the residual concentration (C_{min}) and the therapeutic response is poorly characterized. The proportion of responders (relapse prevention) increases proportionally with the increases in the C_{min} to approximately 0.8–1.0 mmol/L (Sproule 2002). However, these concentrations are close to toxic concentrations (≥1.5 mmol/L) (Simard et al. 1989). A significant proportion of patients exhibit documented response to lithium despite low plasma concentrations (approximately 0.4–0.7 mmol/L) (Coppen et al. 1973). Furthermore, toxicity has been observed in some patients with C_{min} within the recommended therapeutic range (Sproule 2002). Several parameters influencing the C_{min} are likely to influence the therapeutic response (Amdisen 1977; Greil et al. 1985; Swartz 1987). Minimal C_{min} fluctuations can exert significant effects on efficacy and safety (Amdisen 1977; Greil et al. 1985; Swartz 1987). In addition, discrepancies between the plasma and erythrocyte concentrations of lithium may exist in some patients and be a source of toxicity or ineffective treatment (Frazer et al. 1973). Altogether, these data clearly suggest that the study of PK-PD relationships is hampered by the complex architecture of factors influencing lithium's PK and PD, heterogeneity of bipolar disorders with regard to lithium response, and narrow therapeutic range. These results further suggest that the estimations of the PK parameters and the PK-PD relationships of XR lithium based on IR lithium results are highly speculative. Specific PK studies of XR lithium using modern population PK modelling must be performed to identify not only the PK model parameters but also their variability.

9.5 Brain Lithium Pharmacokinetics

Animal studies provide abundant direct and indirect evidence to support regional variations in the brain distribution of lithium. Twenty-four hours after a single subcutaneous lithium injection, heterogeneous lithium distribution was observed among rat brain regions, with the hypothalamus, corpus striatum, and midbrain being the regions with the highest lithium accumulation (Rios and Guzman-Mendez 1990). After the administration of 3 mEq/kg lithium chloride, the hypothalamus exhibited the highest concentration between 0.5 and 8 h, and after 8 h, the highest concentration was observed in the caudate nucleus, followed by the cortex and rest of the diencephalon (Mukherjee et al. 1976). Heterogeneous lithium distribution in rat brain regions at therapeutic dosages after prolonged treatment has also been reported (Ramaprasad 2004). Significant differences were found in vivo among lithium concentration ratios for the various brain regions, using either lithium-7 (^{7}Li) MR spectroscopic imaging technology or in vitro methods (Komoroski et al. 1997).

Not only is lithium's brain distribution heterogeneous, but its activity at a molecular level also exhibits high variability. Lithium appears to be most active in a region stretching from the anterior cingulate cortex and striatum to the caudal midbrain, with the greatest activity in the preoptic area and hypothalamic regions (Ramaprasad et al. 2005). Only some activity was observed in prefrontal cortex, and the lowest molecular activity was observed in the cerebellum and metencephalic brainstem. Accordingly, in rats chronically fed with a diet containing 0.2 % lithium carbonate, myo-inositol monophosphatase-1 activity increased substantially in various brain regions analysed, even doubling in some regions, in the following order: hippocampus > cerebellum > striatum > cerebral cortex > brainstem (Parthasarathy et al. 2003). In addition, the effect of 1, 2 and 4 weeks of lithium treatment (1 mmol/kg/day, intraperitoneal) on the myo-inositol concentrations across several rat brain regions, including the prefrontal, temporal and occipital cortical areas, as well as the hippocampus, was analysed using MRS at 18.8 T (McGrath et al. 2006). However, while other studies suggest brain region-specific alterations in myo-inositol concentrations among bipolar patients, this study found that lithium-induced reduction of myo-inositol is more global.

Excellent opportunities have emerged using MRS at high magnetic fields to image brain lithium-7 with unprecedented sensitivity (Lee et al. 2012; Renshaw et al. 1985). Lithium's distribution in the brain, the mechanisms influencing lithium brain distribution, and lithium's mechanism of molecular action in these different regions are important questions that need to be addressed, and answers to these questions may help to identify sensitive and specific biomarkers of lithium response.

The brain pharmacokinetics of lithium does not follow its plasma pharmacokinetics (time to reach maximal concentration and elimination half-life), suggesting active mechanisms underlying the distribution of lithium through the brain barriers; to date, the role of the blood-brain barrier (BBB) and the blood-cerebrospinal fluid barrier (BCSFB) in brain lithium PK-PD relationships has never been studied. Moreover, these transport mechanisms through the brain barriers, due to their region-specific expression and/or activity, may be responsible for the heterogeneity in lithium's brain distribution. Interestingly, the Na^+-dependent Cl^-/HCO_3^- exchanger (NCBE) encoded by the *SLC4A10* gene is highly expressed and is of major importance in the physiology of choroid plexuses for the production of cerebrospinal fluid (CSF) (Damkier and Praetorius 2012). This Na^+ exchanger likely transports lithium, and, recently, the *SLC4A10* gene has been suggested in a GWAS study as a candidate gene for lithium response in bipolar disorder (Rybakowski 2013).

9.6 Molecular Signature of Lithium Response

Lithium's mechanism of action is rather complex because it exerts inhibitory and facilitatory actions on several cellular signalling pathways. In the brain, lithium acts on pre- and postsynaptic neurons and has been demonstrated to inhibit excitatory transmission and to promote inhibitory transmission. Its modulatory effects on dopamine, glutamate and GABA neurotransmissions have been investigated in animal and clinical studies (Ahluwalia et al. 1981; Vargas et al. 1998; Ferrie et al. 2005, 2006). Lithium inhibition of the solute carrier family 5 member 3 (SLC5A3)-mediated entry of inositol into the cells and direct inhibition of inositol monophosphatase (IMPase) and inositol polyphosphate 1-phosphatase (IPPase) (Gould and Manji 2002; Willmroth et al. 2007) contribute to a decrease in intracellular inositol, therefore inhibiting protein kinase C (PKC)-associated signalling and reducing excitatory neurotransmission (Machado-Vieira et al. 2009; Lenox and Wang 2003; Zarate and Manji 2009; Lenox et al. 1992). Lithium also directly inhibits GSK3β, which in turn activates the Akt pathway (Klein and Melton 1996; Tajes et al. 2009). Another pathway affected by lithium is the cAMP pathway, resulting in an increase in the basal levels of cAMP and adenylate cyclase (AC) (Mann et al. 2009; Marmol 2008; Quiroz et al. 2010). All of these targets and effectors contribute to the neuroprotective, neuroproliferative and mood stabilizing actions of lithium in the brain and complicate the understanding of the lithium prophylactic response.

To understand the variability in responses to lithium, a complementary approach to genetic studies is to study lithium's molecular effects. Lithium has several cellular targets and modulates expression levels at the transcriptomic, proteomic and also miRNA levels in several in vitro and in vivo models.

9.6.1 Animal Models

Available studies on lithium's molecular effects in animal models are summarized in Table 9.2. The short-term effects of lithium on the proteome have been analysed after acute lithium intraperitoneal injection, and an increase was observed in the levels of fourteen proteins in the rat hypothalamus (Lee et al. 2009). Acute lithium has also been demonstrated to increase the mRNA levels of the immediate early gene early growth response 1 (*Egr1*) and the circadian gene period circadian clock 2 (*Per2*) in mice frontal cortices (Kim et al. 2013). In mice, chronic lithium was found to modulate 121 mRNAs in the brain, including *Per2* (McQuillin et al. 2007); these results are in agreement with the observed impact of lithium on circadian rhythms in rodents (Abe et al. 2000; Subramanian et al. 1998; Williams and Jope 1995).

Because lithium is a long-term treatment, most of the reported studies have focused on lithium-induced modulations after chronic treatment. In rats, chronic lithium treatment has been demonstrated to modulate the mRNA levels of several genes in the brain. In line with lithium's effects on the cAMP pathway, the upregulation of AC and downregulation of Giα were observed in the rat cerebral cortex after chronic treatment with lithium (Colin et al. 1991). In the prefrontal cortex, a comparative study of the effects of lithium, valproate and paliperidone found that only the levels of synapsin I (Syn1) were increased by the three drugs; in contrast, lithium decreased specifically Goα, tubulin-β and glutathione peroxidase 4 (Gpx4) protein levels (Corena-McLeod Mdel et al. 2008). Another comparative study found that the expression of two isoforms of phosphodiesterase (PE) (PE4B and PE4E)

Table 9.2 Available studies on lithium's molecular effects in animal models

Species	Brain region	Dose	Duration of treatment	Route of administration	Observed modulations	Reference
Rat	Cerebral cortex	0.24 %	4 weeks	Food	↑ *AC* mRNA ↓ *Giα* mRNA	Colin et al. (1991)
Rat	Hippocampus Temporal cortex Frontal cortex	0.2 %	14 days 28 days	Food	↑ BDNF peptide	Fukumoto et al. (2001)
Rat	Hippocampus Cerebral cortex	0.2 %	4 weeks	Food	↑ IMPase 1 protein	Parthasarathy et al. (2003)
Rat	Choroid plexus	75 mg/kg	28 days	Osmotic pump	↓ *Ttr* mRNA	Pulford et al. (2006)
Rat	Prefrontal cortex	22 mg/kg	28 days	i.p.	↓ Gpx4, Goα, tubulin β proteins ↑ Syn1 protein	Corena-Mcleod Mdel et al. (2008)
Rat	Hypothalamus	0.15 M	1 h and 6 h	i.p.	↑ 14 proteins at 1 h ↑ 32 proteins at 6 h	Lee et al. (2009)
Rat	Hippocampus	2.4 g/kg	4 weeks	Food	↓ Let-7b, let-7c, miR-128a, miR-24a, miR-30c, miR-34a, miR-221 miRNAs ↑ miR-144 miRNAs	Zhou et al. (2009)
Rat	Cortex	120 mg/kg	10 days	i.p.	↑ *Adra2A* mRNA	Cuffi et al. (2010)
Rat	Frontal cortex	50 mg/kg	21 days	i.p.	↓ Pde4ax and Pde4b1 proteins	Fatemi et al. (2010)
Rat	Hippocampus	0.1 %	4 weeks	Food	↑ 104genes (mRNA) ↓ 96 genes (mRNA)	Lee et al. (2015)
Mice	Brain	0.4 %	2 weeks	Food	121 genes (mRNA)	McQuillin et al. (2007)
Mice	Frontal cortex	150 mg/kg	2 h	i.p.	↑ *Per2* and *Egr1* mRNA	Kim et al. (2013)
Mice	Microglia	600 mg/L	2 weeks	Water	↑ 104genes (mRNA) ↓ 126 genes (mRNA)	Yu et al. (2015)

decreased in the rat frontal cortex after lithium, clozapine, haloperidol and valproate chronic treatment, while lithium alone reduced the mRNA levels of PDE8B, suggesting a potential common therapeutic pathway for these psychotropic medications (Fatemi et al. 2010). Comparative transcriptomic effects of lithium and valproate in the rat hippocampus have also been recently studied (Lee et al. 2015). Lithium impact on gene expression was found more limited than that of valproate, and only few genes were found to be regulated by both drugs. In agreement with the observed increase in BDNF peptide and anti-apoptotic genes levels in the peripheral blood of lithium responders (Tramontina et al. 2009; Lowthert et al. 2012; Beech et al. 2014), an increase in the peptide levels of BDNF in the rat hippocampus and temporal cortex was observed after chronic lithium treatment (Fukumoto et al. 2001). Chronic lithium can also regulate the protein level and activity of IMPase 1, as observed in the rat hippocampus and cerebral cortex (Parthasarathy et al. 2003). Pulford and collaborators reported the downregulation of transthyretin (Ttr) mRNA in the choroid plexus (Pulford et al. 2006). Lithium treatment has been demonstrated to upregulate the transcription of the adrenoceptor alpha 2A (AdrA2a) gene in the rat cortex (Cuffi et al. 2010). In mice microglia, lithium has been shown to modulate the expression level of several genes, among these the complement component 3 (C3) was found to be the only gene significantly regulated in the mice microglia and human monocyte-derived dendritic cells (Yu et al. 2015).

A possible mechanism for the regulation of mRNA, and therefore protein, levels is the modulation of miRNA levels (Guarnieri and DiLeone 2008). A study of the impact of chronic lithium on miRNA levels in the rat hippocampus found a decrease in eight miRNAs and an increase in miR-144 (Zhou et al. 2009). This result suggests that lithium could exert some of its molecular effects through this epigenetic mechanism, which opens a new field of research. An effect of lithium on histone modifications in the leptin receptor gene *Lepr* has been recently found in rat hippocampus emphasizing the need to study the epigenetic impacts of this drug (Lee et al. 2015).

Animal models possess the added benefit of allowing the study of lithium effects in targeted areas of the brain. These studies have confirmed some of the observed modulations of protein and mRNA levels in patients and also initiated new research directions. However, the dissimilar routes of administration used (infusion by osmotic pump, intraperitoneal injection, lithium in food pellets, etc.) and the differences in the lithium doses and treatment duration (from 22 to 120 mg/day during 1 h to 28 days) make it difficult to compare the results.

9.6.2 Cellular Models

Several animal and human cellular models have been used to further analyse the molecular response to lithium. Available studies are summarized in Table 9.3. In line with the in vivo results, lithium-induced increases in the *Bcl2* and *Bdnf* mRNA levels were observed in rat cultured astrocytes, cortical or hippocampal neurons (Yasuda et al. 2009; Keshavarz et al. 2013; Dwivedi and Zhang 2014). The decrease

Table 9.3 Available studies on lithium's molecular effects in cellular models

Species	Cellular model	Dose	Duration of treatment	Observed modulations	Reference
Rat	Cortical neurons	1 mM	48 h	↑ Promoter IV *Bdnf* mRNA	Yasuda et al. (2009)
Rat	Astrocytes	1 mM	7 days	↑ *Bcl2* mRNA	Keshavarz et al. (2013)
Rat	Hippocampal neurons	1 mM	48 h	↑ Promoter IV *Bdnf*, *Bcl2*, *Bcl2l1* mRNA ↓*Bad*, *Casp3* mRNA	Dwivedi and Zhang (2015)
Mice	HN33	5 mM	7 days	↓ *Marcks* mRNA	Wang et al. (2001)
Mice	NIH 3T3	20 mM	3 days	↑ *Per2* and *Cry1* mRNA ↓ *Per3*, *Cry2*, *Bmal1*, *E4BP4* and *Rev-Erb-α* mRNAs	Osland et al. (2011)
Human	LCL BP vs control	1 mM	7 days	↓ *ADRA1B*, *CHRNA1*, *PDE4D*, *SPR*, *and RAB7* mRNA	Sun et al. (2004)
Human	LCLs LiR vs unaffected relatives vs controls	1 mM	7 days	↓ BDNF peptide	Tseng et al. (2008)
Human	SK-N-AS	1.5 mM	33 days	↑ 347 genes ↓ 324 genes	Seelan et al. (2008)
Human	LCL BP vs discordant siblings	1 mM	4, 7 and 21 days	↑ miR-221, miR-152, miR-155, miR-24a miRNAs	Chen et al. (2009)
Human	LCL Control ± lithium or val	0.75 mM	7 days	44 genes modulated by lithium and 416 by val (18 in common)	Sugawara et al. (2010)
Human	LCL LiR vs LiNR	1 mM	7 days	Modulation of *SYN2a* and *SYN2b* mRNAs	Cruceanu et al. (2012)
Human	LCL LiR vs LiNR	1 mM	7 days	No significant effects of Li *IGF1* differentially expressed between R and NR	Squassina et al. (2013)

Table 9.3 (continued)

Human	iPSCs BP and control	1 mM	24 h	↑ EMX2 protein	Chen et al. (2014)
Human	LCL BP vs controls	1 mM	3 weeks	↓ *DBP* mRNA	Kittel-Schneider et al. (2015)
Human	LCL LiR vs LiNR	1 mM	7 days	↓ 10 members of Let-7miRNAs family in R	Hunsberger et al. (2015)

in one of the myristoylated alanine-rich PKC substrates (MARCKS) observed in the rat brain was also observed in immortalized hippocampal cells from mice (Wang et al. 2001). In line with the effects of lithium on the circadian rhythms of rodents, the molecular effects of lithium on several circadian genes have been analysed in the fibroblast cell line NIH-3T3, and significant modulations of seven genes were observed (Osland et al. 2011).

Human cellular models have also been used, and despite some inherent limitations resulting from cellular changes induced by transformation with the Epstein-Barr virus, lymphoblastoid cell lines (LCLs) have been extensively used for the identification of genes associated with bipolar disorder. Gene expression analyses in this model have identified differential gene expression in bipolar patients compared with controls or monozygotic asymptomatic twins (Matigian et al. 2007; Kato et al. 2011). In line with the impact of lithium on circadian genes observed in in vivo and cellular animal models, a lithium-induced decrease of the expression level of *DBP* has been shown in LCL of bipolar patients and healthy controls (Kittel-Schneider et al. 2015). The genes regulated by lithium in the LCLs of bipolar patients were found to include alpha1B-adreno-ceptor (*ADRA1B*), acetylcholine receptor protein alpha chain precursor (*CHRNA1*), phosphodiesterase 4D, cAMP-specific (PDE4D), substance P-receptor (SPR), and the ras-related protein RAB7 (Sun et al. 2004). A decrease in the BDNF peptide levels was found in LCLs from lithium-responsive bipolar patients compared with their unaffected relatives (Tseng et al. 2008).

As suggested from animal studies, lithium may exert some of its effects through miRNA modulation. miR-134 plasma levels are reported to be decreased in bipolar patients compared with controls and increased upon the introduction of medication and improvement of mood states (Rong et al. 2011), suggesting that miRNAs could be used as circulating biomarkers of lithium response variability. LCLs have been successfully used as a model to study differential miRNA modulation induced by lithium; four miRNAs (miR-221, miR-24a, miR-152 and miR-155) were found to be differentially regulated in LCLs from affected patients compared with those from their siblings (Chen et al. 2009). Using the same model, the effects of lithium in control subjects have been compared to those of another mood stabilizer valproate. This study identified 44 genes regulated by lithium and 416 by valproate and only 18 genes with mRNA levels regulated by both molecules (Sugawara et al. 2010). The gene most significantly downregulated by lithium was BCL2-associated

X protein (*BAX*, -1, eightfold), and the most significantly upregulated gene was platelet-activating factor acetylhydrolase 1b catalytic subunit 2 (*PAFAH1B2*, twofold).

To examine whether lithium-induced molecular effects differ between responsive and nonresponsive patients, studies using the LCL model were also undertaken. Interestingly, a study focusing on two isoforms of synapsin II (SYN2A and SYN2B) found differential effects of lithium on the synapsin II mRNA levels in LCLs from lithium responders vs nonresponder patients (Cruceanu et al. 2012). Microarray profiling of LCLs indicated that lithium did not differentially regulate any genes in lithium responders compared to nonresponder patients (Squassina et al. 2013). However, this study showed that *IGF1* was significantly downregulated in LCL from nonresponders as compared to responders. Although preliminary, the interactive networks of mRNAs and miRNAs modulated by lithium were recently studied in LCL from responder and nonresponder patients (Hunsberger et al. 2015). MiRNAs and associated pathways differentially regulated by lithium in the two groups were identified.

Other human cellular models are available, but they have been used sparingly in molecular studies. Microarray profiling of human neuronal cells after long-term treatment with lithium identified 347 upregulated and 324 downregulated transcripts (Seelan et al. 2008). More recently, a study using induced pluripotent stem cells generated from bipolar patients indicated that lithium increased the protein level of a gene involved in telencephalic neuronal fate, empty spiracles homeobox 2 (EMX2), in the derived neurons (Chen et al. 2014).

The results obtained in these animal and cellular studies contribute to the in-depth study of the molecular and biochemical regulations underlying the response to lithium in bipolar disorder and, therefore, the variability in lithium treatment responses. The candidate genes and proteins identified will help to provide potential peripheral biomarkers of treatment response in bipolar patients. Promising results have been obtained with serum BDNF levels, which could be correlated with the emotional state of the patients (de Oliveira et al. 2009; Machado-Vieira et al. 2007; Cunha et al. 2006) and also can be modulated upon treatment (Huang et al. 2012; Palomino et al. 2006; Tramontina et al. 2007). Peripheral mononuclear cells (PBMCs) provide readily accessible material to overcome the low accessibility of brain samples for researchers. They present the advantage of being easily isolated from blood samples, allowing the longitudinal follow-up of patients at different stages of the disease and/or the therapy. For genes expressed in the brain and in blood cells, a correlation has been reported between their levels of expression in both tissues (Sullivan et al. 2006). Moreover, a study of miRNA distribution in normal human tissues demonstrated that PBMCs exhibit the highest similarity in miRNA expression patterns relative to the brain (Liang et al. 2007). A recent study reported the differential regulation of genes and DNA methylation in PBMCs from BP patients as compared to control subjects (D'Addario et al. 2012). Interestingly, DNA, mRNA and miRNA can be isolated from the same material to perform a more comprehensive study of the observed gene and protein levels. PBMCs therefore represent a promising material for future studies of the molecular effects of lithium at the periphery.

9.7 Summary

The hope that a biological marker might predict lithium response and tolerance has stimulated decades of research, without the identification of biomarkers of clear clinical utility. Thus, the identification of such predictive biomarkers of prophylactic lithium efficacy remains a major issue for the development of personalized medicine in bipolar disorders. This field of research has benefited from recent developments in modern pharmacokinetic modelling to better understand the lithium PK model parameters and their interindividual variability. PK model(s) and the characterization of blood-brain-barrier transportation will help to elucidate the means through which lithium accesses the brain and is distributed. New neuroimaging techniques may represent interesting tools for these investigations. In particular, excellent opportunities have emerged through the use of high magnetic fields to image lithium-7 with unprecedented sensitivity. Indeed, based on recent animal studies and pilot studies in humans, the concentration and regional distribution of lithium in the brain may be reasonably associated with therapeutic success. Thus, the development of a technique to accurately determine the concentrations and regional distribution of lithium in the brain, as well as to investigate the molecular determinants of lithium diffusion across the different brain barriers, will help to identify pharmacokinetic and imaging-based biomarkers to better predict treatment response. Pharmacogenetic studies and studies of blood biomarkers that correlate with (or predict) lithium response are likely to identify variations in lithium molecular targets. The identification of genes differentially modulated by lithium in cellular models (LCLs or PBMCs) from lithium responder patients in comparison with nonresponders should provide insights into the cellular pathways that are differentially regulated by lithium. The study of lithium-induced DNA methylation or miRNA modulations in these models will also generate new hypotheses about the role or importance of epigenetic regulation in response to lithium. These studies will contribute to research efforts aimed at elucidating lithium's mechanisms of action and propose hypotheses for the cellular basis of the variability in the lithium treatment response. These findings would have a high translational potential for clinical use and could ultimately improve the prediction of therapeutic responses and the management of patient care. For all of these goals to be achieved, the collection and collation of follow-up clinical and biological data with careful assessment of adherence to treatment is essential.

References

Abe M, Herzog ED, Block GD (2000) Lithium lengthens the circadian period of individual suprachiasmatic nucleus neurons. Neuroreport 11(14):3261–3264

Abou-Auda HS, Al-Yamani MJ, Abou-Shaaban RR, Khoshhal SI (2008) A new accurate method for predicting lithium clearance and daily dosage requirements in adult psychiatric patients. Bipolar Disord 10(3):369–376. doi:10.1111/j.1399-5618.2007.00549.x

Ahluwalia P, Grewaal DS, Singhal RL (1981) Brain gabaergic and dopaminergic systems following lithium treatment and withdrawal. Prog Neuropsychopharmacol 5(5–6):527–530

Amdisen A (1977) Serum level monitoring and clinical pharmacokinetics of lithium. Clin Pharmacokinet 2(2):73–92

Backlund L, Ehnvall A, Hetta J, Isacsson G, Agren H (2009) Identifying predictors for good lithium response – a retrospective analysis of 100 patients with bipolar disorder using a life-charting method. Eur Psychiatry J Assoc Eur Psychiatr 24(3):171–177. doi:10.1016/j.eurpsy.2008.12.009

Baldessarini RJ, Tondo L (2000) Does lithium treatment still work? Evidence of stable responses over three decades. Arch Gen Psychiatry 57(2):187–190

Beech RD, Leffert JJ, Lin A, Sylvia LG, Umlauf S, Mane S, Zhao H, Bowden C, Calabrese JR, Friedman ES, Ketter TA, Iosifescu DV, Reilly-Harrington NA, Ostacher M, Thase ME, Nierenberg A (2014) Gene-expression differences in peripheral blood between lithium responders and non-responders in the lithium treatment-moderate dose use study (litmus). Pharmacogenomics J 14(2):182–191. doi:10.1038/tpj.2013.16

Chen C, Veronese L, Yin Y (2000) The effects of lamotrigine on the pharmacokinetics of lithium. Br J Clin Pharmacol 50(3):193–195

Chen H, Wang N, Burmeister M, Mcinnis MG (2009) Microrna expression changes in lymphoblastoid cell lines in response to lithium treatment. Int J Neuropsychopharmacol 12(7):975–981

Chen HM, Delong CJ, Bame M, Rajapakse I, Herron TJ, Mcinnis MG, O'shea KS (2014) Transcripts involved in calcium signaling and telencephalic neuronal fate are altered in induced pluripotent stem cells from bipolar disorder patients. Transl Psychiatry 4:e375

Chiu CC, Shen WW, Chen KP, Lu ML (2007) Application of the cockcroft-gault method to estimate lithium dosage requirement. Psychiatry Clin Neurosci 61(3):269–274. doi:10.1111/j.1440-1819.2007.01654.x

Colin SF, Chang HC, Mollner S, Pfeuffer T, Reed RR, Duman RS, Nestler EJ (1991) Chronic lithium regulates the expression of adenylate cyclase and gi-protein alpha subunit in rat cerebral cortex. Proc Natl Acad Sci U S A 88(23):10634–10637

Coppen A, Peet M, Bailey J, Noguera R, Burns BH, Swani MS, Maggs R, Gardner R (1973) Double-blind and open prospective studies on lithium prophylaxis in affective disorders. Psychiatr Neurol Neurochir 76(6):501–510

Corena-Mcleod Mdel P, Oliveros A, Charlesworth C, Madden B, Liang YQ, Boules M, Shaw A, Williams K, Richelson E (2008) Paliperidone as a mood stabilizer: a pre-frontal cortex synaptoneurosomal proteomics comparison with lithium and valproic acid after chronic treatment reveals similarities in protein expression. Brain Res 1233:8–19. doi:10.1016/j.brainres.2008.07.021

Coryell W, Akiskal H, Leon AC, Turvey C, Solomon D, Endicott J (2000) Family history and symptom levels during treatment for bipolar i affective disorder. Biol Psychiatry 47(12):1034–1042

Cruceanu C, Alda M, Grof P, Rouleau GA, Turecki G (2012) Synapsin ii is involved in the molecular pathway of lithium treatment in bipolar disorder. PLoS ONE 7(2), e32680. doi:10.1371/journal.pone.0032680

Cuffi ML, Artells R, Navarro A, Ciruela F, Carbonell L (2010) Regulation of alpha(2)-adrenoceptor gene expression by chronic lithium treatment in rat brain. Methods Find Exp Clin Pharmacol 32(10):721–725. doi:10.1358/mf.2010.32.10.1545783

Cunha AB, Frey BN, Andreazza AC, Goi JD, Rosa AR, Goncalves CA, Santin A, Kapczinski F (2006) Serum brain-derived neurotrophic factor is decreased in bipolar disorder during depressive and manic episodes. Neurosci Lett 398(3):215–219. doi:10.1016/j.neulet.2005.12.085

D'addario C, Dell'osso B, Palazzo MC, Benatti B, Lietti L, Cattaneo E, Galimberti D, Fenoglio C, Cortini F, Scarpini E, Arosio B, Di Francesco A, Di Benedetto M, Romualdi P, Candeletti S, Mari D, Bergamaschini L, Bresolin N, Maccarrone M, Altamura AC (2012) Selective DNA methylation of bdnf promoter in bipolar disorder: differences among patients with bdi and bdii. Neuropsychopharmacology 37(7):1647–1655. doi:10.1038/npp.2012.10

Damkier HH, Praetorius J (2012) Genetic ablation of slc4a10 alters the expression pattern of transporters involved in solute movement in the mouse choroid plexus. Am J Physiol Cell Physiol 302(10):C1452–C1459. doi:10.1152/ajpcell.00285.2011

Dasso L, Campanini T, De Risio C (1977) Electroencephalographic changes during lithium salt therapy: correlation with the erythrocyte lithium content. Ateneo Parmense Acta Biomed Organo Soc Med Sci Nat Parma 48(5):417–423

De Oliveira GS, Cereser KM, Fernandes BS, Kauer-Sant'anna M, Fries GR, Stertz L, Aguiar B, Pfaffenseller B, Kapczinski F (2009) Decreased brain-derived neurotrophic factor in medicated and drug-free bipolar patients. J Psychiatr Res 43(14):1171–1174

Dols A, Sienaert P, Van Gerven H, Schouws S, Stevens A, Kupka R, Stek ML (2013) The prevalence and management of side effects of lithium and anticonvulsants as mood stabilizers in bipolar disorder from a clinical perspective: a review. Int Clin Psychopharmacol 28(6):287–296. doi:10.1097/YIC.0b013e32836435e2

Duffy A, Horrocks J, Doucette S, Keown-Stoneman C, Mccloskey S, Grof P (2014) The developmental trajectory of bipolar disorder. Br J Psychiatr J Ment Sci 204(2):122–128. doi:10.1192/bjp.bp.113.126706

Dunner DL, Fieve RR (1974) Clinical factors in lithium carbonate prophylaxis failure. Arch Gen Psychiatry 30(2):229–233

Dwivedi T, Zhang H (2014) Lithium-induced neuroprotection is associated with epigenetic modification of specific BDNF gene promoter and altered expression of apoptotic-regulatory proteins. Front Neurosci 8:457. doi:10.3389/fnins.2014.00457

Dwivedi T, Zhang H (2015) Lithium-induced neuroprotection is associated with epigenetic modification of specific BDNF gene promoter and altered expression of apoptotic-regulatory proteins. Front Neurosci 14;8:457

Dysken MW, Comaty JE, Pandey GN, Davis JM (1979) Asterixis associated with a high rbc lithium concentration. Am J Psychiatry 136(12):1610

ElDesoky ES, Kumar V, Alorainy MS, Hamdi MM, Derendorf H (2008) Estimation of lithium clearance from routine clinical data in Egyptian bipolar patients. A population pharmacokinetic approach. Int J Clin Pharmacol Ther 46(12):617–626

Elizur A, Yeret A, Segal Z, Graff E (1982) Lithium and electrolytes plasma/rbc ratio and paradoxical lithium neurotoxicity. Prog Neuro-Psychopharmacol Biol Psychiatry 6(3):235–241

Fatemi SH, Folsom TD, Reutiman TJ, Braun NN, Lavergne LG (2010) Levels of phosphodiesterase 4a and 4b are altered by chronic treatment with psychotropic medications in rat frontal cortex. Synapse 64(7):550–555. doi:10.1002/syn.20762

Ferrie L, Young AH, Mcquade R (2005) Effect of chronic lithium and withdrawal from chronic lithium on presynaptic dopamine function in the rat. J Psychopharmacol 19(3):229–234. doi:10.1177/0269881105051525

Ferrie L, Young AH, Mcquade R (2006) Effect of lithium and lithium withdrawal on potassium-evoked dopamine release and tyrosine hydroxylase expression in the rat. Int J Neuropsychopharmacol 9(6):729–735. doi:10.1017/S1461145705006243

Frazer A, Mendels J, Secunda SK, Cochrane CM, Bianchi CP (1973) The prediction of brain lithium concentrations from plasma or erythrocyte measures. J Psychiatr Res 10(1):1–7

Fukumoto T, Morinobu S, Okamoto Y, Kagaya A, Yamawaki S (2001) Chronic lithium treatment increases the expression of brain-derived neurotrophic factor in the rat brain. Psychopharmacology 158(1):100–106. doi:10.1007/s002130100871

Gaillot J, Steimer JL, Mallet AJ, Thebault JJ, Bieder A (1979) A priori lithium dosage regimen using population characteristics of pharmacokinetic parameters. J Pharmacokinet Biopharm 7(6):579–628

Garnham J, Munro A, Slaney C, Macdougall M, Passmore M, Duffy A, O'donovan C, Teehan A, Alda M (2007) Prophylactic treatment response in bipolar disorder: results of a naturalistic observation study. J Affect Disord 104(1–3):185–190. doi:10.1016/j.jad.2007.03.003

Geddes JR, Burgess S, Hawton K, Jamison K, Goodwin GM (2004) Long-term lithium therapy for bipolar disorder: systematic review and meta-analysis of randomized controlled trials. Am J Psychiatry 161(2):217–222

Gould TD, Manji HK (2002) The wnt signaling pathway in bipolar disorder. Neuroscientist Rev J Bringing Neurobiol Neurol Psychiatry 8(5):497–511

Greil W, Stoltzenburg MC, Mairhofer ML, Haag M (1985) Lithium dosage in the elderly. A study with matched age groups. J Affect Disord 9(1):1–4

Grof P, Alda M, Grof E, Zvolsky P, Walsh M (1994) Lithium response and genetics of affective disorders. J Affect Disord 32(2):85–95

Grof P, Duffy A, Alda M, Hajek T (2009) Lithium response across generations. Acta Psychiatr Scand 120(5):378–385. doi:10.1111/j.1600-0447.2009.01454.x

Guarnieri DJ, Dileone RJ (2008) Micrornas: a new class of gene regulators. Ann Med 40(3):197–208. doi:10.1080/07853890701771823

Hewick DS, Murray N (1976) Red-blood-cell levels and lithium toxicity. Lancet 2(7983):473

Higuchi S, Fukuoka R, Aoyama T, Horioka M (1988) Predicting serum lithium concentration using bayesian method: a comparison with other methods. J Pharmacobiodyn 11(3):158–174

Hoegberg LC, Jurgens G, Zederkof VW, Holgersson B, Andersson JE, Dalhoff KP, Larsen EB, Angelo HR (2012) A computerised sampling strategy for therapeutic drug monitoring of lithium provides precise estimates and significantly reduces dose-finding time. Basic Clin Pharmacol Toxicol 110(3):259–263. doi:10.1111/j.1742-7843.2011.00800.x

Huang HC, Chang YL, Lan TH, Chiu HJ, Liu WM, Lee TJ (2008) Prediction of optimal lithium doses for Taiwanese psychiatric patients. J Clin Pharm Ther 33(2):115–121. doi:10.1111/j.1365-2710.2007.00888.x

Huang TL, Hung YY, Lee CT, Chen RF (2012) Serum protein levels of brain-derived neurotrophic factor and tropomyosin-related kinase b in bipolar disorder: effects of mood stabilizers. Neuropsychobiology 65(2):65–69. doi:10.1159/000328991

Hunsberger JG, Chibane FL, Elkahloun AG, Henderson R, Singh R, Lawson J, Cruceanu C, Nagarajan V, Turecki G, Squassina A, Medeiros CD, Del Zompo M, Rouleau GA, Alda M, Chuang DM (2015) Novel integrative genomic tool for interrogating lithium response in bipolar disorder. Transl Psychiatry 5, e504. doi:10.1038/tp.2014.139

Hunter R (1988) Steady-state pharmacokinetics of lithium carbonate in healthy subjects. Br J Clin Pharmacol 25(3):375–380

Jermain DM, Crismon ML, Martin ES 3rd (1991) Population pharmacokinetics of lithium. Clin Pharm 10(5):376–381

Johnston BB, Dick EG, Naylor GJ, Dick DA (1979) Lithium side effects in a routine lithium clinic. Br J Psychiatry J Ment Sci 134:482–487

Kato T, Shioiri T, Inubushi T, Takahashi S (1993) Brain lithium concentrations measured with lithium-7 magnetic resonance spectroscopy in patients with affective disorders: relationship to erythrocyte and serum concentrations. Biol Psychiatry 33(3):147–152

Kato T, Fujii K, Shioiri T, Inubushi T, Takahashi S (1996) Lithium side effects in relation to brain lithium concentration measured by lithium-7 magnetic resonance spectroscopy. Prog Neuro-Psychopharmacol Biol Psychiatry 20(1):87–97

Kato T, Inubushi T, Kato N (2000) Prediction of lithium response by 31p-mrs in bipolar disorder. Int J Neuropsychopharmacol 3(1):83–85. doi:10.1017/S1461145799001674

Kato T, Hayashi-Takagi A, Toyota T, Yoshikawa T, Iwamoto K (2011) Gene expression analysis in lymphoblastoid cells as a potential biomarker of bipolar disorder. J Hum Genet 56(11):779–783. doi:10.1038/jhg.2011.101

Keshavarz M, Emamghoreishi M, Nekooeian AA, Warsh J, Zare HR (2013) Increased bcl-2 protein levels in rat primary astrocyte culture following chronic lithium treatment. Iran J Med Sci 38(3):255–262

Kim SH, Yu HS, Park HG, Ahn YM, Kim YS, Lee YH, Ha K, Shin SY (2013) Egr1 regulates lithium-induced transcription of the period 2 (per2) gene. Biochim Biophys Acta 1832(12):1969–1979. doi:10.1016/j.bbadis.2013.06.010

Kittel-Schneider S, Schreck S, Ziegler C, Weissflog L, Hilscher M, Schwarz R, Schnetzler L, Neuner M, Reif A (2015) Lithium-induced clock gene expression in lymphoblastoid cells of bipolar affective patients. Pharmacopsychiatry 48(4–5):145–149. doi:10.1055/s-0035-1549946

Klein PS, Melton DA (1996) A molecular mechanism for the effect of lithium on development. Proc Natl Acad Sci U S A 93(16):8455–8459

Kleindienst N, Engel R, Greil W (2005) Which clinical factors predict response to prophylactic lithium? A systematic review for bipolar disorders. Bipolar Disord 7(5):404–417. doi:10.1111/j.1399-5618.2005.00244.x

Komoroski RA, Pearce JM, Newton JE (1997) The distribution of lithium in rat brain and muscle in vivo by 7lithium nmr imaging. Magn Reson Med Off J Soc Magn Reson Med Soc Magn Reson Med 38(2):275–278

Kristoff CA, Hayes PE, Barr WH, Small RE, Townsend RJ, Ettigi PG (1986) Effect of ibuprofen on lithium plasma and red blood cell concentrations. Clin Pharm 5(1):51–55

Lee JY, Lee JH, Moon YW, Chun BG, Jahng JW (2009) Proteomic analysis of lithium-induced gene expression in the rat hypothalamus. Int J Neurosc 119(9):1267–1281

Lee JH, Adler C, Norris M, Chu WJ, Fugate EM, Strakowski SM, Komoroski RA (2012) 4-t 7lithium 3d mr spectroscopy imaging in the brains of bipolar disorder subjects. Magn Reson Med Off J Soc Magn Reson Med Soc Magn Reson Med 68(2):363–368. doi:10.1002/mrm.24361

Lee RS, Pirooznia M, Guintivano J, Ly M, Ewald ER, Tamashiro KL, Gould TD, Moran TH, Potash JB (2015) Search for common targets of lithium and valproic acid identifies novel epigenetic effects of lithium on the rat leptin receptor gene. Transl Psychiatry 5, e600. doi:10.1038/tp.2015.90

Lehmann W, Kanarkowski R, Matkowski K, Rybakowski J (1988) Studies of lithium pharmacokinetics in patients with affective illness. Pol J Pharmacol Pharm 40(1):47–54

Lenox RH, Wang L (2003) Molecular basis of lithium action: integration of lithium-responsive signaling and gene expression networks. Mol Psychiatry 8(2):135–144. doi:10.1038/sj.mp.4001306

Lenox RH, Watson DG, Patel J, Ellis J (1992) Chronic lithium administration alters a prominent pkc substrate in rat hippocampus. Brain Res 570(1–2):333–340

Liang Y, Ridzon D, Wong L, Chen C (2007) Characterization of microrna expression profiles in normal human tissues. BMC Genomics 8:166. doi:10.1186/1471-2164-8-166

Lowthert L, Leffert J, Lin A, Umlauf S, Maloney K, Muralidharan A, Lorberg B, Mane S, Zhao H, Sinha R, Bhagwagar Z, Beech R (2012) Increased ratio of anti-apoptotic to pro-apoptotic bcl2 gene-family members in lithium-responders one month after treatment initiation. Biol Mood Anxiety Disord 2(1):15. doi:10.1186/2045-5380-2-15

Machado-Vieira R, Dietrich MO, Leke R, Cereser VH, Zanatto V, Kapczinski F, Souza DO, Portela LV, Gentil V (2007) Decreased plasma brain derived neurotrophic factor levels in unmedicated bipolar patients during manic episode. Biol Psychiatry 61(2):142–144. doi:10.1016/j.biopsych.2006.03.070

Machado-Vieira R, Manji HK, Zarate CA Jr (2009) The role of lithium in the treatment of bipolar disorder: convergent evidence for neurotrophic effects as a unifying hypothesis. Bipolar Disord 11(Suppl 2):92–109. doi:10.1111/j.1399-5618.2009.00714.x

Maj M, Starace F, Nolfe G, Kemalithium D (1986) Minimum plasma lithium levels required for effective prophylaxis in dsm iii bipolar disorder: a prospective study. Pharmacopsychiatry 19(6):420–423. doi:10.1055/s-2007-1017280

Maj M, Pirozzi R, Magliano L (1995) Nonresponse to reinstituted lithium prophylaxis in previously responsive bipolar patients: prevalence and predictors. Am J Psychiatry 152(12):1810–1811

Mann L, Heldman E, Bersudsky Y, Vatner SF, Ishikawa Y, Almog O, Belmaker RH, Agam G (2009) Inhibition of specific adenylyl cyclase isoforms by lithium and carbamazepine, but not valproate, may be related to their antidepressant effect. Bipolar Disord 11(8):885–896. doi:10.1111/j.1399-5618.2009.00762.x

Marmol F (2008) Lithium: bipolar disorder and neurodegenerative diseases possible cellular mechanisms of the therapeutic effects of lithium. Prog Neuro-Psychopharmacol Biol Psychiatry 32(8):1761–1771. doi:10.1016/j.pnpbp.2008.08.012

Matigian N, Windus L, Smith H, Filippich C, Pantelis C, Mcgrath J, Mowry B, Hayward N (2007) Expression profiling in monozygotic twins discordant for bipolar disorder reveals dysregulation of the wnt signalling pathway. Mol Psychiatry 12(9):815–825. doi:10.1038/sj.mp.4001998

Mcgrath BM, Greenshaw AJ, Mckay R, Slupsky CM, Silverstone PH (2006) Lithium alters regional rat brain myo-inositol at 2 and 4 weeks: an ex-vivo magnetic resonance spectroscopy study at 18.8 t. Neuroreport 17(12):1323–1326. doi:10.1097/01.wnr.0000230501.40349.41

McKnew DH, Cytryn L, Buchsbaum MS, Hamovit J, Lamour M, Rapoport JL, Gershon ES (1981) Lithium in children of lithium-responding parents. Psychiatry Res 4(2):171–180

Mcquillin A, Rizig M, Gurling HM (2007) A microarray gene expression study of the molecular pharmacology of lithium carbonate on mouse brain mrna to understand the neurobiology of mood stabilization and treatment of bipolar affective disorder. Pharmacogenet Genomics 17(8):605–617. doi:10.1097/FPC.0b013e328011b5b2

Mendels J, Frazer A (1973) Intracellular lithium concentration and clinical response: towards a membrane theory of depression. J Psychiatr Res 10(1):9–18

Mukherjee BP, Bailey PT, Pradhan SN (1976) Temporal and regional differences in brain concentrations of lithium in rats. Psychopharmacology 48(1):119–121

Okuma T (1993) Effects of carbamazepine and lithium on affective disorders. Neuropsychobiology 27(3):138–145, doi:118969

Olie JP, Gay C, Loo H, Zarifian E, Poirier-Littre MF, Deniker P (1982) The value of the measurement of intra-erythrocytic lithium in the bio-clinical surveillance of lithium therapy (author's transl). Ann Biol Clin 40(1):3–10

Ong YL (1983) Lithium levels in white blood cells and their relationship to observed side effects. Br J Psychiatry J Ment Sci 143:36–39

Osland TM, Ferno J, Havik B, Heuch I, Ruoff P, Laerum OD, Steen VM (2011) Lithium differentially affects clock gene expression in serum-shocked nih-3 t3 cells. J Psychopharmacol 25(7):924–933. doi:10.1177/0269881110379508

Palomino A, Vallejo-Illarramendi A, Gonzalez-Pinto A, Aldama A, Gonzalez-Gomez C, Mosquera F, Gonzalez-Garcia G, Matute C (2006) Decreased levels of plasma bdnf in first-episode schizophrenia and bipolar disorder patients. Schizophr Res 86(1–3):321–322. doi:10.1016/j.schres.2006.05.028

Parthasarathy LK, Seelan RS, Wilson MA, Vadnal RE, Parthasarathy RN (2003) Regional changes in rat brain inositol monophosphatase 1 (impase 1) activity with chronic lithium treatment. Prog Neuro-Psychopharmacol Biol Psychiatry 27(1):55–60

Passmore MJ, Garnham J, Duffy A, Macdougall M, Munro A, Slaney C, Teehan A, Alda M (2003) Phenotypic spectra of bipolar disorder in responders to lithium versus lamotrigine. Bipolar Disord 5(2):110–114

Price LH, Heninger GR (1994) Lithium in the treatment of mood disorders. N Engl J Med 331(9):591–598. doi:10.1056/NEJM199409013310907

Pulford DJ, Adams F, Henry B, Mallinson DJ, Reid IC, Stewart CA (2006) Chronic lithium administration down regulates transthyretin mrna expression in rat choroid plexus. Neuropsychiatr Dis Treat 2(4):549–555

Quiroz JA, Machado-Vieira R, Zarate CA Jr, Manji HK (2010) Novel insights into lithium's mechanism of action: neurotrophic and neuroprotective effects. Neuropsychobiology 62(1):50–60. doi:10.1159/000314310

Ramaprasad S (2004) Lithium spectroscopic imaging of rat brain at therapeutic doses. Magn Reson Imaging 22(5):727–734. doi:10.1016/j.mri.2004.01.063

Ramaprasad S, Newton JE, Cardwell D, Fowler AH, Komoroski RA (1992) In vivo 7lithium nmr imaging and localized spectroscopy of rat brain. Magn Reson Med Off J Soc Magn Reson Med Soc Magn Reson Med 25(2):308–318

Ramaprasad S, Ripp E, Pi J, Lyon M (2005) Pharmacokinetics of lithium in rat brain regions by spectroscopic imaging. Magn Reson Imaging 23(8):859–863. doi:10.1016/j.mri.2005.07.007

Renshaw PF, Haselgrove JC, Leigh JS, Chance B (1985) In vivo nuclear magnetic resonance imaging of lithium. Magn Reson Med Off J Soc Magn Reson Med Soc Magn Reson Med 2(5):512–516

Rios C, Guzman-Mendez R (1990) Determination of lithium in rat brain regions and synaptosomes by graphite furnace atomic absorption spectrophotometry. J Pharmacol Methods 24(4):327–332

Rohayem J, Bayle JF, Richa S (2008) Predictors of prophylactic response to lithium. L'Encéphale 34(4):394–399. doi:10.1016/j.encep.2007.05.002

Rong H, Liu TB, Yang KJ, Yang HC, Wu DH, Liao CP, Hong F, Yang HZ, Wan F, Ye XY, Xu D, Zhang X, Chao CA, Shen QJ (2011) Microrna-134 plasma levels before and after treatment for bipolar mania. J Psychiatr Res 45(1):92–95

Rybakowski JK (2013) Genetic influences on response to mood stabilizers in bipolar disorder: current status of knowledge. CNS Drugs 27(3):165–173. doi:10.1007/s40263-013-0040-7

Rybakowski JK, Suwalska A, Skibinska M, Dmitrzak-Weglarz M, Leszczynska-Rodziewicz A, Hauser J (2007) Response to lithium prophylaxis: interaction between serotonin transporter and bdnf genes. Am J Med Genet B Neuropsychiatr Genet 144B(6):820–823. doi:10.1002/ajmg.b.30420

Sarantidis D, Waters B (1981) Predictors of lithium prophylaxis effectiveness. Prog Neuropsychopharmacol 5(5–6):507–510

Seelan RS, Khalyfa A, Lakshmanan J, Casanova MF, Parthasarathy RN (2008) Deciphering the lithium transcriptome: microarray profiling of lithium-modulated gene expression in human neuronal cells. Neuroscience 151(4):1184–1197. doi:10.1016/j.neuroscience.2007.10.045

Simard M, Gumbiner B, Lee A, Lewis H, Norman D (1989) Lithium carbonate intoxication. A case report and review of the literature. Arch Intern Med 149(1):36–46

Solomon DA, Keitner GI, Miller IW, Shea MT, Keller MB (1995) Course of illness and maintenance treatments for patients with bipolar disorder. J Clin Psychiatry 56(1):5–13

Sproule B (2002) Lithium in bipolar disorder: can drug concentrations predict therapeutic effect? Clin Pharmacokinet 41(9):639–660. doi:10.2165/00003088-200241090-00002

Squassina A, Costa M, Congiu D, Manchia M, Angius A, Deiana V, Ardau R, Chillotti C, Severino G, Calza S, Del Zompo M (2013) Insulin-like growth factor 1 (igf-1) expression is up-regulated in lymphoblastoid cell lines of lithium responsive bipolar disorder patients. Pharmacol Res Off J Ital Pharmacol Soc 73:1–7. doi:10.1016/j.phrs.2013.04.004

Subramanian P, Menon VP, Arokiam FV, Rajakrishnan V, Balamurugan E (1998) Lithium modulates biochemical circadian rhythms in wistar rats. Chronobiol Int 15(1):29–38

Sugawara H, Iwamoto K, Bundo M, Ishiwata M, Ueda J, Kakiuchi C, Ishigooka J, Kato T (2010) Effect of mood stabilizers on gene expression in lymphoblastoid cells. J Neural Transm 117(2):155–164. doi:10.1007/s00702-009-0340-8

Sullivan PF, Fan C, Perou CM (2006) Evaluating the comparability of gene expression in blood and brain. Am J Med Genet B Neuropsychiatr Genet 141(3):261–268

Sun X, Young LT, Wang JF, Grof P, Turecki G, Rouleau GA, Alda M (2004) Identification of lithium-regulated genes in cultured lymphoblasts of lithium responsive subjects with bipolar disorder. Neuropsychopharmacology 29(4):799–804. doi:10.1038/sj.npp.1300383

Swartz CM (1987) Correction of lithium levels for dose and blood sampling times. J Clin Psychiatry 48(2):60–64

Tajes M, Yeste-Velasco M, Zhu X, Chou SP, Smith MA, Pallas M, Camins A, Casadesus G (2009) Activation of akt by lithium: pro-survival pathways in aging. Mech Ageing Dev 130(4):253–261. doi:10.1016/j.mad.2008.12.006

Terao T, Okuno K, Okuno T, Nakano H, Shinkai T, Ohmori O, Yoshimura R, Suzuki T, Abe K (1999) A simpler and more accurate equation to predict daily lithium dose. J Clin Psychopharmacol 19(4):336–340

Tondo L, Baldessarini RJ, Floris G (2001) Long-term clinical effectiveness of lithium maintenance treatment in types i and ii bipolar disorders. Br J Psychiatry J Ment Sci 178(Suppl 41):S184–S190

Tramontina J, Frey BN, Andreazza AC, Zandona M, Santin A, Kapczinski F (2007) Val66met polymorphism and serum brain-derived neurotrophic factor levels in bipolar disorder. Mol Psychiatry 12(3):230–231. doi:10.1038/sj.mp.4001941

Tramontina JF, Andreazza AC, Kauer-Sant'anna M, Stertz L, Goi J, Chiarani F, Kapczinski F (2009) Brain-derived neurotrophic factor serum levels before and after treatment for acute mania. Neurosci Lett 452(2):111–113

Tseng M, Alda M, Xu L, Sun X, Wang JF, Grof P, Turecki G, Rouleau G, Young LT (2008) Bdnf protein levels are decreased in transformed lymphoblasts from lithium-responsive patients with bipolar disorder. J Psychiatry Neurosci JPN 33(5):449–453

Turck D, Heinzel G, Luik G (2000) Steady-state pharmacokinetics of lithium in healthy volunteers receiving concomitant meloxicam. Br J Clin Pharmacol 50(3):197–204

Vargas C, Tannhauser M, Barros HM (1998) Dissimilar effects of lithium and valproic acid on gaba and glutamine concentrations in rat cerebrospinal fluid. Gen Pharmacol 30(4):601–604

Wang L, Liu X, Lenox RH (2001) Transcriptional down-regulation of marcks gene expression in immortalized hippocampal cells by lithium. J Neurochem 79(4):816–825

Williams MB, Jope RS (1995) Circadian variation in rat brain ap-1 DNA binding activity after cholinergic stimulation: modulation by lithium. Psychopharmacology 122(4):363–368

Willmroth F, Drieling T, Lamla U, Marcushen M, Wark HJ, Van Calker D (2007) Sodium-myo-inositol co-transporter (smit-1) mrna is increased in neutrophils of patients with bipolar I disorder and down-regulated under treatment with mood stabilizers. Int J Neuropsychopharmacol 10(1):63–71. doi:10.1017/S1461145705006371

Wright R, Crismon ML (2000) Comparison of three a priori methods and one empirical method in predicting lithium dosage requirements. Am J Health Syst Pharm AJHP Off J Am Soc Health Syst Pharm 57(18):1698–1702

Yang YY (1985) Prophylactic efficacy of lithium and its effective plasma levels in Chinese bipolar patients. Acta Psychiatr Scand 71(2):171–175

Yang YY, Yeh EK, Chang SS, Deng HC, Lee CF (1991) Maintenance lithium levels could be lowered: based on Taiwanese and Danish studies. J Formos Med Assoc Taiwan Yi Zhi 90(5):509–513

Yasuda S, Liang MH, Marinova Z, Yahyavi A, Chuang DM (2009) The mood stabilizers lithium and valproate selectively activate the promoter iv of brain-derived neurotrophic factor in neurons. Mol Psychiatry 14(1):51–59. doi:10.1038/sj.mp.4002099

Yazici O, Kora K, Ucok A, Tunalithium D, Turan N (1999) Predictors of lithium prophylaxis in bipolar patients. J Affect Disord 55(2–3):133–142

Yildiz A, Vieta E, Leucht S, Baldessarini RJ (2011) Efficacy of antimanic treatments: meta-analysis of randomized, controlled trials. Neuropsychopharmacology 36(2):375–389. doi:10.1038/npp.2010.192

Yu Z, Ono C, Aiba S, Kikuchi Y, Sora I, Matsuoka H, Tomita H (2015) Therapeutic concentration of lithium stimulates complement c3 production in dendritic cells and microglia via gsk-3 inhibition. Glia 63(2):257–270. doi:10.1002/glia.22749

Yukawa E, Nomiyama N, Higuchi S, Aoyama T (1993) Lithium population pharmacokinetics from routine clinical data: role of patient characteristics for estimating dosing regimens. Ther Drug Monit 15(2):75–82

Zakowska-Dabrowska T, Rybakowski J (1973) Lithium-induced eeg changes: relation to lithium levels in serum and red blood cells. Acta Psychiatr Scand 49(4):457–465

Zarate CA, Manji HK (2009) Protein kinase c inhibitors: rationale for use and potential in the treatment of bipolar disorder. CNS Drugs 23(7):569–582. doi:10.2165/00023210-200923070-00003

Zhou R, Yuan P, Wang Y, Hunsberger JG, Elkahloun A, Wei Y, Damschroder-Williams P, Du J, Chen G, Manji HK (2009) Evidence for selective micrornas and their effectors as common long-term targets for the actions of mood stabilizers. Neuropsychopharmacology 34(6): 1395–1405

Part II

Lithium in Practice

The History of Lithium in Medicine and Psychiatry

10

Marc-Louis Bourgeois and Marc Masson

Abstract

This chapter outlines the main turning points in the history of the use of lithium in mental disorders. It is a history that begins in the eighteenth century with the study of gout, a condition that was endemic at that time. Gout was associated with melancholia, and its cause was widely attributed to an excess of uric acid. Thus when Garrod erroneously concluded in 1859 that lithium carbonate could dissolve uric acid in vivo, he became the first to specify the role of uric acid in affective disorders and to advise the medicinal use of lithium as a remedy. Hence, in the second half of the nineteenth century, in alignment with this hypothesis, Carl and Fritz Lange were the first to implement lithium therapy in patients to prevent depression. Consequently, they are regarded as the founding fathers of lithium therapy. However, the use of lithium was relatively short lived, and, at the turn of the century, lithium therapy for affective disorders was dismissed as nonsense by Christiansen (among others) and was almost completely forgotten for half a century. This is all the more remarkable given the scarcity of alternatives.

In 1949, John Cade, whose work aimed to identify the toxic factors of psychosis, discovered by chance what he believed to be a sedative effect of lithium and experimented with its use on manic patients. This treatment, though effective, led to the death of a patient, which prompted Cade to abandon lithium therapy. However, other psychiatrists began using it, and in 1954 Mogens Schou eventually confirmed the therapeutic effect of lithium. Nowadays, lithium is considered

M.-L. Bourgeois, MD, Psycho D (✉)
Hôpital Charles Perrens, Université de Bordeaux 2, IPSO CHS Ch. Perrens, 121, rue de la Béchade, 33076 Bordeaux, France
e-mail: ipso.bourgeois@u-bordeaux2.fr

M. Masson
Clinique du Château, 11bis rue de la Porte Jaune, 92380 Garches, France
e-mail: marc.masson@clinique-garches.com

G.S. Malhi et al. (eds.), *The Science and Practice of Lithium Therapy*,
DOI 10.1007/978-3-319-45923-3_10

to be the most (and probably only) effective mood stabilizer, and it is one of many examples of how prolific the middle of the twentieth century was in terms of new treatments for mental disorders.

Key Points

- This chapter outlines the main turning points in the history of the use of lithium in mental disorders.
- In 1859, Garrod, working on a cure for gout, was the first to focus on the role of uric acid in affective disorders and advocate the use of lithium.
- In the second half of the nineteenth century, Carl and Fritz Lange were the first to use lithium therapy on a wide range of patients to prevent depression.
- In 1949, John Cade discovered, by chance, what he believed to be the sedative effect of lithium and experimented with it successfully on manic patients.
- In 1954, Mogens Schou eventually confirmed the therapeutic effect of lithium, which is now considered to be the most effective and perhaps only true mood stabilizer.

10.1 Introduction

Lithium re-established psychiatry as the medical speciality it is
Goodwin and Ghaemi (1999)

The history of lithium in medicine and psychiatry has been extensively documented and published most recently by Professor Johan Schioldann (first edition 2001; second expanded edition 2009).[1] The case of lithium illustrates the role that serendipity can play in the development of medical treatments (with John Cade's discovery). It also demonstrates that important discoveries can be based upon an incorrect hypothesis (the role of gout and the uric acid diathesis in affective disorders).

Three periods can be distinguished for the use of lithium in the treatment of mental disorders:

1. During the nineteenth century, the treatment of recurrent depression with lithium was proposed by the brothers Carl and Fritz Lange (1874–1907).
2. Cade's experiment on guinea pigs (1949) and his seminal report of ten cases of lithium-based treatment of mania.
3. Finally, Schou and Baastrup's work (1967) facilitating safe and efficient lithium therapy in manic and depressive patients: the first studies to include a control group and the beginning of plasma-level dosing.

[1] Born in Aalborg (Denmark) in 1941, Schioldann graduated in Copenhagen, specialized in psychiatry and took up residency, then professorship, in Adelaïde (Australia).

10.2 The Use of Lithium in the Eighteenth and Nineteenth Centuries

Lithium is as old as the Earth, but its story in relation to manic-depressive illness begins in the eighteenth and nineteenth centuries.

10.2.1 Lithium and Gout

Gout was a common illness in Europe in the eighteenth century, and 'melancholia was considered as the inseparable companion of gout' (*Morbus Dominorum and Domina Morborum*), known as melancholia arthritic ('goutte mélancolique'). For Lorry (1789): 'la goutte se fixe au cerveau, pour déterminer le délire, ou plutôt la mélancolie et le délire'. Uric acid was discovered in urinary calculi in 1775 and later on in gouty concretions of the joints. In 1847, the renowned British physician Alfred Baring Garrod confirmed the hypothesis that the origin was excess uric acid (urate of soda) in the blood of gouty patients. However, Garrod extended himself and extrapolated too far by linking affective disorders to this uric acid diathesis. His famous book on the topic (Garrod 1859) was translated into several languages, including German (1861) and French (1867), and Jean-Martin Charcot (Charcot 1866) himself—the founder of modern neurology—adopted Garrod's recommendation of lithium treatment (carbonate of lithia, 30–45 grains over 24 h) for gout especially for 'folie goutteuse' (gouty insanity).

10.2.2 Carl and Fritz Lange and the Treatment of 'Periodical Depression'

According to Schioldann (2009), Frederick (Fritz) Lange (1842–1907) and Carl Lange (1834–1900), his older brother, are the true 'founding fathers of lithium therapy', building on the theory that uric acid diathesis was the cause of 'folie goutteuse' and could be effectively treated with the use of lithium. Carl Lange published a monograph in 1886 entitled *On Periodical Depressions and Their Pathogenesis* (Lange 1886). In this he explained that lithium could be used to prevent the recurrence of 'periodic depression', which he believed was also caused by uric acid diathesis.

Carl Lange was a noted Danish neurologist and a pathologist, famous for his theory on emotions (the James-Lange theory). In private practice, he saw 700–800 cases of periodical depression, over approximately 12 years: 'my experience shows beyond any doubt that the condition, at least in that century, is extremely common' wrote Carl Lange (Lange 1886; quoted in Schioldann 2009 p. 37), and in 1895 he noted that in his clinical experience, he never observed among 'approximately 2 000 patients suffering from periodical depression "the slightest sign of elevated periods"' (Lange 1895; quoted in Schioldann p. 42). Fritz Lange worked as superintendent psychiatrist in a large mental asylum in Middelfart on the island of Funen,

Denmark, from 1874 to 1907. According to Schioldann, 'Fritz Lange's text book of psychiatry, The most important group of insanity (1884), possibly contains the first explicit mention of systematic lithium treatment of periodical depression' (Schioldann 2009, p. 46). His experience with lithium prescription inspired his brother Carl to resort to using the drug to treat private patients. Between them, from 1874 to 1907, the two Lange brothers gained considerable experience in treating both private patients and patients at the Middelfart asylum using lithium carbonate.

10.2.3 The 'Old Danish Lithium Treatment' Denied and Forgotten for Half a Century

The Lange theory of depression and its treatment with lithium subsequently became discredited by Viggo Christiansen (1867–1939), who held the first chair of neurology in Denmark. He conducted over 500 experiments in which he injected rabbits with urine from psychotic patients to test the Lange theory of depression and eventually dismissed it. Similarly, in his thesis, Hans Jacob Schou (1886–1952), Christiansen's student, also did not support the 'old Danish lithium treatment' (1938). Likewise, Emil Kraepelin also found that the uric acid hypothesis made little sense, and so the old Danish treatment was abandoned.

10.3 Lithium Treatment of Mania: John Cade's Discovery (1949)

John Cade (1912–1980) was born in Victoria, Australia, the son of a medical practitioner. He studied medicine at Melbourne University. Interested in biological treatments, he was an enthusiastic proponent and prescriber of insulin coma therapy (also known as the 'Sakel cure') in schizophrenia. In 1934 Cade enrolled in the Australian Army Medical Corps. He rose to the rank of captain, in 1940, and then became a major. From February 1942 to September 1945, he was held prisoner of war by the Japanese, along with 2,500 Australian soldiers, most of whom died because of deficiency diseases such as beriberi and pellagra, alongside encephalopathy and other illnesses. In the Changi camps in Singapore, Cade took care of sick and wounded soldiers. His work as a psychiatrist and his taste for limericks, malapropisms and petty yarns earned him the nickname of the 'mad major'.

When Cade returned to Australia, he was a 'walking skeleton' weighing just 40 kg. Remarkably, he had survived and then recovered to the point that he could function once again as a psychiatrist. Once his health was restored, he promptly resumed his experiments. Cade, who was both a psychiatrist and a neurologist, saw biological psychiatry as an antidote to the growing influence of psychoanalysis. Such was his commitment to his work that, according to his son Jack, John Cade used to frighten and dismay his wife by experimenting with drugs on himself and keeping urine samples in their kitchen refrigerator.

Cade was convinced that toxic factors were the cause of psychosis. He injected patients' urine into the peritoneal cavities of guinea pigs. Almost all the animals died in *status epilepticus*. Cade injected the guinea pigs with lithium salts, choosing lithium because it was a good solvent of uric acid (and not because of the fallacious concept of uric acid diathesis). Surprisingly, the guinea pigs Cade injected with lithium salts became very calm: 'The animals became thoughtful and preoccupied' (Hartigan 1959). Then came the creative leap: prompted by the sedative effect of lithium salts, Cade tried using them in cases of mania. However, the sought-for anticonvulsant action did not emerge. Thus Cade viewed the introduction of lithium salts into medicine as the consequence of a misconception, whereas Samuel Gershon describes the 'delivery' of lithium into the world of manic-depressive illness as 'serendipity seems to have been the midwife for lithium'—emphasizing the undeniable role that chance has played (Gershon 1968).

From 29 March 1948 to 11 February 1949, Cade gave lithium citrate or carbonate to ten manic patients: three suffered from chronic mania, seven recurrent mania, six schizophrenia and three melancholia. Cade published his trial in his famous paper of 1949 (Cade 1949). With the very first manic patient, WB (one of those suffering from chronic mania), treatment with lithium salts (1948–1949) was initially successful, but then a little while later, WB began to vomit and suffered nocturnal enuresis. Ultimately, despite an initial 'remarkable improvement', he relapsed and eventually died because of lithium intoxication. Thereupon Cade, alarmed by the toxic effects of lithium, renounced its use for therapeutic purposes. Interestingly, a previous report of the use of lithium in the treatment of mania had been made, in 1881, by Hammond, from Bellevue Hospital, New York, in his famous *Treatise on Diseases of the Nervous System* (Hammond 1871). However, in his later works, Hammond no longer mentioned the use of lithium, probably because he initially used high doses of lithium bromide, which is likely to have caused severe toxicity (Schioldann 2009, 29–30). It is unclear whether Cade was aware of these early uses.

Even though Cade himself stopped promoting the use of lithium and became very cautious because of its potential to cause serious harm, others in Australia (Noack and Trautner) and then worldwide continued to experiment with it (1951). For instance, in France, soon after Cade's publication, there were very early clinical reports of the effectiveness of lithium in mania (Despinoy and Romeuf in 1951, Carrere and Pochard in 1954, and Chanoit and Sivadon in 1955 (Masson 2013, p. 317–335)), and thus interest in exploring the effects of lithium flourished (Malhi and Gershon 2009).

10.4 Mogens Schou: The Safer and Efficacious Use of Lithium Therapy

In Denmark, Eric Strömgren (Aarhus) drew the attention of Mogens Schou (Hans Jacob Schou's son) to the Australian reports of the apparent effects of lithium. Using an old instrument, they monitored and controlled serum lithium levels in patients and published their results in 1954 (Schou et al. 1954) confirming Cade's hypothesis about the therapeutic effects of lithium. Further corroboration also came from Denmark (Baastrup and Schou 1967), England, France and Australia. However, Blackwell and Shepherd (1968) were critical of the data of earlier studies and regarded the supposed prophylactic effects of lithium to be yet another therapeutic myth, a criticism that has been echoed more recently in a re-examination of placebo-controlled trials (Moncrieff 1995). However, at the time, Mogens Schou's research (1954) confirmed Cade's discovery, which had heralded the dawn of a pharmacological revolution.

The use of lithium was eventually approved by the US Food and Drug Administration (FDA) in 1970–1972, and it is now considered to be the gold standard for maintenance therapy and, in particular, the prevention of depression and mania in bipolar disorder. Arguably, it is the first and perhaps only true mood stabilizer (Bauer et al. 2006; Goodwin and Malhi 2007). Many individuals played a key role in establishing lithium therapy (see Fig. 10.1), and among these, Mogens Schou warrants particular mention. He deservedly received the Albert Lasker Clinical Medical Research Award and championed lithium throughout his life. He died in 2005.

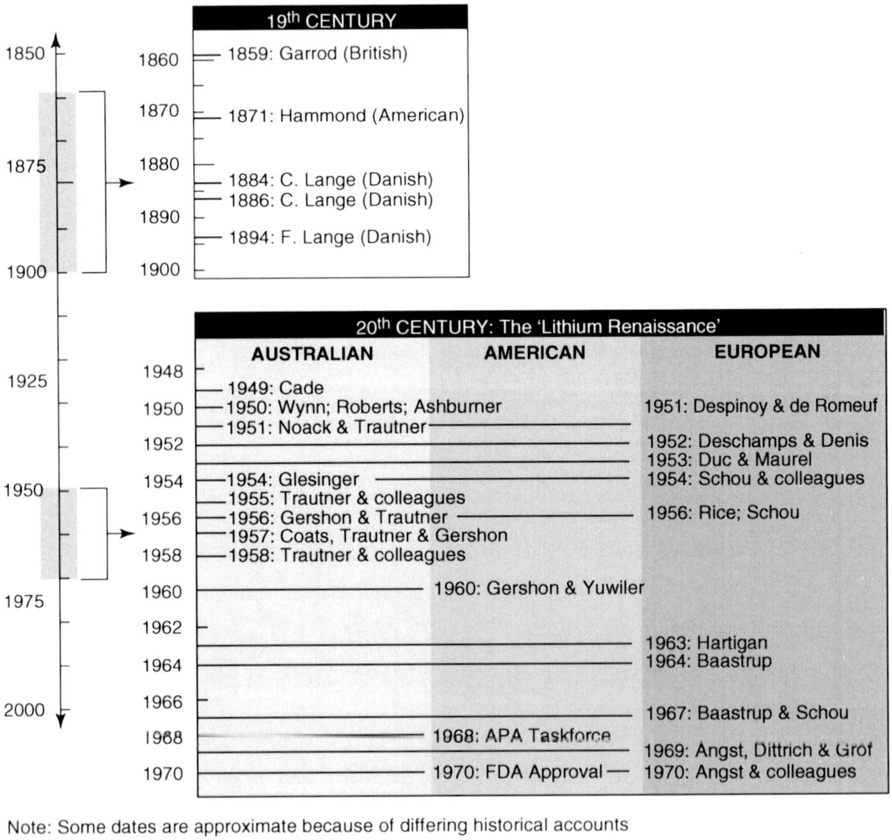

Fig. 10.1 Timeline of key figures in relation to lithium (From Malhi and Gershon 2009)

10.5 Summary

The 1950s was a prodigious decade for modern psychopharmacology. The principal efficient molecules were discovered over the course of just a few short years:

- Isoniazide, iproniazide, imipramine, etc., for depressive disorders
- Neuroleptics (major tranquillizers): phenothiazines (chlorpromazine, thorazine, etc.) and buthirophénones, for psychosis
- Anxiolytics (tranquillizers), chlordiazepoxide, diazepam, etc.
- And, of course, lithium as the first psychotropic drug (1949) and the very first (and arguably only true) mood stabilizer

Lithium was the first psychotropic drug ever discovered in psychopharmacology and is still used as a first-line treatment in bipolar disorders.

Garrod, Hammond, the brothers Lange, Cade and Schou (Malhi 2009) have all linked their names forever to this revolution in the world of mental health:

- Garrod was the first to focus on the role of uric acid in affective disorders and to advocate the use of lithium.
- Hammond reported on his use of lithium bromide in treating mania in 1881.
- Carl and Fritz Lange proposed the use of lithium to treat recurrent depression in the nineteenth century.
- Cade discovered by chance what he believed to be the sedative effect of lithium and experimented with it successfully on manic patients.
- In the 1950s and 1960s, Schou and Baastrup confirmed the therapeutic effect of lithium and facilitated its safe and efficient use by developing plasma-level testing.

References

Baastrup PC, Schou M (1967) Lithium as a prophylactic agent. Its effect against recurrent depressions and manic-depressive psychosis. Arch Gen Psychiatry 16(2):162–72

Bauer M, Grof P, Müller-Oerlinghause B (2006) Lithium in neuropsychiatry. The comprehensive guide. IGSLI Informa UK, Abingdon, 532 pages

Blackwell B, Shepherd M (1968) Prophylactic lithium: another therapeutic myth? An examination of evidence. Lancet I:968–971

Cade JF (1949) Lithium salts in the treatment of psychotic excitement. Med J Austr 36(2):349–352

Garrod AB (1859) The nature and treatment of gout and rheumatic gout, 1st edn. Walton & Maberly, London

Gershon S (1968) Use of lithium salts in psychiatric disorders. Dis Nerv Syst 29(1):51–55

Goodwin FK, Ghaemi SN (1999) The impact of the discovery of lithium on psychiatry thought and practice in the USA and Europe. In: Mitchell PB, Hadzi-Pavlovic D, Manji HK (eds) Fifty years of treatments for bipolar disorder. A celebration of John Cade's discovery. Aust NZ J Psychiatry, (Oxford UK) 33, pp S54–S64

Goodwin GM, Malhi GS (2007) What is a mood stabilizer? Psychol Med 37(5):609–614

Hammond WA (1871) Treatise on diseases of the nervous system. Appleton, New York, pp 358–399 ("Mania"), pp. 380–381 ("Treatment")
Hartigan GP (1959) Experiences of treatment with lithium salts. In: Johnson (Brit J, London UK) (1984). pp 183–187
Lange C (1886) Om Periodiske Depressionstilstande og deres patogenese. Lunds Forlag, Copenhagen
Lorry AL (1789) De praecipius morborum conversionibus. Paris. 280 pages
Malhi GS, Gershon S (2009) Ion men and their mettle. Aust NZ J Psychiatry 43(12):1091–1095
Masson M (2013) 24 textes fondateurs de la psychiatrie introduits et commentés par la Société Médico-Psychologiques. Armand Colin, Paris, 380 pages
Moncrieff J (1995) Lithium revisited. A re-examination of the placebo-controlled trials of lithium prophylaxis in manic-depressive disorder. Br J Psychiatry 167:569–574
Schioldann JA (2001) The Lange theory of "periodical depressions". A landmark in the History of Lithium Therapy. In: Commemoration of the centenary of the death of Carl Lange. Adelaide Academic Press, Adelaïde (Australia) 200 pages
Schioldann JA (2009) History of the introduction of Lithium into Medicine and Psychiatry. Birth of modern psychopharmacology 1949. Academic Press, Adelaïde
Schou M, Juel-Nielsen N, Stromgren E, Voldby H (1954) The treatment of manic psychoses by the administration of lithium salts. J Neurol Neurosurg Psychiatry 17(4):250–260

11 Recommendations in International Clinical Practice Guidelines for Lithium Therapy of Bipolar Disorder

Gin S. Malhi, Danielle Gessler, Kristina Fritz, Christine Allwang, Tim Outhred, and Pritha Das

Abstract

Lithium is the gold-standard treatment for prophylaxis in bipolar disorder and has remained in pole position for the past 60 years in international guidelines. However, in practice, the perception that lithium is complicated to prescribe and associated with acute and long-term side effects constitutes a barrier to its effective prescription and management. In this chapter, we examine the positioning of lithium within current clinical practice guidelines for the treatment of bipolar disorder and describe how it is effective in quelling the acute symptoms of mania, as well as sustaining full and functional remission in prophylaxis, alongside conferring antisuicidal and neuroprotective benefits. We then discuss the specificity of lithium as a mood stabiliser and the recognition of an ideal lithium responder. The chapter also includes clinical advice regarding the prescription of lithium—emphasising the role of monitoring and measuring lithium plasma levels and using the 'lithiumeter' to optimise therapeutic levels and avoid toxicity.

G.S. Malhi, MBChB BSc FRCPsych FRANZCP MD (✉) • D. Gessler • K. Fritz
T. Outhred • P. Das
Academic Department of Psychiatry, Sydney Medical School, The University of Sydney and CADE Clinic, Royal North Shore Hospital, Sydney, Australia
e-mail: gin.malhi@sydney.edu.au

C. Allwang
Clinic for Psychosomatic Medicine and Psychotherapy, Klinikum Rechts der Isar der Technischen Universitaet Muenchen, Langerstrasse 3, Munich 81675, Germany

G.S. Malhi et al. (eds.), *The Science and Practice of Lithium Therapy*,
DOI 10.1007/978-3-319-45923-3_11

Key Points

- Lithium therapy is a recommended treatment for bipolar disorder across all international guidelines.
- Lithium is an effective mood stabiliser and possesses additional antisuicidal and neuroprotective benefits.
- Characterising patient subgroups within future guidelines will aid the management of those best suited to lithium prophylaxis.
- Clinically, an ideal lithium responder has a classic episodic pattern of illness consisting of *recurrent, recognisable* mood episodes that are separated by periods of *remission.*
- Early commencement of lithium therapy ensures the best long-term outcome, and at the time of initiation, it is important to optimise serum lithium levels.
- Refinement of lithium therapy depends upon both effectiveness as well as tolerability and can be achieved using the 'lithiumeter'.

11.1 Introduction

Since its rediscovery by John Cade over 60 years ago (Cade 1949), lithium—the original and perhaps only 'real' mood stabiliser—has prevailed amongst an ever-expanding arsenal of supposed bipolar medications. Its track record for continually and effectively treating patients with mood disorders has positioned it securely as a key long-term option, despite the absence of Big Pharma marketing. Its clinical usefulness is reflected by the fact that it features prominently across all international guidelines for the treatment of bipolar disorder, particularly for prophylaxis (Malhi et al. 2015). Concurrently, during this period of more than half a century, psychiatry has undergone a vast transformation and adopted a far more pragmatic approach to the diagnosis of psychiatric disorders and their treatment, with greater investment in the development of newer therapeutic agents—especially those that address existing diagnostic categories. During this metamorphosis, lithium has gradually lost scientific ground and clinical popularity amongst researchers and clinicians, respectively, partly because of exaggerated fears regarding its potential long-term toxicity and side-effect profile but also because of a lack of diagnostic specificity, a consequence of manic-depressive illness being supplanted diagnostically by bipolar disorder. Such misguided beliefs have undoubtedly marred the reputation of lithium and curtailed its use, but interestingly it has maintained its role in the treatment of mood disorders and its pole position in international clinical practice guidelines. However, not all clinicians refer to guidelines (Malhi et al. 2008), and, in reality, guidelines are not intended to be prescriptive. Instead, they provide advice (guidance), and, even when clinicians are aware of a particular recommendation, they may still choose to pursue alternatives based on their judgement and personal experience. A key problem that has limited the use of lithium is that, despite its molecular simplicity, it is a complicated medication to prescribe and manage as compared to other treatments and is often associated with both acute and long-term side effects, both troublesome and serious. Therefore, even though lithium demonstrates robust efficacy, it is regarded overall as having only modest effectiveness.

11.1.1 Acute Mania and Bipolar Depression

The treatment of acute mania usually requires rapid containment of behavioural disturbance and stabilisation of mood, and though lithium is antimanic, a clinically meaningful response often takes considerable time (days rather than hours). Therefore, in practice, its use is usually trumped by more potent, rapidly acting agents such as antipsychotics and benzodiazepines. Ideally, this should not preclude its use alongside these agents, but here too the need for lithium to be taken orally precludes its administration in patients who are behaviourally disturbed, uncooperative and lacking insight. In the treatment of acute bipolar depression, the efficacy of lithium is, at best, on par with other agents and takes time to take effect. In other words, its efficacy is modest and delayed. Hence, lithium is seldom used as monotherapy for first-line treatment of bipolar depression and instead is often added when transitioning to maintenance and prophylaxis. Indeed, it is only in this phase of bipolar disorder that lithium excels.

11.1.2 Maintenance and Prophylaxis

The BALANCE study reaffirmed lithium's status as a mood stabiliser and as an effective agent in maintenance therapy of bipolar disorder and its prophylaxis, demonstrating importantly an advantage over valproate (BALANCE investigators 2010). However, concerns persist regarding the long-term use of lithium with respect to tolerability and potential toxicity. In practice, lithium levels require regular monitoring, and often a delicate balance is needed to achieve and maintain efficacy whilst, at the same time, avoid unnecessary side effects and diminish the likelihood of long-term serious consequences. The latter include thyroid and parathyroid dysfunction, leading to hypothyroidism and hypercalcaemia, respectively. However, the concern that is perhaps paramount and somewhat exaggerated is that of renal failure. Lithium can, and does, compromise renal function, and recent research has considered the risk factors of nephrotoxicity in long-term lithium users (Davis et al. 2015). Although prolonged exposure to lithium treatment has been associated with greater risk of decline in renal function (Aiff et al. 2015; Shine et al. 2015; Close et al. 2014), recent research in maintenance lithium therapy has found, when adjusting for confounds, that duration of exposure to lithium treatment and mean plasma concentration level of lithium were not predictors of renal decline

(Clos et al. 2015). Rather, decline was attributed to age, baseline renal function, comorbidity, combination neurotrophic treatment and episodes of lithium toxicity. Such findings emphasise the importance of achieving stable lithium maintenance therapy, and recent research has pointed to the necessity of moderating lithium levels to within the therapeutic range in the long term. In doing so, the associated risks can be minimised and with regular monitoring perhaps avoided altogether (Clos et al. 2015; Davis et al. 2015; Shine et al. 2015). Therefore, whilst such concerns are legitimate, a more sophisticated and more tailored approach to monitoring lithium therapy may well avoid many of the long-term problems and reduce the risk of severe complications, thereby ensuring maximal benefit for those patients suited to lithium therapy (Malhi 2015).

11.1.3 Clinical Practice Guidelines

Clinical practice guidelines attempt to synthesise the available pharmacotherapeutic evidence and make overarching recommendations for the effective management of major psychiatric disorders. Whilst the formulation of clinical guidelines is based on internally valid evidence-based research informed by clinical trial findings, the implementation of guidelines and their clinical uptake are generally poor (Rosenman et al. 2008). The length and density of many clinical practice guideline documents perhaps reduce their accessibility to clinicians. In addition, they usually have an emphasis on findings regarding patients who fit neatly into current diagnostic nosology, thus losing some generalisability to real-world patients and scenarios. The known intricacies of illness course, as well as the role of clinical judgement in assessing these intricacies, are generally not well captured (Malhi and Adams 2009). As such, to best inform clinical practice, guidelines must be considered alongside clinical practice recommendations. Hence, this chapter reviews where lithium is placed within current international guidelines for bipolar disorder by considering the recommendations made by recognised organisations such as the American Psychiatric Association (APA; American Psychiatric Association 2002), the British Association of Psychopharmacology (BAP; Goodwin et al. 2016), the Canadian Network for Mood and Anxiety Treatments (CANMAT; Yatham et al. 2013), the National Institute for Health and Care Excellence (NICE; NICE 2014), the Royal Australian and New Zealand College of Psychiatrists (RANZCP; Malhi et al. 2015), the World Federation of Societies of Biological Psychiatry (WFSBP; Grunze et al. 2013), the German Association for Bipolar Disorders (DGBS e.V.) and the German Association for Psychiatry, Psychotherapy and Psychosomatics (DGPPN e.V.; DGBS, DGPPN 2012).

11.2 Lithium in Current Guidelines

Across a whole host of guidelines, lithium is the gold standard for the long-term treatment of bipolar disorder. Recommendations for its use in acute mania, acute bipolar depression and maintenance/prophylaxis are presented in Table 11.1 and discussed further below.

Table 11.1 Lithium in current international guidelines

	APA (2002)	BAP (2016)	CANMAT (2013)	DGBS, DGPPN (2012)	NICE 2014	RANZCP (2015)	WFSBP (2013)	Summary
Acute mania	First-line combination therapy for severe or mixed episode; lithium monotherapy for less ill patients	Lithium for mild acute mania	First-line monotherapy; second-line combination therapy	Yes		First line	PNES CE A[a] PES CE B	Strong support in monotherapy and combination therapy for severe acute episodes
Acute bipolar depression	First line	Antimanic combination therapy if bipolar I	First-line monotherapy; second-line combination therapy			Yes	PNES CE B[b] PES CE B	Modest support for the use of lithium in bipolar depression, predominantly because of delay in onset of action
Prophylaxis	First-line monotherapy	First line if mania is predominant feature; second line if depression is predominant feature	First-line monotherapy; second-line combination therapy	First line	First line	Yes	PNES CE A[c] PES CE B	All guidelines recommend lithium as first-line treatment in prophylaxis. Primacy re-established recently by virtue of BALANCE study[d]

[a]Lithium received a Category A (full evidence for treatment in acute mania for efficacy in preventing treatment-emergent episodes and separately manic episodes in non-enriched study samples) and Category B (limited positive evidence for efficacy in preventing treatment-emergent episodes of any polarity and separately manic episodes in study samples enriched for acute response and/or acute tolerability of this medication)
[b]Lithium received a Category B (limited positive evidence for treatment in acute bipolar depression for efficacy in preventing treatment-emergent episodes and separately depressive episodes in non-enriched study samples) and Category B (limited positive evidence for efficacy in preventing treatment-emergent episodes of any polarity and separately depressive episodes in study samples enriched for acute response and/or acute tolerability of this medication)
[c]Lithium received a Category A (full evidence for treatment in prophylaxis for efficacy in preventing treatment-emergent episodes and separately in non-enriched study samples) and Category B (limited positive evidence for efficacy in preventing treatment-emergent episodes of any polarity and separately episodes in study samples enriched for acute response and/or acute tolerability of this medication)
[d]BALANCE investigators (2010)

11.2.1 Maintenance Treatment and Prophylaxis

Presently, lithium monotherapy is recommended as a first-line treatment for prophylaxis by all key bipolar disorder guidelines (RANZCP, APA, BAP, NICE, CANMAT, WFSBP, DGBS and DGPPN). The long-term mood-stabilisation efficacy of lithium and its potential to reduce suicide/suicidality rates have informed a number of guidelines (RANZCP, BAP, DGBS and DGPPN) and contributed to an overall rating of grade 1 by the WFSBP for the management of bipolar disorder (Grunze et al. 2013; see Table 11.1).

11.2.2 Treatment of Acute Mania

In treating acute mania/hypomania, lithium monotherapy is supported as first-line treatment in RANZCP and CANMAT guidelines and, in less severe patients, as first-line treatment for mild acute mania in BAP and APA guidelines. As a component of combination pharmacotherapy, lithium is recommended as first-line treatment for acute mania in APA guidelines and as second-line treatment in CANMAT guidelines. WFSBP guidelines also recommend lithium for acute mania but with some additional concerns (Table 11.1). DGBS and DGPPN guidelines also recommend lithium for the treatment of acute mania but again with some limitations that stem from a narrow therapeutic index.

11.2.3 Treatment of Acute Bipolar Depression

In recommending treatment for acute bipolar depression, lithium is favoured as first-line monotherapy treatment in RANZCP, APA and CANMAT guidelines and receives some limited support (Category B) within the WFSBP guidelines. However, the DGBS and DGPPN guidelines do not recommend lithium monotherapy for the treatment of acute bipolar depression, and where the diagnosis is bipolar I disorder, BAP recommends combination therapy along with antimanic therapy.

11.2.4 Lithium Plasma Concentration

Monitoring of plasma lithium concentration is recommended for lithium therapy by APA, BAP, CANMAT, DGBS, DGPPN, NICE, RANZCP and WFSBP guidelines, with therapeutic plasma levels stipulated for both the initiation and maintenance of lithium therapy. The NICE guidelines recommend levels of 0.6–0.8 mmol/L for the initiation of treatment, whereas APA allows for a broader interval of 0.5–1.2 mmol/L within which the initial dose can be titrated. In prophylaxis, guidelines vary in recommended therapeutic intervals. The APA guidelines propose a range of 0.4–1.0 mmol/L, whereas in the BAP guidelines the

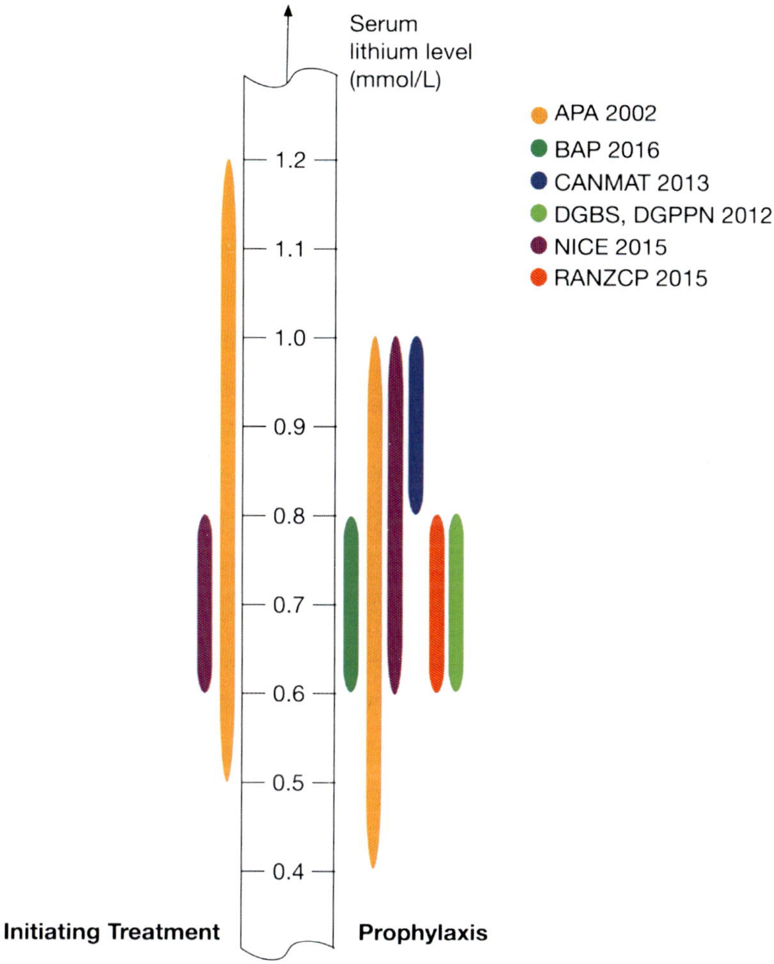

Fig. 11.1 Recommended lithium levels for initiating treatment and prophylaxis (Adapted from Malhi and Tanious 2011). The figure shows the recommended plasma lithium concentrations for initiation of treatment and prophylaxis in patients with bipolar disorder, according to APA, BAP, CANMAT, DGBS, DGPPN, NICE and RANZCP guidelines. When initiating treatment, NICE recommends plasma lithium levels at 0.6–0.8 mmol/L, whereas APA suggests an interval from 0.5 to 1.2 mmol/L. In prophylaxis, BAP recommends lithium plasma levels stabilised between 0.6 and 0.8 mmol/L, a wider range is suggested by APA at 0.4–1.0 mmol/L, NICE suggests a therapeutic interval ranging from 0.6 to 1.0 mmol/L and CANMAT from 0.8 to 1.0 mmol/L, and DGBS, DGPPN and RANZCP propose that lithium plasma levels be stabilised at 0.6–0.8 mmol/L

recommended range is between 0.6 and 0.8 mmol/L (Goodwin et al. 2016), and the NICE guidelines suggest a therapeutic interval of 0.6–1.0 mmol/L. In comparison, CANMAT (0.8–1.0 mmol/L), DGBS, DGPPN and RANZCP (0.6–0.8 mmol/L) recommend narrower intervals, and these are depicted alongside existing guidelines in Fig. 11.1.

11.3 Specificity of Response: To Whom Should Lithium Be Prescribed?

Lithium has demonstrated efficacy in acute mania and prophylaxis in bipolar disorder, as well as in recurrent unipolar depression (Gershon et al. 2009). Specifically, first-line lithium treatment clinical trials have shown lithium to be effective in the treatment of mania, as compared to valproate (Bowden et al. 2010; Geddes et al. 2004; Cipriani et al. 2011; Yildiz et al. 2011), and placebo (Cipriani et al. 2005; Gershon et al. 2009; Geddes et al. 2004; Grunze et al. 2009; Malhi et al. 2009), as well as antipsychotics such as olanzapine and risperidone (Segal et al. 1998; Shafti 2010; Yildiz et al. 2011). Lithium is also a superior antimanic agent relative to asenapine, carbamazepine, ziprasidone, valproate, lamotrigine, topiramate and gabapentin (Cipriani et al. 2011). Whilst treatment approaches to severe acute episodes of mania have relied on the efficacy of antipsychotics, it has been duly noted that effective uptake in the acute phase of bipolar illness might not indicate the best choice for long-term treatment, where prophylaxis is imperative (Cipriani et al. 2011). As such, although lithium achieves a delayed antidepressant effect, the strength of this agent lies in prophylactic treatment, where recurrent episodes of depression can be effectively prevented (Gershon et al. 2009).

In the maintenance phase of therapy, lithium monotherapy has demonstrated superiority in the prophylaxis of both mania and depression (Malhi et al. 2009), and although the delayed antidepressant effect of this agent might direct recommendations towards faster-acting agents in the short term, commencing this prophylactic agent early in the course of the illness provides the best opportunity for good long-term outcomes (Freeman and Freeman 2006; Geddes et al. 2004). In addition to this, lithium has specificity within bipolar disorder, as well as antisuicidal and neuroprotective properties (Berk 2009; Gershon et al. 2009; Malhi et al. 2012b). Prophylactic benefit has also been shown to be optimal when used to treat classic manic depression (Berk and Malhi 2011), where bipolar family history, commencement of maintenance treatment and episodic clinical course are predictors of ideal lithium response (Frye et al. 1998; Grof 2010). Thus, recognising the predictors of ideal lithium response in clinical practice holds great promise for remission within this disorder, where selective administration is key. These finer considerations are largely missing from most guidelines, even though in practice they are essential for the effective treatment of patients with bipolar disorder.

By identifying 'excellent lithium responders', clinicians can tailor bipolar disorder treatment and provide individualised care. Long-term mood stability with lithium characterises bipolar patients as responders versus nonresponders, and the clinical profile of those likely to respond to prophylaxis has been defined empirically (Grof 2010). Recent research in the developmental trajectory of bipolar disorder has also demonstrated differential clinical staging in the offspring of lithium responders and lithium nonresponders, such that the offspring of lithium responders appear to develop episodic mood disorders, whereas the offspring of lithium nonresponders tend to develop non-episodic/non-fully remitting disorders (Duffy et al. 2014). This divergent psychopathology indicates a genetic basis and phenotype for lithium response (Duffy et al. 2014).

The mostly prophylactic actions of lithium make it evident that lithium response is predicated on episodic illness course rather than direct and rapid resolution of illness episodes. It appears that the 'classic' manic-depressive illness course, as originally described by Kraepelin (1921), with recurrent episodes of mania and depression alternating with periods of remission, is the presentation most responsive to lithium therapy (see Fig. 11.2; Grof 2010; Rybakowski et al. 2001). With evolution of diagnostic categories over time, the definition of this 'classic' bipolar illness has been lost, and thus guidelines and clinical trials have not been well placed to incorporate such intricacies. This is important to note, as lithium responders comprise approximately one-third of lithium-treated patients, and, in this subgroup, it is possible to achieve mood stability, near-complete prophylaxis, functional remission and preservation of cognitive function over the long term (Rybakowski and Suwalska 2010), with many patients achieving this over extended periods of time—often more than 10 years (Grof 2010). Other indicators of this 'classic' presentation are a family history of episodic bipolar disorder, as well as an absence of psychiatric comorbidity. With further investigation, a 'classic' illness course specifier could be introduced simply because of its potential utility in identifying common and characteristic illness course patterns that are most likely to respond to mood-stabilising treatment.

In sum, it is important to identify lithium responders in clinical practice by recognising the classic episodic pattern of illness: recurrent, recognisable mood episodes (predominantly depressive but manic as well) separated by periods of complete symptom-free remission (as shown in Fig. 11.2). A key difficulty in translating guidelines into clinical practice is that identifying lithium response requires a greater degree of individualised care, and this is seldom achieved to the standard necessary. This is not surprising given the reduced taxonomic emphasis on a 'classic' bipolar subtype, which means that clinical trials and guidelines fail to take this key consideration into account. Promisingly, the phenotypic signature of lithium response (Gershon et al. 2009) appears to have a genetic basis (Duffy et al. 2014), but identifying these patients as suitable candidates for lithium therapy requires long-term surveillance, and this once again remains a significant challenge in practice. Perhaps further research investigating ways to identify suitable candidates for lithium therapy from the onset will enable us to better address this problem.

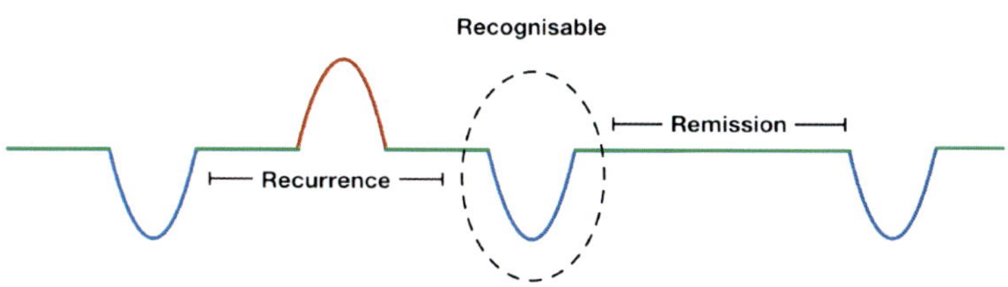

Fig. 11.2 Recognising lithium responders. Lithium *responders* have a classic episodic pattern of illness: *recurrent*, *recognisable* mood episodes (predominantly depressive) separated by periods of complete symptom-free *remission*. Recurrent depressive episodes are represented in blue, separated by periods of remission and manic episodes in *red* (Reproduced with permission from Gershon, Chengappa and Malhi (2009))

11.4 How Should Lithium Be Prescribed?

Effective pharmacotherapy of bipolar disorder involves, first, rapid amelioration of the acute symptoms of mania and depression and, then, prevention of relapse and sustained remission. The latter aim of maintaining remission requires the early commencement of prophylactic treatment allowing therapeutic actions to be set in motion as the acute symptoms diminish. Commencement of treatment is then followed by a period of careful fine-tuning to ensure maximal effectiveness.

11.4.1 Initial Considerations

Many clinicians perceive prescribing lithium as problematic because of the need for monitoring of plasma levels and the risk of opening the door to a myriad of additional health concerns. However, when lithium is prescribed to a suitable candidate and managed judiciously, the therapeutic return is worthwhile and far outweighs the effort expended in initial assessment and the refinement of dosage and administration. Prior to the initiation of lithium, a number of factors must be considered to assess a patient's suitability for treatment. These concern the psychiatric profile, episodic course and presentation of bipolar disorder, all of which must be weighed up in a probabilistic approach to determine the likelihood of lithium responsiveness (see Fig. 11.3). Once lithium has been selected as the therapy of choice, appropriate medical screening must be conducted to detect potential contraindications, inform the clinician of likely side effects and schedule subsequent monitoring (Malhi et al. 2009). In particular, this screening should address renal, thyroid, cardiac and metabolic functioning, all of which might affect, or be affected by, lithium treatment (American Psychiatric Association 2002; NICE 2014). At the commencement of treatment, baseline measures should include an assessment of cardiovascular functioning by ECG, as well as complete blood count, to inform the interpretation of subsequent monitoring. In

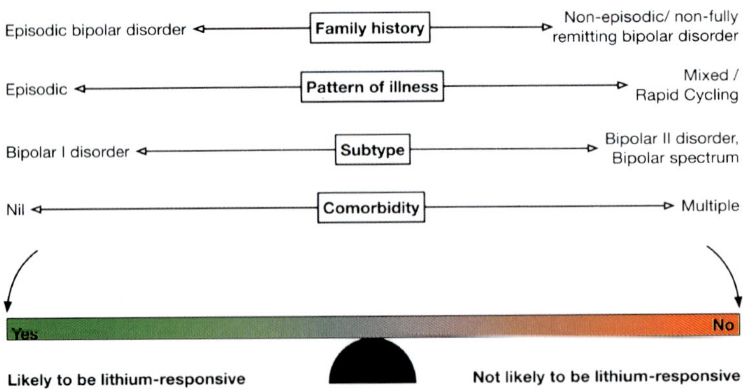

Fig. 11.3 A probabilistic approach to prescribing lithium. Likelihood to be lithium responsive is predicated on family history, pattern of illness, bipolar disorder subtype and psychiatric comorbidity. A probabilistic approach to prescribing lithium takes into account features that predict lithium response is likely and 'weighs' them against features that predict that they are not likely to respond

addition, thyroid function evaluation, pregnancy testing, physical examination and tests for urea and electrolytes are required to determine appropriate dosage and possibility of contraindications (American Psychiatric Association 2002; Malhi et al. 2009). Precautions must be communicated to the patient with regard to maintaining a normal diet with adequate salt and fluid intake and avoiding excessive sweating that may result for example from heavy exercise (Malhi et al. 2009). Patient education is essential in emphasising the importance of treatment adherence and administration of lithium, and the patient must be made aware of managing the potential side effects and risk factors of long-term lithium therapy.

11.4.2 Dosage and Administration

Both the narrow therapeutic index of lithium and its associated risk of toxicity indicate the imperative for appropriate dosing and careful monitoring of serum levels (Malhi et al. 2012b). The appropriate dose should be informed by illness course and severity, and the aim should be to achieve the lowest level whilst balancing effectiveness and adverse effects. At the commencement of treatment, higher lithium plasma levels can be used to treat acute symptoms, and lithium plasma levels should be measured weekly and eventually stabilised at 0.6–0.8 mmol/L (NICE 2014; Malhi et al. 2011, 2012b).

Lithium is best administered early in the course of illness (Kessing et al. 2014), but this also means that it may be administered to patients who do not present with a classic bipolar I disorder pattern. It is no surprise that lithium is much less effective when used in patients who have significant personality problems or additional psychiatric pathology, but, unfortunately, this is often misinterpreted as lithium 'lacking efficacy', or occurring because it is difficult to prescribe. Therefore it is essential to identify those patients who are likely to be lithium responsive (see Fig. 11.3; Malhi and Geddes 2014). Small, divided daily doses are recommended in APA guidelines to minimise side effects in the early course of the treatment, gradually titrating serum levels to achieve the highest dose tolerated without side effects, ranging between 0.5 and 1.2 mmol/L (American Psychiatric Association 2002). However, if tolerated, a single daily dose may be preferred. Compared to a multiple dose, a single-dose regimen induces less polyuria and thus reduces the long-term risk of renal damage (Malhi et al. 2012b). If tolerated, a single dose taken at night can be trialled (American Psychiatric Association 2002; Goodwin et al. 2016), and in some cases this dose can be reduced by up to 25 % in order to limit side effects, such as fatigue, and increase compliance (Letica-Crepulja et al. 2008; Malhi et al. 2009, 2012b).

11.5 The Role of Monitoring and of Measuring Lithium Plasma Levels

Dosage, acute and chronic tolerability and ongoing changes in illness severity must be considered when monitoring lithium therapy of bipolar disorder.

11.5.1 Optimal Levels and the Lithiumeter

To optimise the effectiveness of lithium treatment, it is important to appropriately manage its potential for adverse side effects and toxicity. Therefore, it is necessary to ensure that the correct blood level concentrations are maintained, and to achieve this lithium plasma concentration, testing must be conducted upon initiation of therapy and at regular intervals subsequently (Malhi and Tanious 2011). This is particularly important for those who respond to lithium, and it is maintained long term for prophylaxis. By instituting adequate monitoring and ensuring that optimal levels are maintained (tailored to the needs of the individual), the likelihood of recurrences will be greatly diminished, and side effects can be promptly managed if they arise. The 'lithiumeter' (Fig. 11.4) depicts indications and risks associated with lithium therapy according to its blood plasma levels. Optimal levels for treatment initiation and maintenance are between 0.6 and 0.8 mmol/L and should be individually tailored on the basis of recurrences and polarity of symptoms (Malhi et al. 2015). In prophylaxis, serum lithium levels can be increased up to 1.0 mmol/L for mania and decreased to 0.4 mmol/L for bipolar depression (Malhi et al. 2011, 2012b). Maintaining optimal plasma levels in lithium prophylaxis is central to the effective maintenance of remission and the prevention of future episodes, as well as minimising side effects and ensuring treatment adherence.

11.5.2 Tolerability

Lithium is associated with both acute and chronic side effects that can limit its tolerability. Nonetheless, correct monitoring practice can enable patients to achieve remission with the fewest side effects possible. At the initiation of lithium treatment, it is common for a rapid increase of plasma lithium concentration to cause acute side effects that include tremor, general fatigue, diarrhoea, thirst, polyuria, nausea, headache and vomiting (Fyrö et al. 1970; Lydiard and Gelenberg 1982). In practice, these normally subside after a day or so and are rarely troublesome especially if patients are forewarned.

Over the long term, clinical consideration must be given to renal and endocrine function, obesity and metabolic syndrome, cardiovascular disease, neuroprogression and cognitive impairment. Although modest in most cases, renal function, and in particular renal concentrating ability, can be impaired by long-term lithium use, leading to polyuria and secondary thirst (Grandjean and Aubry 2009). The risk of renal failure with lithium therapy is often at the forefront of clinicians' minds and no doubt in many cases prevents or limits its use. In actuality, end-stage renal failure only occurs in 0.53 % of lithium-treated individuals, compared to 0.2 % in the general population (Coresh et al. 2003). The vast majority of patients on long-term lithium therapy experience a gradual decrease in glomerular function that is comparable to that of normal ageing (Gitlin 1999; Silverstone 2000). Nonetheless, routine examinations of urea, creatinine and electrolytes are recommended both upon initiation of lithium therapy and during ongoing maintenance treatment (see Table 11.2). Similarly, because

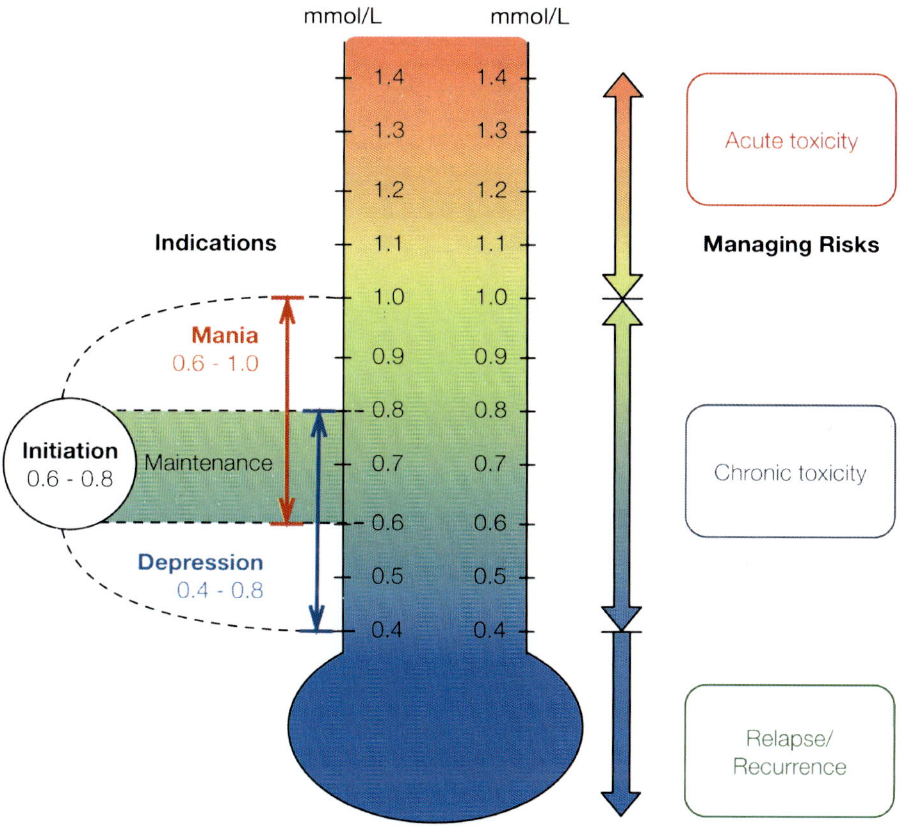

Fig. 11.4 The lithiumeter (Adapted from Malhi and Berk 2012). The lithiumeter depicts indications and risks associated with lithium according to its blood plasma levels. Optimal plasma levels for treatment initiation are between 0.6 and 0.8 mmol/L and should be individually tailored on the basis of recurrences and polarity of symptoms. *Indications*: this depicts suitable plasma levels for initiation and maintenance therapy in relation to the phases of illness, which are shown on the *left*. Prophylaxis for depression (shown in *blue*) might be achieved satisfactorily at lower levels than for mania (shown in *red*). The risk of acute toxicity is increasingly likely at levels above 1.0 mmol/L, but, over time, even much lower therapeutic levels of lithium can lead to chronic toxicity, especially in elderly individuals. However, relapse/recurrence is more likely at plasma lithium levels lower than 0.4 mmol/L

clinically significant hypothyroidism is observed in 8–19 % of patients receiving lithium therapy, endocrine monitoring is essential to detect and manage thyroid abnormalities, especially since hypothyroidism is associated with depression and rapid cycling (Yatham et al. 2005). Furthermore, patients with hypothyroidism whilst on lithium therapy should be prescribed thyroxine (Freeman and Freeman 2006). Another endocrine problem that patients on long-term lithium therapy are prone to develop is that of hypercalcaemia secondary to elevated parathyroid hormone levels, which is

Table 11.2 Lithium therapy monitoring recommendations

		Suggested timetable for assessments during lithium therapy (in months)			
		Initiation	Maintenance		
Investigations/examinations[a]		0	6	12	24
Renal	Urea	◆	◆	◆	◆
	Creatinine	◆	◆	◆	◆
	Electrolytes	◆		◆	◆
Endocrine	TSH	◆	◆	◆	◆
	Serum Ca^{2+}	◆		◆	◆
Physical	Waist circumference	◆		◆	◆
	BMI	◆		◆	◆

BMI body mass index, *TSH* thyroid-stimulating hormone
[a]Abnormal findings should prompt investigation and more frequent monitoring. Any significant change in mood state should prompt a repeat of investigations. Women of childbearing age should be advised that the risk of congenital malformation whilst taking lithium therapy is unknown. A baseline ECG should be considered in patients over 40 years, or if otherwise indicated

also associated with a decrease in bone resorption (Grandjean and Aubry 2009). Therefore, careful monitoring of the approximate endocrine measures is necessary to inform long-term management—involving adjustments in lithium dose to mitigate potential adverse effects. Regular monitoring (see Table 11.2) should also include assessments of obesity, metabolic dysfunction and cardiovascular illness, given that these illnesses are associated with bipolar disorder and have shared pathogenetic pathways (Outhred et al. 2016; Vancampfort et al. 2013).

11.6 Additional Considerations

11.6.1 Suicidality

Lithium has a recognised antisuicidal effect in long-term administration, such that suicidal behaviour and risk are significantly reduced with therapy (Gershon et al. 2009; Goodwin et al. 2016; Guzzetta et al. 2007; Malhi et al. 2009). This study supports the specific use of lithium especially in treating bipolar disorders, given that the risk of suicide and suicide attempts in this illness is approximately ten times higher than in the general population (Baldessarini et al. 2006). It is suspected that the anti-impulsive and anti-aggressive properties of lithium confer the antisuicidal effects of lithium treatment (Grunze et al. 2013; Kovacsics et al. 2009), and it has been shown clinically that lithium reduces the risk of death by suicide by 60% in this group and the risk of self-harm by 70% (Cipriani et al. 2005; Goodwin et al. 2003). In a meta-analysis of 32 randomised control trials assessing suicide and all-cause mortality rates in patients with mood disorders (Cipriani et al. 2005), it was concluded that patients receiving lithium were much less likely to die from suicide or from any cause than were patients given an alternative to lithium. These data support its use as first-line therapy in bipolar disorder especially in those at risk of suicidal behaviour (Schaffer et al. 2015).

11.6.2 Neuroprotective Benefits

Bipolar disorder is not considered a neurodegenerative disorder per se; however neuroprogression and functional deterioration likely occur over the course of the illness because of oxidative dysfunction and nitrosative damage within inflammatory and neurotrophic expression pathways (Berk 2009; Berk et al. 2011). In particular, neurotransmitter excitation of dopamine and glutamate, alongside diminished inhibition from underactive pathways in γ-aminobutyric acid (GABA), accounts for oxidative and nitrosative damage in postsynaptic cells (Berk et al. 2011). Whilst in lithium treatment the mechanisms of action remain unknown, lithium is thought to modify these mechanisms and potentially ameliorate some of these—addressing the underlying pathophysiology of bipolar disorder (see Chap. 3, Malhi et al. 2017; Malhi and Outhred 2016). Lithium protects against the excitatory neurotoxicity that results from increased neural glutamate by inactivating *N*-methyl-D-aspartic acid (NMDA) receptors (Nonaka et al. 1998) and increasing the release of GABA, which mediates excitatory neurotransmission and potentially prevents neurotoxicity (Brunello and Tascedda 2003; Malhi et al. 2013). Improved neuronal function has also been associated with increased levels of *N*-acetyl-asparate (NAA) with lithium treatment, neuroprotective proteins such as brain-derived neurotrophic factor (BDNF) and B-cell lymphoma 2 (Bcl-2), as well as decreases in pro-apoptotic enzymes such as glycogen synthase kinase-3 (GSK-3) (Marmol 2008). Despite these well-characterised neuroprotective actions, we are lacking a complete understanding of which of these pathways is important for achieving prophylaxis in bipolar disorder with lithium administration and whether specific or adjunctive drug targets along these pathways can be exploited to achieve maximal benefit. With promising leads already being followed (Chiu et al. 2013; Chuang and Priller 2006; Marmol 2008), future investigations along these lines may also yield novel indications for lithium in diseases that share common pathophysiology, including Parkinson's disease, Alzheimer's disease and Huntington's disease.

11.7 Guidelines for Real-World Presentations of Bipolar Disorder

Returning to the development of clinical practice guidelines, the majority of which are, in turn, guided by evidence-based clinical trial findings, it is pertinent to note that a one-size-fits-all approach to treating bipolar disorder may not be appropriate, due in part to what is known about ideal lithium response but also because building recommendations on the basis of treating pristine presentations of bipolar disorder fails to address the many complex manifestations of bipolar disorder encountered in day-to-day practice (Malhi et al. 2012a).

Most guidelines are based on empirical research pertaining to bipolar I disorder; however this research is used to inform the treatment of variations of the illness, including bipolar II disorder, subsyndromal bipolar disorder, mixed mood states and rapid cycling. This bias towards bipolar I disorder makes it challenging to interpret most guidelines when treating such real-world complexities of bipolar disorder (Malhi et al. 2012a). Therefore, there is a need for a stratified approach to the treatment of bipolar disorders with selective administration of lithium that recognises the complexities of the illness in terms of its varied manifestations and course trajectory (see Fig. 11.5). The diversification and expansion of the diagnosis of bipolar disorder in DSM-5 fail to adopt a stratified approach, instead collapsing a myriad of presentations and conflating bipolar II diagnosis, resulting in misdiagnosis and a lack of treatment specificity for bipolar I disorder (Malhi and Porter 2014). Moreover, in defining

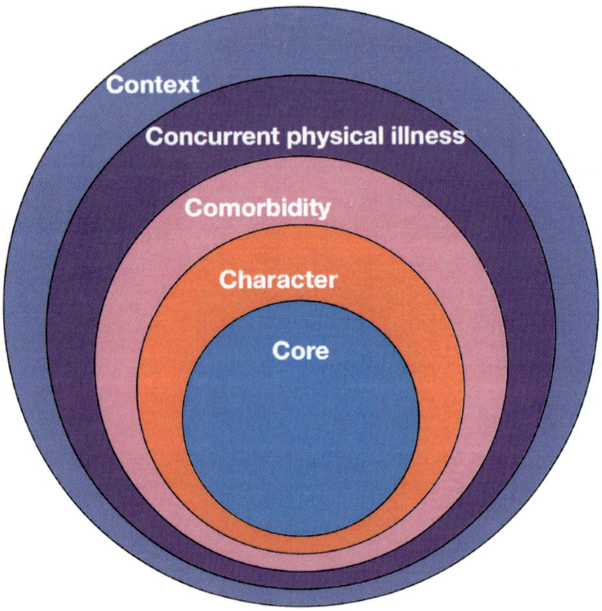

Fig. 11.5 Stratified model of bipolar disorder complexity (Adapted from Malhi et al. 2012c). Bipolar complexity can be conceptualised as a stratified model in which successive layers of complexity build upon core features of the illness. The core comprises the defining mood symptoms that manifest as a variety of syndromes (e.g. bipolar I disorder, bipolar II disorder). Personality factors that constitute character are also central to bipolar complexity and hence form the next layer. Concomitant symptoms of anxiety and associated substance abuse are common comorbidities that form the middle layer of the stratified model. Alongside these, concurrent physical illness provides the penultimate layer of complexity. The final, outermost layer is the context in which bipolar disorder arises. This is an important consideration and one that often complicates management. Lithium is very much dependent on context to determine treatment response, where lithium response is predicated on family history, pattern of illness, bipolar disorder subtype and psychiatric comorbidity

diagnostic categories based on features of polarity, DSM-5 fails to capture the episodic course of the illness, in particular the 'classic' manic-depressive subtype—described by Kraepelin (1921) which was recurrent by definition (Malhi and Berk 2014). Despite this limitation, which results in clinical trial findings over-representing treatment for bipolar I disorder (Swartz and Thase 2011), lithium monotherapy can be used to treat bipolar II disorder and rapid cycling bipolar disorder and can be of benefit in both youth and geriatric populations. Lithium may also be of some benefit in the treatment of mixed states, when used in combination with other medications (Malhi et al. 2012a). However, in general, these presentations or features are not typically indicative of suitability to lithium therapy (Fig. 11.3).

11.8 Individualised Care

It is clear from the considerations regarding specificity of lithium response and the importance of monitoring and measuring lithium plasma levels that the therapeutic benefit of lithium treatment is likely to vary from patient to patient. This underscores the importance of individualised care, and it is proposed that the 'lithiumeter' (Fig. 11.4) be used to determine the optimal plasma levels required, from initiation to maintenance, by also taking into account fluctuations in mood state. In a routine monitoring schedule in long-term prophylactic treatment, the measurement of lithium plasma levels would accompany renal, endocrine and physical examinations (Table 11.2), all of which would inform dosing and management of lithium treatment. Such monitoring practices are typically conducted by the managing physician; however there is a place for outpatient specialised psychiatric care, in the form of lithium clinics (Ahrens et al. 1995; Bey et al. 1972; Courtney et al. 1995; Kallner et al. 2000; Licht et al. 2001; Osher et al. 2010; Schweitzer et al. 1999). Through the implementation of best-practice psychoeducation and monitoring, such clinics optimise the therapeutic benefit of lithium treatment whilst reducing the burden of management that may be too intensive and cumbersome in many cases (Berk et al. 2004; Outhred et al. 2016).

11.9 Summary

Lithium's pole position across international clinical practice guidelines for the treatment of bipolar disorder indicates unequivocal support for this agent that clearly demonstrates efficacy in quelling acute symptoms and maintaining long-term remission whilst providing antisuicidal and potentially neuroprotective benefits in an ideal responder type. Managed judiciously, acute and chronic side effects of lithium can be minimised and mitigated over the course of illness, resulting in long-term efficacy of the agent. This chapter has emphasised the variance in individual suitability towards this agent, the narrow therapeutic window of lithium and the requirement of individualised monitoring practices, all of which represent challenges in the administration of lithium. Areas in which the clinical practice of lithium for bipolar disorder requires greater emphasis and evidence base are (1) identification of, and lithium use in, 'classic' bipolar illness course presentations, as these are most likely to be 'excellent lithium responders', and (2) approaches of assessment and monitoring that are more individualised and thus effective but at the same time aim to minimise the healthcare burden. The recommendations contained within this chapter provide a synthesis of the existing clinical practice guidelines, whilst recognising that the real-world presentations of bipolar disorder are far from pristine and impose additional challenges (see also appendices 1 and 2 for assessment of cognition using lithium battery – clinical and research). There is clearly a need for the further development of management approaches to recognise this and further optimise lithium therapy.

References

Ahrens B, Grof P, Moeller H-J, Müller-Oerlinghausen B, Wolf T (1995) Extended survival of patients on long-term lithium treatment. Can J Psychiatry 40(5):241–246

Aiff H, Attman P-O, Aurell M, Bendz H, Ramsauer B, Schön S, Svedlund J (2015) Effects of 10 to 30 years of lithium treatment on kidney function. J Psychopharmacol 29:608–614, 0269881115573808

American Psychiatric Association (2002) Practice guideline for the treatment of patients with bipolar disorder (revision). Am J Psychiatry 159(4 Suppl):1

BALANCE investigators (2010) Lithium plus valproate combination therapy versus monotherapy for relapse prevention in bipolar I disorder (BALANCE): a randomised open-label trial. Lancet 375(9712):385–395

Baldessarini RJ, Tondo L, Davis P, Pompili M, Goodwin FK, Hennen J (2006) Decreased risk of suicides and attempts during long-term lithium treatment: a meta-analytic review. Bipolar Disord 8(5p2):625–639

Berk M (2009) Neuroprogression: pathways to progressive brain changes in bipolar disorder. Int J Neuropsychopharmacol 12(4):441–445

Berk M, Malhi GS (2011) Should antipsychotics take pole position in mania treatment? Lancet 378(9799):1279–1281

Berk M, Berk L, Castle D (2004) A collaborative approach to the treatment alliance in bipolar disorder. Bipolar Disord 6(6):504–518

Berk M, Kapczinski F, Andreazza A, Dean O, Giorlando F, Maes M, Yücel M, Gama C, Dodd S, Dean B (2011) Pathways underlying neuroprogression in bipolar disorder: focus on inflammation, oxidative stress and neurotrophic factors. Neurosci Biobehav Rev 35(3):804–817

Bey D, Chapman R, Tornquist K (1972) A lithium clinic. Am J Psychiatry 129(4):468–470

Bowden CL, Mosolov S, Hranov L, Chen E, Habil H, Kongsakon R, Manfredi R, Lin H-N (2010) Efficacy of valproate versus lithium in mania or mixed mania: a randomized, open 12-week trial. Int Clin Psychopharmacol 25(2):60–67

Brunello N, Tascedda F (2003) Cellular mechanisms and second messengers: relevance to the psychopharmacology of bipolar disorders. Int J Neuropsychopharmacol 6(02):181–189

Cade JF (1949) Lithium salts in the treatment of psychotic excitement. Med J Aust 2:349–352

Chiu CT, Wang Z, Hunsberger JG, Chuang DM (2013) Therapeutic potential of mood stabilizers lithium and valproic acid: beyond bipolar disorder. Pharmacol Rev 65(1):105–142. doi:10.1124/pr.111.005512

Chuang D-M, Priller J (2006) Potential use of lithium in neurodegenerative disorders. In: Lithium in neuropsychiatry: the comprehensive guide. Taylor & Francis Books, London, 381–397

Cipriani A, Pretty H, Hawton K, Geddes JR (2005) Lithium in the prevention of suicidal behavior and all-cause mortality in patients with mood disorders: a systematic review of randomized trials. Am J Psychiatry 162:1805–1819

Cipriani A, Barbui C, Salanti G, Rendell J, Brown R, Stockton S, Purgato M, Spineli LM, Goodwin GM, Geddes JR (2011) Comparative efficacy and acceptability of antimanic drugs in acute mania: a multiple-treatments meta-analysis. Lancet 378(9799):1306–1315. doi:10.1016/s0140-6736(11)60873-8

Clos S, Rauchhaus P, Severn A, Cochrane L, Donnan PT (2015) Long-term effect of lithium maintenance therapy on estimated glomerular filtration rate in patients with affective disorders: a population-based cohort study. Lancet Psychiatry 2(12):1075–1083

Close H, Reilly J, Mason JM, Kripalani M, Wilson D, Main J, Hungin A (2014) Renal failure in lithium-treated bipolar disorder: a retrospective cohort study. PLoS ONE 9(3):e90169

Coresh J, Astor BC, Greene T, Eknoyan G, Levey AS (2003) Prevalence of chronic kidney disease and decreased kidney function in the adult US population: Third National Health and Nutrition Examination Survey. Am J Kidney D 41(1):1–12

Courtney ME, Acomb JA, Lovatt V (1995) A pharmacy-controlled lithium clinic. Psychiatrist 19(1):15–17

Davis JM, Rosenbaum A, Shahinian V, Brosius FC (2015) Prevention of lithium-associated renal failure: recent evidence. Lancet Psychiatry 2(12):1045–1047

DGBS, DGPPN (2012) S3-Leitlinie zur Diagnostik und Therapie Bipolarer Störungen. Langversion. Available from: http://www.leitlinie-bipolar.de

Duffy A, Horrocks J, Doucette S, Keown-Stoneman C, McCloskey S, Grof P (2014) The developmental trajectory of bipolar disorder. Br J Psychiatry 204(2):122–128. doi:10.1192/bjp.bp.113.126706

Freeman MP, Freeman SA (2006) Lithium: clinical considerations in internal medicine. Am J Med 119(6):478–481

Frye MA, Ketter TA, Altshuler LL, Denicoff K, Dunn RT, Kimbrell TA, Corá-Locatelli G, Post RM (1998) Clozapine in bipolar disorder: treatment implications for other atypical antipsychotics. J Affect Disord 48(2):91–104

Fyrö B, Pettersson U, Sedvall G (1970) Serum lithium levels and side effects during administration of lithium carbonate and two slow release lithium preparations to human volunteers. Pharmacologia Clin 2(4):236–240

Geddes JR, Burgess S, Hawton K, Jamison K, Goodwin GM (2004) Long-term lithium therapy for bipolar disorder: systematic review and meta-analysis of randomized controlled trials. Am J Psychiatry 161(2):217–222

Gershon S, Chengappa K, Malhi GS (2009) Lithium specificity in bipolar illness: a classic agent for the classic disorder. Bipolar Disord 11(s2):34–44

Gitlin M (1999) Lithium and the kidney. Drug Saf 20(3):231–243

Goodwin FK, Fireman B, Simon GE, Hunkeler EM, Lee J, Revicki D (2003) Suicide risk in bipolar disorder during treatment with lithium and divalproex. JAMA 290(11):1467–1473

Goodwin GM, Haddad PM, Ferrier IN, Aronson JK, Barnes T, Cipriani A (2016) Evidence-based guidelines for treating bipolar disorder: revised third edition recommendations from the British Association for Psychopharmacology. J Psychopharmacol 30(6):495–553

Grandjean EM, Aubry J-M (2009) Lithium: updated human knowledge using an evidence-based approach: part III: clinical safety. CNS Drugs 23(5):397

Grof P (2010) Sixty years of lithium responders. Neuropsychobiology 62(1):8–16. doi:10.1159/000314305

Grunze H, Vieta E, Goodwin GM, Bowden C, Licht RW, Moeller H-J, Kasper S, WTFoTGfB Disorders, Grunze H, Vieta E (2009) The World Federation of Societies of Biological Psychiatry (WFSBP) guidelines for the biological treatment of bipolar disorders: update 2009 on the treatment of acute mania. World J Biol Psychiatry 10(2):85–116

Grunze H, Vieta E, Goodwin GM, Bowden C, Licht RW, Moller HJ, Kasper S, WTFoTGfB Disorders (2013) The World Federation of Societies of Biological Psychiatry (WFSBP) guidelines for the biological treatment of bipolar disorders: update 2012 on the long-term treatment of bipolar disorder. World J Biol Psychiatry 14(3):154–219. doi:10.3109/15622975.2013.770551

Guzzetta F, Tondo L, Centorrino F, Baldessarini RJ (2007) Lithium treatment reduces suicide risk in recurrent major depressive disorder. J Clin Psychiatry 68(3):380–383

Kallner G, Lindelius R, Petterson U, Stockman O, Tham A (2000) Mortality in 497 patients with affective disorders attending a lithium clinic or after having left it. Pharmacopsychiatry 33(1):8–13

Kessing LV, Vradi E, Andersen PK (2014) Starting lithium prophylaxis early v. late in bipolar disorder. Br J Psychiatry 205(3):214–220

Kovacsics CE, Gottesman II, Gould TD (2009) Lithium's antisuicidal efficacy: elucidation of neurobiological targets using endophenotype strategies. Annu Rev Pharmacol Toxicol 49: 175–198

Kraepelin E (1921) Manic depressive insanity and paranoia. J Nerv Ment Dis 53(4):350

Letica-Crepulja M, Bistrovic IL, Ljubicic R (2008) Lithium treatments: single and multiple daily dosing. Can J Psychiatry 53(5):323

Licht R, Vestergaard P, Rasmussen NA, Jepsen K, Brodersen A, Hansen PEB (2001) A lithium clinic for bipolar patients: 2-year outcome of the first 148 patients. Acta Psychiatr Scand 104(5):387–390

Lydiard RB, Gelenberg A (1982) Hazards and adverse effects of lithium. Annu Rev Med 33(1):327–344

Malhi GS (2015) Lithium therapy in bipolar disorder: a balancing act? Lancet 386(9992):415–416. doi:10.1016/s0140-6736(14)62123-1

Malhi G, Adams D (2009) Are guidelines in need of CPR? The development of clinical practice recommendations (CPR). Acta Psychiatr Scand 119(s439):5–7

Malhi GS, Berk M (2012) Is the safety of lithium no longer in the balance? Lancet 379:690–692

Malhi GS, Berk M (2014) Diagnosing bipolar disorder: defining thresholds and setting boundaries. Aust N Z J Psychiatry 48(6):500–504

Malhi GS, Geddes JR (2014) Carving bipolarity using a lithium sword. Br J Psychiatry 205(5):337–339

Malhi GS, Outhred T (2016) Therapeutic Mechanisms of Lithium in Bipolar Disorder: Recent Advances and Current Understanding. CNS Drugs 30:931–949. http://doi.org/10.1007/s40263-016-0380-1

Malhi GS, Outhred T, Das P (2017) Lithium: Neurotransmission and Cellular Mechanism Pathways Underlying Neuroprogression in Bipolar Disorder. In: Malhi GS, Masson M, Bellivier F (ed) Science and Practice of Lithium Therapy. Springer, Berlin

Malhi GS, Porter RJ (2014) Are 'buy-polar' forces and 'try-polar' thinking expanding bipolarity? Aust N Z J Psychiatry 48(8):697–700

Malhi GS, Tanious M (2011) Optimal frequency of lithium administration in the treatment of bipolar disorder. CNS Drugs 25(4):289–298

Malhi G, Adams D, O'Connor N (2008) Comment: clinical practice guidelines–guiding or goading? Australas Psychiatry 16(1):5–7

Malhi GS, Adams D, Berk M (2009) Is lithium in a class of its own? A brief profile of its clinical use. Aust N Z J Psychiatry 43(12):1096–1104

Malhi GS, Tanious M, Gershon S (2011) The lithiumeter: a measured approach. Bipolar Disord 13(3):219–226

Malhi GS, Bargh DM, Cashman E, Frye MA, Gitlin M (2012a) The clinical management of bipolar disorder complexity using a stratified model. Bipolar Disord 14(s2):66–89

Malhi GS, Tanious M, Das P, Berk M (2012b) The science and practice of lithium therapy. Aust N Z J Psychiatry 46(3):192–211

Malhi GS, Bargh DM, McIntyre R, Gitlin M, Frye MA, Bauer M, Berk M (2012c) Balanced efficacy, safety, and tolerability recommendations for the clinical management of bipolar disorder. Bipolar Disord 14(Suppl 2):1–21. doi:10.1111/j.1399-5618.2012.00989.x

Malhi GS, Tanious M, Das P, Coulston CM, Berk M (2013) Potential mechanisms of action of lithium in bipolar disorder. CNS Drugs 27(2):135–153

Malhi GS, Bassett D, Boyce P, Bryant R, Fitzgerald PB, Fritz K, Hopwood M, Lyndon B, Mulder R, Murray G (2015) Royal Australian and New Zealand College of Psychiatrists clinical practice guidelines for mood disorders. Aust N Z J Psychiatry 49(12):1087–1206

Marmol F (2008) Lithium: bipolar disorder and neurodegenerative diseases possible cellular mechanisms of the therapeutic effects of lithium. Prog Neuropsychopharmacol Biol Psychiatry 32(8):1761–1771

NICE (2014) Bipolar disorder: the assessment and management of bipolar disorder in adults, children and young people in primary and secondary care, Clinical guideline 185. National Institute for Health and Clinical Excellence, London

Nonaka S, Hough CJ, Chuang D-M (1998) Chronic lithium treatment robustly protects neurons in the central nervous system against excitotoxicity by inhibiting N-methyl-D-aspartate receptor-mediated calcium influx. Proc Natl Acad Sci 95(5):2642–2647

Osher Y, Bersudsky Y, Belmaker R (2010) The new lithium clinic. Neuropsychobiology 62(1):17–26

Outhred T, McAulay C, Gessler D, Malhi G (2016) Monitoring for metabolic dysfunction and cardiovascular disease in bipolar disorder: a shared illness process approach. In: Baune BT, Tully PJ (eds) Cardiovascular diseases and depression – treatment and prevention in psychocardiology. Springer International Publishing, Switzerland (pp 333–350)

Rosenman S, Christensen H, Griffiths K (2008) What is to become of the college clinical practice guidelines? Australas Psychiatry 16(1):1–4

Rybakowski JK, Suwalska A (2010) Excellent lithium responders have normal cognitive functions and plasma BDNF levels. Int J Neuropsychopharmacol 13(5):617–622

Rybakowski JK, Chlopocka-Wozniak M, Suwalska A (2001) The prophylactic effect of long-term lithium administration in bipolar patients entering treatment in the 1970s and 1980s. Bipolar Disord 3(2):63–67

Schaffer A, Isometsä ET, Azorin J-M, Cassidy F, Goldstein T, Rihmer Z et al (2015) A review of factors associated with greater likelihood of suicide attempts and suicide deaths in bipolar disorder: part II of a report of the International Society for Bipolar Disorders Task Force on Suicide in Bipolar Disorder. Aust N Z J Psychiatry 49(11):1006–1020

Schweitzer I, Davies B, Burrows G, Branton L, Turecek L, Tiller J (1999) The Royal Melbourne Hospital Lithium Clinic. Aust N Z J Psychiatry 33(s1):S35–S38

Segal J, Berk M, Brook S (1998) Risperidone compared with both lithium and haloperidol in mania: a double-blind randomized controlled trial. Clin Neuropharmacol 21(3):176–180

Shafti SS (2010) Olanzapine vs. lithium in management of acute mania. J Affect Disord 122(3):273–276

Shine B, McKnight RF, Leaver L, Geddes JR (2015) Long-term effects of lithium on renal, thyroid, and parathyroid function: a retrospective analysis of laboratory data. Lancet 386(9992):461–468. doi:10.1016/s0140-6736(14)61842-0

Silverstone T (2000) Is lithium still the maintenance treatment of choice for bipolar disorder? CNS Drugs 14(2):81–94

Swartz HA, Thase ME (2011) Pharmacotherapy for the treatment of acute bipolar II depression: current evidence. J Clin Psychiatry 72(3):356–366

Vancampfort D, Vansteelandt K, Correll CU, Mitchell AJ, De Herdt A, Sienaert P, Probst M, De Hert M (2013) Metabolic syndrome and metabolic abnormalities in bipolar disorder: a meta-analysis of prevalence rates and moderators. Am J Psychiatry 170(3):265–274

Yatham LN, Kennedy SH, O'Donovan C, Parikh S, MacQueen G, McIntyre R, Sharma V, Silverstone P, Alda M, Baruch P (2005) Canadian Network for Mood and Anxiety Treatments (CANMAT) guidelines for the management of patients with bipolar disorder: consensus and controversies. Bipolar Disord 7(s3):5–69

Yatham LN, Kennedy SH, Parikh SV, Schaffer A, Beaulieu S, Alda M, O'Donovan C, MacQueen G, McIntyre RS, Sharma V (2013) Canadian Network for Mood and Anxiety Treatments (CANMAT) and International Society for Bipolar Disorders (ISBD) collaborative update of CANMAT guidelines for the management of patients with bipolar disorder: update 2013. Bipolar Disord 15(1):1–44

Yildiz A, Vieta E, Leucht S, Baldessarini RJ (2011) Efficacy of antimanic treatments: meta-analysis of randomized, controlled trials. Neuropsychopharmacology 36(2):375–389

12 Lithium in Acute and Maintenance Treatment of Bipolar Disorders

Chantal Henry, Aroldo A. Dargél, and Jan Scott

Abstract

Following the first efficacy trial of lithium prophylaxis in recurrent mood disorders in 1967, lithium has been considered the gold standard for bipolar disorders. However, its use has declined over time as newer drugs have been identified as mood stabilizers. This chapter offers a selective review of studies, focusing on the usefulness of lithium in acute episodes of bipolar disorders and in maintenance treatment. The efficacy of lithium in acute episodes and as a prophylactic treatment has been tested against various antipsychotics in comparative, high-quality clinical trials. These clinical trials demonstrate the efficacy of lithium and confirm that it should still be considered as a major mood stabilizer. We then consider the prevalence of partial or non-adherence with lithium and the reasons why individuals might stop taking a potentially beneficial treatment. The chapter highlights

C. Henry (✉)
Université Paris-Est, UMR_S955, UPEC, 94000 Créteil, France

Inserm U955, Translational Psychiatry Laboratory – Team 15, 94000 Créteil, France

AP-HP, Hôpital H. Mondor – A. Chenevier, Pôle de psychiatrie, 94000 Créteil, France

Hôpital A. Chenevier, Pavillon Hartmann, Pôle de psychiatrie,
40, rue de Mesly, 94000 Créteil, France
e-mail: chantal.henry@inserm.fr

A.A. Dargél
Institut Pasteur, Unité Perception et Mémoire, F-75015 Paris, France
e-mail: aroldodargel@gmail.com

J. Scott
Academic Psychiatry, Institute of Neuroscience, Newcastle University,
England NE6 4Bf, UK

Centre for Affective Disorders, Institute of Psychiatry, London, UK
e-mail: jan.scott@newcastle.ac.uk

G.S. Malhi et al. (eds.), *The Science and Practice of Lithium Therapy*,
DOI 10.1007/978-3-319-45923-3_12

that adherence can be predicted by considering patients' beliefs about health and exploring their views about the necessity for treatment versus their concerns about medication. We offer a framework for intervention that is more reliable and valid than previous, overly simplistic, predictor models that failed to take into account dynamic changes in adherence over time within and between individuals.

Key Points

- Lithium has held pole position for more than half a century and is especially effective in treating acute mania and providing long-term prophylaxis.
- Lithium is the only true mood stabilizer in the management of bipolar disorder.
- Lithium has profound antisuicidal properties in bipolar disorder.
- The mechanisms of the neuroprotective properties of lithium are not fully understood; however, lithium appears to preserve grey matter volume in patients with bipolar disorder.
- Lithium possesses a complex set of actions involving neurotransmission and cellular signalling pathways that translate to its many clinical effects involving mood and cognition.
- In the presence of bipolar disorder, lithium possibly preserves cognitive functioning.

12.1 Introduction

Bipolar disorders are characterized by recurrences of mood episodes swinging from mania (or hypomania) to depression. Beyond the classical subcategories defined in current classification – bipolar I, II and NOS (not otherwise specified) – there is a high heterogeneity in clinical presentation when considering predominant polarity, psychiatric and somatic comorbidities and quality of functioning during inter-crisis periods. Treatment of such a complex disorder needs active intervention for manic and depressive episodes but also requires a preventive effect.

As described in Chap. 10, lithium was first used to treat patients with gout, since lithium had been shown to dissolve uric acid crystals (Schou et al. 1954). Because of the cyclicity of the painful periods in gout, some theories originating in the eighteenth century linked gout and recurrent psychiatric disorders. But it was not until the middle of the twentieth century, after observing that lithium salt acted as a tranquilizer in rodents, that Cade decided to use it to treat patients with mania (Cade 1949). However, after this, nearly 20 years passed before the first proof of lithium's prophylactic efficacy in recurrent mood disorders with the work of Baastrup and Schou (1967). Since this observation, lithium has been considered the gold standard for bipolar disorders, but over time its use has declined as newer drugs have been introduced. In the USA, in 1994, 80% of new mood stabilizer prescriptions in bipolar patients were for lithium, but in 2002 lithium was only prescribed as the initial psychotropic monotherapy in 7.5 % of cases – much less than the prescription of anticonvulsants (17.1 %, with 8.3 % of this being for divalproex sodium), antipsychotic medications (10.7 %) and antidepressants (49.8 %) (Baldessarini et al. 2007).

This chapter offers a selective review of studies that provide insights into the usefulness of lithium in acute episodes and in the maintenance treatment of bipolar disorders (Yatham et al. 2009). The goal of acute treatment is to improve symptoms to the point of remission for the two polarities: depressive and manic. Maintenance treatment refers to post-acute treatment as a preventive treatment to avoid relapse (re-emerging symptoms of the index episode), recurrence (re-emergence of symptoms after recovery) and emergent affective switches defined as an episode of opposite polarity within a continuation phase. Note that we do not distinguish between the terms 'maintenance', 'prophylactic', 'preventive' and 'long-term treatment' when referring to continuation treatment used to protect patients from any episode.

12.2 Lithium Efficacy in the Treatment of Acute Mania

The primary goals of treatment of a manic episode are the rapid control of symptoms such as agitation, impulsivity or dangerous behaviour, to allow a return to appropriate psychosocial functioning and to avoid a depressive switch (Grunze et al. 2010).

To date, about 30 studies have evaluated the acute antimanic efficacy of lithium, starting with those by Shou (Schou et al. 1954). However, here we consider only the most recent studies that meet current methodological standards for randomized clinical trials (RCTs).

The antimanic efficacy of lithium has been tested versus various antipsychotics (chlorpromazine, haldol, olanzapine, risperidone) (Grunze et al. 2003) and anticonvulsants (lamotrigine, carbamazepine, valproate) (Ichim et al. 2000; Placidi et al. 1986; Bowden et al. 1994).

In randomized studies, the response rates given for lithium range from 32 % to 94 %. However, lithium seems less effective than classic antipsychotics in a subgroup of severely agitated patients. Lower efficacy in severe mania was not linked to the presence of psychotic symptoms, but seems due to the severity of the manic episode. Slower onset of action relative to the need for regular plasma level checks to avoid toxicity, and the low level of sedation, could explain the lower efficacy in severe mania. This is why, even though most guidelines recommend monotherapy as the initial treatment for manic phases, in routine practice, monotherapy with classic mood stabilizers (lithium and anticonvulsants) is not sufficiently effective (Malhi et al. 2015). In a recent review, we investigated the combination of classic mood stabilizers with antipsychotics. It appears that combination therapy is superior to monotherapy in terms of efficacy, but side effects are more frequent and discontinuation rates due to adverse events are higher (Geoffroy et al. 2012), raising questions about the continuation of the combination after the manic phase has resolved.

Concerning mixed episodes, associating manic and depressive syndromes, lithium has long been considered as less efficient than anticonvulsants or antipsychotics. In most guidelines differentiating treatment for manic and mixed episodes, lithium is not considered as a first-line monotherapy option, but rather as an add-on treatment (Grunze et Azorin 2014). However, this subgroup represents a generally more difficult phase of the illness to treat with *any* medication. Increasing knowledge about lithium's other beneficial effects, including protection against suicide, should lead us to reconsider this position (Muzina 2009).

12.3 Lithium Efficacy in the Treatment of Bipolar Depression

Although mania is considered to be the hallmark of bipolar disorders, bipolar depression and depressive symptoms lead to a higher burden, as patients spend an average of three times longer with depressive rather than manic symptoms (Judd et al. 2002). Moreover, bipolar depressions are more difficult to treat than unipolar depression as there is poorer response to pharmacological treatments and the risk of treatment-emergent affective switches (Tohen et al. 2009. Beyond major depressive episodes, residual depressive symptoms are extremely common, impair the functioning of individuals (Judd et al. 2002) and are predictors of relapse. The goal of treatment is thus to achieve total remission of depressive episodes, which are also the major risk factor for suicide (Grunze et al. 2010). Another challenge for treating bipolar depression is its clinical heterogeneity. The DSM-5 recognizes depression with mixed characteristics, including manic symptoms (Malhi 2013), which are particularly frequent in bipolar subjects and which present a greater risk of switch with antidepressant treatment (Goldberg et al. 2009). Recommendations for the treatment of bipolar depression report the most conflicting data concerning lithium monotherapy, because there is a very limited number of RCTs versus placebo, and those that have been done are methodologically questionable (Grunze et al. 2003). Among the most recent and robust trials, those comparing the efficacy and tolerability of quetiapine or lithium versus placebo showed that lithium did not significantly differ from placebo on the main measures of efficacy (Young et al. 2010). As a consequence, the inclusion of this negative study ranges lithium to a low level of evidence in the recommendation of the World Federation of Societies of Biological Psychiatry (WFSBP) (Grunze et al. 2010), whilst the Canadian Network for Mood and Anxiety Treatments (CANMAT), which ignores negative studies, ranges lithium as a first-line treatment (2009).

More generally, in current international guidelines, there is a shift from traditional mood stabilizers (lithium and valproate) to atypical antipsychotics and a clear regression of antidepressants in the recommendation for treatment of bipolar depression, even if antidepressants are commonly used in day-to-day practice (Malhi et al. 2015).

Even if some recommendations contest lithium as a first-line treatment for this indication, the rule remains to optimize the dosage when the patient is already on lithium. However, a review (Malhi & Tanious 2011) suggests that a low serum range of lithium (0.4–0.8 mmol/L) would be more effective in the treatment of bipolar depression.

Very little evidence exists of effective strategies for patients who do not respond to first-line treatment to lithium. Recently, a study has investigated the efficacy of lurasidone, a novel antipsychotic agent, as adjunctive therapy with lithium or valproate for the treatment of patients with bipolar I depression who have had an insufficient response to monotherapy. The results showed that treatment with lurasidone adjunctive to lithium or valproate significantly improved depressive symptoms and was generally well tolerated (Loebel et al. 2014). Lurasidone was approved for the treatment of bipolar I depression as a monotherapy or as an adjunct to lithium or divalproex by the US Food and Drug Administration. Evidence also supports efficacy for lamotrigine associated with lithium and suggests efficacy for ketamine for a short period of time (McIntyre et al. 2013).

12.4 Maintenance Treatment of Bipolar Disorders

In 1967, Baastrup and Schou's study (1967) reported a significant decrease in the frequency of episodes in patients with bipolar and unipolar disorders receiving lithium treatment. A subsequent number of placebo-controlled clinical trials (RCTs) followed this first observation, establishing the long-term efficacy of lithium in bipolar disorders. Some of these early clinical trials do not fulfil today's methodological criteria and may lead to an overestimation of the efficacy of lithium. However, a systematic review and meta-analysis of randomized controlled trials comparing lithium with placebo in the long-term treatment of bipolar disorders, which included five randomized controlled trials (770 participants), has reported that lithium was more effective than placebo in preventing all relapses (random effects relative risk = 0.65, 95 % confidence interval (CI) = 0.50–0.84) (Geddes et al. 2004). Manic relapses were reduced by 38 % (relative risk = 0.62, 95 % CI = 0.40–0.95) and depressive relapses by 28 % (relative risk = 0.72, 95 % CI = 0.49–1.07). On the other hand, the BALANCE trial showed that lithium was better than valproate in the prevention of mood episodes in patients followed up for up to 24 months (Geddes et al. 2010).

Other results come from continuation trials for antipsychotics, where lithium is used as comparator. Thus a continuation of quetiapine versus switching to placebo or lithium for maintenance treatment of bipolar I disorder showed that time to recurrence was significantly longer for patients receiving quetiapine or lithium versus placebo (Weisler et al. 2011).

Beyond its efficacy, lithium is also associated with other potential benefits, including a decreased rate of suicide and a neuroprotective effect. Recently, a network meta-analysis, including 48 randomized controlled trials (6,674 participants, 15 comparisons), has demonstrated that lithium was more effective than placebo in reducing the number of suicides (odds ratio (OR) = 0.13, 95 % CI = 0.03–0.66) and deaths from any cause (Cipriani et al. 2013). No clear benefits were observed for lithium compared with placebo in preventing deliberate self-harm (Cipriani et al. 2013). Despite its efficacy, some subtypes of patients could be less responsive to lithium, such as patients without a positive family history of bipolar disorder in first-degree relatives, those with rapid cycling and those with a pattern of illness with depression followed by mania (as opposed to patients presenting with manic episodes followed by depression) (Severus et al. 2014; Malhi and Geddes 2014).

There are also two forms of non-response more specifically linked to lithium that may develop over the course of treatment: discontinuation-related refractoriness and tolerance to long-term treatment by lithium (Kendall et al. 2014). A recent study by Post and colleagues (2012) reviewed this phenomenon and indicates that refractoriness may emerge in some 10–12 % of patients (Post 2012). However, the main problem is a high level of relapses during discontinuation. In a study by Suppes et al. (1991), 50 % of patients relapsed within the first 5 months off treatment and 85–90 % within the first year (Suppes et al. 1991).

Development of tolerance is defined by the recurrence of thymic episodes with an increased severity, higher frequency or longer duration after an extended period of excellent responsiveness and despite the same lithium dose (Kendall et al. 2014). In various studies, the rate of tolerance ranges from 9 % (Koukopoulos et al. 1995) to 34 % (Post et al. 1993), but it is difficult to take into account all the potential confounding factors.

Another crucial point that may explain effectiveness is the timing of onset of treatment. It seems that early initiation of lithium may increase the chance of a good response. Thus, a review (Kleindienst et al. 2005) found that a higher number of prior admissions to hospital are related to poor response to lithium. Another recent Danish study, using national registers, clearly suggests that early prophylactic intervention with lithium following first hospitalization, or the first manic/mixed episode, is associated with improved long-term response to lithium (Kessing et al. 2014).

12.5 The Efficacy–Effectiveness Gap

The above review highlights that, despite the availability of an increased number of pharmacological agents for treating patients with bipolar disorders, lithium remains a major treatment (Licht 2012). However, the acute response, remission and reductions in relapse rates achieved in clinical practice often fall below those reported under controlled conditions in research trials (Tacchi and Scott 2005). Non-adherence rates for long-term prophylaxis in bipolar disorders range from 20 % to 66 %, with a median prevalence of about 40 %. About 30 % of individuals receiving prophylactic lithium were non-adherent within 6 months (Scott and Pope 2002).

The consequences of non-adherence are significant. For example, a retrospective study by Svarstad et al. (2001) reported that 33 % of individuals with bipolar disorders were irregular users of medications; these individuals were more likely to be hospitalized and to have longer admissions than were regular medication users, leading to highly significant differences in treatment costs ($9,701 for irregular users compared with $1,657 for regular users). There are relatively few prospective studies of lithium adherence, but Scott and Pope (2002) explored the relationship between lithium adherence, plasma levels and psychiatric hospitalizations in 98 individuals. Over 18 months, 27 out of 92 individuals experienced one or more admissions; however, the cumulative probability of admission was about 80 % in individuals with partial adherence and subtherapeutic plasma levels, compared with those reporting high adherence and therapeutic plasma levels (about 10 %). There are no definitive data showing any association between adherence rates and mortality, but Muller-Oerlinghausen and colleagues (2005) have published a series of papers that show that adequate long-term lithium treatment significantly reduced, and even normalized, the excess mortality rates found in patients with mood disorders.

Given that the gap between research efficacy and clinical effectiveness is largely explained by problems with adherence and that the consequences are potentially severe, it is important to understand the underlying reasons. Early research often identified associations such as being in the first year of lithium treatment, having a past history of non-adherence, being younger and/or male, complaints of 'missing highs', etc. (Duffy et al. 2014). However, these studies were atheoretical and took a rather simplistic view of the association between adherence and individual characteristics. Importantly, they failed to take into account that adherence status is not 'fixed': although it may vary between individuals, it may also change within individuals over time. As a consequence, the findings are unreliable and often contradictory. Recent research has taken a more sophisticated approach, attempting to use health-beliefs models to identify the underlying reasons for lithium non-adherence and to explore what combination of variables explains an individual patient's 'adherence behaviour'.

12.6 Health Beliefs and Lithium Adherence

The use of a health-beliefs model (which is commonly employed in the psychological management of chronic physical disorders such as diabetes) allows a clinician to explore patient's beliefs about bipolar disorder and their attitudes towards lithium treatment. Tacchi and Scott (2005) advocate the use of a modified version of the 'cognitive representation of illness model', which describes (a) how an individual constructs an internal representation of what is happening to them when they experience physical or mental symptoms and (b) how they react to this. It is an example of a self-regulatory model with three core elements as follows:

1. A cognitive representation, which reflects the meaning of the health 'threat' to the individual. This can be activated by internal cues (e.g. prodromal symptoms) or external cues (e.g. information in the media).
2. A coping response that is developed and instigated by the individual to deal with the threat.
3. The individual's appraisal of the outcome of the coping strategy and their decision to continue with this approach or change their coping response.

The elements of this model can be drawn as a flow diagram that shows how all these factors may be associated with adherence behaviour (Fig. 12.1). It is suggested that, no matter what the nature of the symptoms, most individuals organize their thinking about any health threat around five key themes (Scott and Tacchi 2002). These are as follows: What is it (identity)? Why has it happened (cause)? How long will it last and/or will I recover and then will it recur (timeline)? What

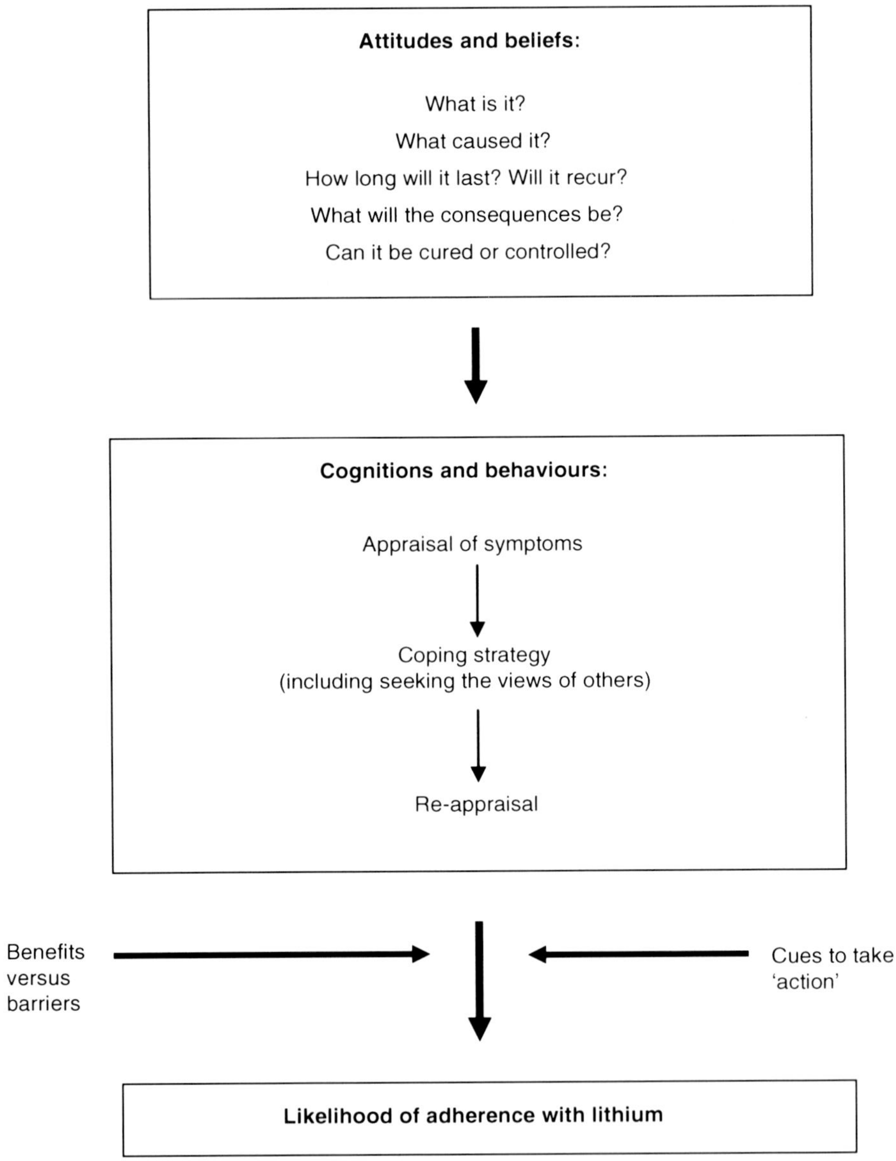

Fig. 12.1 Flow diagram of factors influencing the likelihood of adherence

effects will it have (consequences)? What can I do to make it go away (cure/control)? These five elements of the cognitive representation are stable structures in the model, but the content of each answer may be highly individualized and can be influenced by past experiences or the views of significant others.

The second component of the model suggests that, if symptoms occur, individuals will make some attempt to cope with them. Crucially, their choice of a particular coping strategy to cure or control the problem (e.g. taking lithium or using alcohol to contain the symptoms) will be influenced by whether that seems to be a logical step given their views (about the identity, cause, timeline, controllability and consequences of their symptoms) (Colom et al. 2005).

The individual will next appraise their coping strategy and come to a decision about how effective it has been. They will then continue to use this strategy or modify it accordingly. So a patient may refuse medication initially but later accept the need for lithium (or vice versa). Finally, it should be noted that individuals who perceive 'coherence' between (i) their experience of symptoms, (ii) the meaning they give to the symptoms (their interpretations) and (iii) the explanation offered to them by a significant other (which might be a clinician or a family member) are more likely to adhere to medication.

When reviewing the identity and consequences of the disorder, individuals will appraise the threat posed by their symptoms in terms of the severity of the illness and their susceptibility to relapses. If they see the threat as being real, they may be more inclined to accept the necessity of taking a mood stabilizer (see Clatworthy et al. 2009). Similarly, when considering whether the problem can be cured or controlled, adherence with lithium will be influenced by any concerns the individual has regarding taking medication (e.g. fear of 'addiction'). Clatworthy et al. (2009) demonstrated that 30 % of 223 individuals with bipolar disorders had low levels of adherence and that adherence status was predicted by lower necessity-for-treatment beliefs (OR = 0.50; 95 % CI = 0.31–0.82) and more concerns about potential adverse effects (OR = 2.00; 95 % CI: 1.20–3.34). These predictors were independent of current mood state, illness and demographic characteristics, and, very importantly, it was noted that concerns about, rather than actual experience of, adverse events predicted lower levels of adherence. In addition, as an individual's mental state improves, their beliefs about the necessity of taking lithium may diminish, whilst any concerns remain constant (or sometimes become more prominent in their thinking) – explaining why adherence status often changes over time.

12.7 Conclusions

Lithium remains an important treatment option for the acute and maintenance phases of bipolar disorder, but to optimize its benefits, clinicians need a clear framework for assessing and managing the risk of partial or total non-adherence in the short and long term. The relationship between necessity beliefs, concerns and adherence offers a target for clinical interventions to help patients to resolve their ambivalence about taking lithium. However, clinicians need a greater understanding of the origins of necessity beliefs and concerns about lithium. For example, research suggests that patients' concerns may arise from 'social representations' (beliefs that are common among the general public about the dangers of treatments), so there is a need for further publicity about the benefits of lithium. Also, it appears that adherence behaviour is undermined by anticipated, rather than actual, experiences of adverse effects, so it is critical to teach patients how to manage specific side effects (and/or understand their significance) and not just to provide information about which side effects may occur. This helps to explain why basic education is less successful than more sophisticated psychoeducation or behavioural strategies in improving adherence (Tacchi and Scott 2005).

12.8 Summary

Following the first efficacy trial of lithium prophylaxis in recurrent mood disorders in 1967, lithium became the gold standard for bipolar disorders, but its use has declined over time as newer drugs have been labelled as mood stabilizers. This chapter has offered a selective review of studies, focusing on the usefulness of lithium in acute episodes of bipolar disorders and in maintenance treatment. The efficacy of lithium in acute episodes and as a prophylactic treatment has been tested against various antipsychotics in comparative, high-quality clinical trials. These trials have demonstrated the efficacy of lithium and confirmed that it is still the key mood stabilizer. Critically, we have considered the prevalence of partial or non-adherence with lithium and the reasons for individuals discontinuing a potentially beneficial treatment. We have described how using health beliefs and exploring patients' views about the necessity for treatment, versus their concerns about medication, can predict and enhance adherence and offer a framework for intervention.

References

Baastrup PC, Schou M (1967) Lithium as a prophylactic agent. Its effect against recurrent depressions and manic-depressive psychosis. Arch Gen Psychiatry 16(2):162–172

Baldessarini RJ, Leahy L, Arcona S, Gause D, Zhang W, Hennen J (2007) Patterns of psychotropic drug prescription for U.S. patients with diagnoses of bipolar disorders. Psychiatr Serv 58(1):85–91

Bowden CL, Brugger AM, Swann AC, Calabrese JR, Janicak PG, Petty F, Dilsaver SC, Davis JM, Rush AJ, Small JG et al (1994) Efficacy of divalproex vs lithium and placebo in the treatment of mania. The Depakote Mania Study Group. JAMA 271(12):918–924

Cade JF (1949) Lithium salts in the treatment of psychotic excitement. Med J Aust 36(2):349–352

Cipriani A, Hawton K, Stockton S, Geddes JR (2013) Lithium in the prevention of suicide in mood disorders: updated systematic review and meta-analysis. BMJ 346:f3646

Clatworthy J, Bowskill R, Parham R, Rank T, Scott J, Horne R (2009) Understanding medication non-adherence in bipolar disorders using a necessity-concerns framework. J Affect Disord 116(1–2):51–55

Colom F, Vieta E, Sánchez-Moreno J, Martínez-Arán A, Reinares M, Goikolea JM, Scott J (2005) Stabilizing the stabilizer: group psychoeducation enhances the stability of serum lithium levels. Bipolar Disord 7(Suppl 5):32–36

Duffy A, Horrocks J, Doucette S, Keown-Stoneman C, McCloskey S, Grof P (2014) The developmental trajectory of bipolar disorder. Br J Psychiatry 204(2):122–128. doi:10.1192/bjp.bp.113.126706, Epub 2013 Nov 21

Geddes JR, Burgess S, Hawton K, Jamison K, Goodwin GM (2004) Long-term lithium therapy for bipolar disorder: systematic review and meta-analysis of randomized controlled trials. Am J Psychiatry 161(2):217–222

Geddes JR, Goodwin GM, Rendell J, Azorin JM, Cipriani A, Ostacher MJ, Morriss R, Alder N, Juszczak E, BALANCE investigators and collaborators (2010) Lithium plus valproate combination therapy versus monotherapy for relapse prevention in bipolar I disorder (BALANCE): a randomised open-label trial. Lancet 375(9712):385–395

Geoffroy PA, Etain B, Henry C, Bellivier F (2012) Combination therapy for manic phases: a critical review of a common practice. CNS Neurosci Ther 18(12):957–964

Goldberg JF, Perlis RH, Bowden CL, Thase ME, Miklowitz DJ, Marangell LB, Calabrese JR, Nierenberg AA, Sachs GS (2009) Manic symptoms during depressive episodes in 1,380 patients with bipolar disorder: findings from the STEP-BD. Am J Psychiatry 166(2):173–181

Grunze H, Azorin JM (2014) Clinical decision making in the treatment of mixed states. World J Biol Psychiatry 15(5):355–368

Grunze H, Kasper S, Goodwin G, Bowden C, Baldwin D, Licht RW, Vieta E, Möller HJ, WFSBP Task Force on Treatment Guidelines for Bipolar Disorders (2003) The World Federation of Societies of Biological Psychiatry (WFSBP) guidelines for the biological treatment of bipolar disorders, part II: treatment of mania. World J Biol Psychiatry 4(1):5–13, Review

Grunze H, Vieta E, Goodwin GM, Bowden C, Licht RW, Möller HJ, Kasper S, WFSBP Task Force On Treatment Guidelines For Bipolar Disorders (2010) The World Federation of Societies of Biological Psychiatry (WFSBP) guidelines for the biological treatment of bipolar disorders: update 2010 on the treatment of acute bipolar depression. World J Biol Psychiatry 11(2):81–109

Ichim L, Berk M, Brook S (2000) Lamotrigine compared with lithium in mania: a double-blind randomized controlled trial. Ann Clin Psychiatry 12(1):5–10

Judd LL, Akiskal HS, Schettler PJ, Endicott J, Maser J, Solomon DA, Leon AC, Rice JA, Keller MB (2002) The long-term natural history of the weekly symptomatic status of bipolar I disorder. Arch Gen Psychiatry 59(6):530–537

Kendall T, Morriss R, Mayo-Wilson E, Marcus E, Guideline Development Group (2014) Assessment and management of bipolar disorder: summary of updated NICE guidance. BMJ 349:g5673. doi:10.1136/bmj.g5673

Kessing LV, Vradi E, Andersen PK (2014) Starting lithium prophylaxis early v. late in bipolar disorder. Br J Psychiatry 205(3):214–220

Kleindienst N, Engel R, Greil W (2005) Which clinical factors predict response to prophylactic lithium? A systematic review for bipolar disorders. Bipolar Disord 7(5):404–417, Review

Koukopoulos A, Reginaldi D, Minnai G, Serra G, Pani L, Johnson FN (1995) The long term prophylaxis of affective disorders. Adv Biochem Psychopharmacol 49:127–147

Licht RW (2012) Lithium: still a major option in the management of bipolar disorder. CNS Neurosci Ther 18(3):219–226. doi:10.1111/j.1755-5949.2011.00260.x, Epub 2011 Jun 23

Loebel A, Cucchiaro J, Silva R, Kroger H, Hsu J, Sarma K, Sachs G (2014) Lurasidone monotherapy in the treatment of bipolar I depression: a randomized, double-blind, placebo-controlled study. Am J Psychiatry 171(2):160–168

Malhi GS (2013) Diagnosis of bipolar disorder: who is in a mixed state? Lancet 381(9878):1599–1600. doi:10.1016/S0140-6736(13)60893-4

Malhi GS, Geddes JR (2014) Carving bipolarity using lithium sword. Br J Psychiatry 205(5):337–339. doi:10.1192/bjp.bp.114.148072

Malhi GS, Tanious M (2011) Optimal frequency of lithium administration in the treatment of bipolar disorder: clinical and dosing considerations. CNS Drugs 25(4):289–298

Malhi GS, Bassett D, Boyce P, Bryant R, Fitzgerald PB, Fritz K, Hopwood M, Lyndon B, Mulder R, Murray G, Porter R (2015) Singh AB Royal Australian and New Zealand College of Psychiatrists clinical practice guidelines for mood disorders. Aust N Z J Psychiatry 49:1087–1206

McIntyre RS, Cha DS, Kim RD, Mansur RB (2013) A review of FDA-approved treatment options in bipolar depression. CNS Spectr 18(Suppl 1):4–20

Müller-Oerlinghausen B, Felber W, Berghöfer A, Lauterbach E, Ahrens B (2005) The impact of lithium long-term medication on suicidal behavior and mortality of bipolar patients. Arch Suicide Res 9(3):307–319

Muzina DJ (2009) Pharmacologic treatment of rapid cycling and mixed states in bipolar disorder: an argument for the use of lithium. Bipolar Disord 11(Suppl 2):84–91

Placidi GF, Lenzi A, Lazzerini F, Cassano GB, Akiskal HS (1986) The comparative efficacy and safety of carbamazepine versus lithium: a randomized, double-blind 3-year trial in 83 patients. J Clin Psychiatry 47(10):490–494

Post RM (2012) Acquired lithium resistance revisited: discontinuation-induced refractoriness versus tolerance. J Affect Disord 140(1):6–13. doi:10.1016/j.jad.2011.09.021, Epub 2011 Dec 7

Post RM, Ketter TA, Pazzaglia PJ, George MS, Marangell L, Denicoff K (1993) New developments in the use of anticonvulsants as mood stabilizers. Neuropsychobiology 27(3):132–137

Schou M, Juel-Nielsen N, Stromgren E, Voldby H (1954) The treatment of manic psychoses by the administration of lithium salts. J Neurol Neurosurg Psychiatry 17(4):250–260

Scott J, Pope M (2002) Self-reported adherence to treatment with mood stabilizers, plasma levels, and psychiatric hospitalization. Am J Psychiatry 159(11):1927–1929

Scott J, Tacchi MJ (2002) A pilot study of concordance therapy for individuals with bipolar disorders who are non-adherent with lithium prophylaxis. Bipolar Disord 4(6):386–392

Severus E, Taylor MJ, Sauer C, Pfennig A, Ritter P, Bauer M, Geddes JR (2014) Lithium for prevention of mood episodes in bipolar disorders: systematic review and meta-analysis. Int J Bipolar Disord 2:15. doi:10.1186/s40345-014-0015-8, eCollection 2014

Suppes T, Baldessarini RJ, Faedda GL, Tohen M (1991) Risk of recurrence following discontinuation of lithium treatment in bipolar disorder. Arch Gen Psychiatry 48(12):1082–1088

Svarstad BL, Shireman TI, Sweeney JK (2001) Using drug claims data to assess the relationship of medication adherence with hospitalization and costs. Psychiatr Serv 52(6):805–811

Tacchi MJ, Scott J (2005) Improving medication adherence in schizophrenia and bipolar disorders. Wiley, London, pp 16–22

Tohen M, Frank E, Bowden CL et al (2009) The International Society for Bipolar Disorders (ISBD) task force report on the nomenclature of course and outcome in bipolar disorders. Bipolar Disord 11(5):453–473

Weisler RH, Nolen WA, Neijber A, Hellqvist A, Paulsson B, Trial 144 Study Investigators (2011) Continuation of quetiapine versus switching to placebo or lithium for maintenance treatment of bipolar I disorder (trial 144: a randomized controlled study). J Clin Psychiatry 72(11):1452–1464

Yatham LN, Kennedy SH, Schaffer A, Parikh SV, Beaulieu S, O'Donovan C, MacQueen G, McIntyre RS, Sharma V, Ravindran A, Young LT, Young AH, Alda M, Milev R, Vieta E, Calabrese JR, Berk M, Ha K, Kapczinski F (2009) Canadian Network for Mood and Anxiety Treatments (CANMAT) and International Society for Bipolar Disorders (ISBD) collaborative update of CANMAT guidelines for the management of patients with bipolar disorder: update 2009. Bipolar Disord 11(3):225–255

Young AH, McElroy SL, Bauer M, Philips N, Chang W, Olausson B, Paulsson B, Brecher M, EMBOLDEN I (Trial 001) Investigators (2010) A double-blind, placebo-controlled study of quetiapine and lithium monotherapy in adults in the acute phase of bipolar depression (EMBOLDEN I). J Clin Psychiatry 71(2):150–162

Lithium and Suicide Prevention

13

Helen Jones, John Geddes, and Andrea Cipriani

It appears that the introduction of lithium treatment in the 1970s was the most recent important breakthrough in the prevention of suicide in bipolar disorder.

(Latalova et al., 2014, p. 113)

Abstract

Over the past 40 years, a growing evidence base has established an association between lithium treatment and reduced rates of suicide in patients with mood disorders. This chapter examines the evidence surrounding this link, from early observational reports to randomized controlled trials. Methodological challenges in conducting this research are discussed. We also illustrate how the use of meta-analysis can overcome some of these limitations and demonstrate a convincing link between lithium treatment and reduced suicide rates. Finally, possible mechanisms of action by which lithium may offer suicide protective effects are explored, indicating the interesting possibility of an independent suicide-preventative action.

H. Jones • J. Geddes
Psychiatry Department, University of Oxford, Oxford, UK
e-mail: helen.jones@oxfordhealth.nhs.uk; john.geddes@psych.ox.ac.uk

A. Cipriani (✉)
Psychiatry Department, Warneford Hospital, University of Oxford, Warneford Lane, Headington, Oxford OX3 7JX, UK
e-mail: andrea.cipriani@psych.ox.ac.uk

G.S. Malhi et al. (eds.), *The Science and Practice of Lithium Therapy*,
DOI 10.1007/978-3-319-45923-3_13

Key Points

- Suicide is common and the second leading cause of death in young people below the age of 40 years.
- Suicide occurs predominantly in individuals with psychiatric illness.
- Lithium does have anti-suicidal properties, but its mechanism of action remains poorly understood.
- More research is needed into the specific suicide-prevention effects of lithium and how precisely these are achieved.

13.1 Introduction

Every year, nearly one million people die from suicide, including nearly 58,000 within the European Union. Suicide is among the top 20 leading causes of death globally for all ages, and in 15–39-year-olds, suicide is the second leading cause of death after accidents (WHO 2014).

It has been estimated that 90 % of completed suicides occur in people with diagnosable psychiatric disorders (Cavanagh et al. 2003). Bipolar disorder and severe recurrent major depression have the highest rates of suicide of any psychiatric condition (Harris & Barraclough 1997). For patients with affective disorders, the lifetime risk of committing suicide is estimated to be between 6 % and 10 % (Harris & Barraclough 1997) or even up to 19 % (Latalova et al. 2014). This is ten times higher than in the nonpsychiatric population. In an analysis of 28 studies involving 823 suicides among 21,484 patients with bipolar disorder at risk for an average of about 10 years, Tondo and colleagues showed that the pooled, weighted mean annual incidence of suicide was 0.39 % (390/100,000), approximately 28 times higher than the international rate for suicide in the general population (which is about 14/100,000/year, i.e. 0.014 %) (Tondo et al. 2003). The higher risk of suicide in patients with bipolar disorder is also supported by a review of the medical records of 2,826 patients with any major mood disorder according to DSM-IV, recruited in Sardinia (Italy) (Tondo and Baldessarini 2007). The authors found on average a greater risk of suicide among patients with bipolar rather than unipolar disorder. Among the 843 bipolar patients, the annual risk of suicide was 150/100,000, more than 15 times higher than in the general population and three times greater than in patients with recurrent major depressive disorder.

Suicide attempts, defined to be any "intentional act of self-inflicted poisoning, injury or self-harm which may or may not have a fatal intent or outcome" (WHO 2014 pg. 12), are much more difficult to quantify than completed suicides. Rates of suicide attempts are however also known to be higher among people diagnosed with bipolar disorder than in the general population. It has been estimated that 25–50 % of bipolar disorder sufferers will at some point make an attempt on their lives (Latalova et al. 2014). Lethality of suicide attempts varies by a number of factors, for instance, age—the ratio of suicide attempts to suicide death in youth is estimated to be approximately 25:1, compared to about 4:1 in the elderly (www.afsp.org/understanding-suicide/facts-and-figures). A history of attempted suicide is reported to be the biggest predictor of later completing suicide (WHO 2014). Moreover, lethality is suggested to be increased in some psychiatric populations; for example, in a study by Baldessarini and colleagues published in 2006, the ratio of attempted suicides versus completed suicides (an index often used to assess lethality of suicidal behaviour) was 8.6 to 1 in bipolar patients and 9.6 to 1 in patients with major depressive disorder (Baldessarini et al. 2006), which, if compared to an estimated risk ratio of 20–30 to 1 in the general population (Kessler et al. 2005), indicates higher lethality of suicide attempts among mood disorder patients. This suggests that patients with bipolar disorder are both more likely to attempt suicide and more likely to complete. Bipolar sufferers who are male, or who have a first-degree family member who has completed suicide, are thought to be at increased risk of suicide (Schaffer et al. 2015). The rate of completed suicides has been reported to be as high as 26 % in men admitted to psychiatric hospital with bipolar disorder and a history of deliberate self-harm (Nordentoft et al. 2011).

Notwithstanding the known risk of suicide among those suffering from mental health problems, a recent retrospective study reported that 60 % of suicide completers were not receiving mental health treatment at the time of their suicide, despite a previous psychiatric diagnosis and history of suicide attempts (Bakst et al. 2014). Such findings demonstrate a real need to increase the awareness of risk profiles for suicide and to use effective treatments to reduce suicide rates.

Suicide and suicidal behaviour have serious social and economic implications. Reducing suicidal ideation and activity is an essential starting point in the recovery of people suffering from affective disorders. The prevention of suicide and treatment of suicidal behaviour in patients with mood disorders has therefore become the focus of therapeutic activities at the individual and the social level. Establishing effective treatments that can reduce the risk of suicide (or possibly prevent it) is nowadays one of the main priorities for the health system (http://www.nhs.uk/Conditions/Suicide/Pages/Prevention.aspx) and for society in general (http://www.nrls.npsa.nhs.uk/resources/?entryid45=65297). There are, however, currently very few treatments that effectively prevent suicide, and the treatment of patients with suicidal behaviour is one of the most challenging tasks for healthcare professionals. The high mortality, morbidity and costs related to attempted suicide mean that the development of treatment and prevention strategies for suicidal behaviour has been the focus for much of the research on suicidality. In their recent report on suicide prevention, the World Health Organization (WHO) has urgently called for more research into effective therapeutic strategies for treating suicidal behaviour (WHO 2014).

Medication currently plays a relatively minor role in most suicide-prevention strategies (Department of Health 2002; US Department of Health Human Services 2001), although its place may have been underestimated (Saunders and Hawton 2009). Interestingly, however, since the early 1970s, a series of international studies (mainly observational) have suggested the anti-suicidal effect of lithium.

13.2 Anti-suicidal Effects of Lithium: From Observational Data to Randomized Evidence

One of the earliest suggestions of lithium's effect on suicidal behaviour was reported in 1972 (Barraclough 1972). In a subsequent report of his retrospective investigation of 100 suicide cases, Barraclough showed that a high proportion of the suicides had occurred in patients with long histories of treatment for psychiatric illness and that 40 % of the sample had previous episodes of affective disorder. Reviewing these affective disorder cases individually and looking at the chronology of illness and treatment regimes, he suggested that 25 % of these cases had been prescribed ineffective treatments at the time of their death. He reported that lithium treatment may have prevented relapse in at least half of these cases and concluded that, with the use of lithium, one fifth of the 100 cases of suicide may have been prevented (Barraclough et al. 1974).

More recent examples of observational reports of lithium's anti-suicidal effects come from a study based on data collected by a health maintenance organisation in the USA, using a sample of 20,638 patients with diagnosed bipolar disorder. The report identified a 2.7 times greater risk of suicide for patients prescribed valproate compared to those prescribed lithium (Goodwin et al. 2003). A similar study by Collins and McFarland (2008) found comparable results in a sample of 12,662 patients with a diagnosis of bipolar disorder. By the end of the study, there had been 11 completed suicides and 79 reported suicide attempts. Results suggested lithium to be superior in reducing rates of both suicide attempts and completed suicides, in comparison to the other anticonvulsants studied (divalproex, gabapentin and carbamazepine) (Collins and McFarland 2008). Consistent findings continue to be reported in the literature; for instance, in a 2014 report of post-hospitalization suicide rates in Finland, Isometsä and colleagues (2014) reported that of three types of pharmacotherapy investigated (lithium, valproate and antidepressants), only lithium had a significant effect in reducing the risk of suicide. Also, in 2015, similar findings were reported in a study following, for 3.5 years, a naturalist cohort of 826 bipolar patients who had previously been hospitalized as a result of a suicide attempt (Toffol et al. 2015). The study authors explored the effect of treatment with a variety of drugs on suicide and all-cause mortality in the sample. Results suggested that several medications were associated with an increase in suicidal attempts: valproic acid (relative risk (RR) = 1.53, 95 % confidence interval (CI) 1.26–1.85, $p=0.001$), antidepressants (RR 1.49, 95 % CI 1.23–1.80, $p=0.001$) and benzodiazepines (RR 1.65, 95 % CI 1.37–1.99, $p<0.001$). Lithium, in contrast, was associated with fewer suicide attempts, although this difference was not significant (RR 0.82, 95 % CI 0.65–1.02, $p=0.09$). Lithium treatment was however, associated with significantly reduced numbers of completed suicides (RR 0.39, 95 % CI 0.17–0.93, $p=0.03$) and decreased all-cause mortality by 49 %. The authors concluded that, among such high-risk bipolar patients, lithium should be considered a suitable treatment to help to prevent suicide.

Trying to make use of all available evidence, meta-analyses have been carried out to combine data and provide pooled estimates from studies from a range of different sources. Baldessarini and Tondo in 2006 analysed data from 31 studies (eight of which were randomized controlled trials) of participants with bipolar disorder and other major mood disorders and found suicide-preventative effects of lithium (Baldessarini et al. 2006). This meta-analysis reported a fivefold reduction in suicidal acts for participants treated with lithium in comparison to those without lithium treatment (RR 4.91, 95 % CI 3.82–6.31, $p<0.0001$) and a risk reduction of suicide by approximately 80 % following treatment with lithium. Subsequent studies by the same team of authors found less positive, but still clearly favourable, results for lithium. In a second meta-analysis of six studies, including more than 30,000 patients who were treated with lithium or with an anticonvulsant, Baldessarini and colleagues observed a nearly threefold superiority of lithium over other mood-stabilizing anticonvulsants (carbamazepine, divalproex or lamotrigine) in terms of reducing suicidal behaviour (RR = 2.86, CI: 2.29–3.57, $p<0.0001$) (Baldessarini et al., 2009).

Although these studies provide a very positive indication of a link between lithium treatment and reduced rates of suicide, their findings lack proper scientific rigour due to their observational nature. Nonrandomized studies, especially those based on lithium clinic samples, are likely to produce biassed findings. Any beneficial effect might reflect compliance with medication being greater in patients attending a clinic. There may also be selective prescribing to patients at risk. Furthermore, deaths from other causes are often not considered in such studies. Theoretically, given the potential physical side effects of lithium on thyroid and renal function, it is possible that a benefit of lithium in preventing suicide might be offset by an increased risk of death from physical disorders. More rigorous research methods are thus needed to provide more reliable data about lithium's preventative effect on suicide.

Randomized controlled trials (RCTs) represent the gold standard in research methodology. Just 1 year after Barraclough's initial observational report, two of the first RCTs investigating the anti-suicidal efficacy of lithium were published in the scientific literature (Prien et al. 1973a, b). In the first of these reports, an RCT with a sample of patients with bipolar disorder ($n=44$) and unipolar depression ($n=78$), participants were randomly assigned treatment with imipramine, lithium or placebo (Prien et al. 1973a). One suicide was reported in the placebo group. In the second report of a placebo-controlled, randomized control trial, allocating patients with bipolar disorder ($n=208$) treatment with either lithium or placebo, the author observed no completed suicides in the lithium-treatment group, in comparison to two completed suicides in the placebo group (Prien et al. 1973b).

The authors concluded that there was a suicide-preventative effect of lithium treatment.

During the 1990s, Greil and colleagues contributed three further RCTs to the literature surrounding lithium's anti-suicidal effects (1996, 1997a, b). All three of these trials involved a relatively long involvement period, following patients for 2.5 years, as well as having reasonably large sample sizes (81, 144 and 90 participants, respectively). Moreover, the open-label design of these trials allows for a more pragmatic result, producing findings that are potentially more relevant to everyday clinical practice. In the first trial, Greil and colleagues compared lithium treatment to amitriptyline in a sample of 81 participants with unipolar depression. One suicide was reported in the amitriptyline group, in contrast to no completed suicides in the lithium group (Greil et al. 1996). Similar effects were reported in a paper in 1997, comparing lithium treatment to carbamazepine, in patients with bipolar disorder ($n = 144$) (Greil et al. 1997a). Again, no suicides were observed in the lithium group, in comparison to one completed and one attempted suicide in the carbamazepine group. In a separate study comparing the same treatments (lithium and carbamazepine), but with a sample of participants with schizoaffective disorder ($n = 90$), the same team reported no suicides in the lithium group, in comparison to four attempted suicides in the carbamazepine group (Greil et al. 1997b).

RCTs comparing suicide rates following treatment with lithium to other mood stabilizers continued to appear in the literature in the following years. In 2000, Bowen published an RCT comparing lithium treatment to valproate and placebo. In a 1-year follow-up, no completed or attempted suicides were reported in any group. In 2003, two randomized controlled trials of lithium in comparison to lamotrigine were published. In an open-label, randomized, placebo-controlled trial of 463 patients with bipolar I disorder, Calabrese and colleagues (2003) compared treatment with lamotrigine, lithium or placebo in monotherapy for a period of 18 months. Here one death was seen in the lamotrigine group, in comparison to no deaths in the placebo group or in the lithium treated group. Similarly, in a double-blind placebo-controlled trial of 175 patients allocated to treatment with lamotrigine, lithium or placebo, the same authors again suggested a suicide-preventative effect of lithium treatment, reporting one suicide attempt in the lamotrigine group, versus none in the lithium group. Additionally, more evidence of suicidal ideation (reported on the Hamilton Depression Rating Scale) was observed in the lamotrigine group in comparison to the lithium group (Bowden et al. 2003).

More recently, in a study of relapse prevention in bipolar disorder, Geddes and colleagues added to these findings by looking at the benefits of using a combination treatment of both lithium and valproate. In this open-label randomized trial, 330 patients were randomly allocated to treatment with lithium, or valproate in monotherapy, or a combination of both lithium and valproate. Participants were followed for a total of 24 months. Findings from this study suggested that using the treatments in combination therapy resulted in the best prophylactic effects. There were two patient deaths from all causes in a lithium-treated group, whilst there were three patient deaths in the valproate group (Geddes et al. 2010).

All these studies suggested a possible effect of lithium in terms of suicide prevention; however, considering that suicide is a rare event, individual studies did not have enough power to assess whether this association was statistically significant. Pooling data through the use of meta-analyses allows for limitations such as low event rates to be minimized. In 2005, the first systematic review investigating evidence on possible anti-suicidal effects of lithium, which included only randomized studies (32 RCTs), was added to the literature (Cipriani et al. 2005). This review combined results from RCTs lasting at least 3 months, in patients with mood disorders, in which there was comparison of lithium versus placebo or versus any active drugs. These included 1,377 patients randomized to lithium and 2,052 to other compounds. Nineteen trials compared lithium with placebo, nine with carbamazepine, three with lamotrigine and one with divalproex. The other trials were comparisons with antidepressants, namely, imipramine, amitriptyline, maprotiline and mianserin (multiple comparisons were included in some trials). Seven trials reported suicide as an outcome and an event occurred in at least one arm of the trial. In these trials 2 out of 503 people prescribed lithium died by suicide, compared with 11 out of 601 patients prescribed placebo or comparator drugs (Peto OR 0.26, 95 % CI 0.09–0.77). Thus there was evidence of a significant reduction of suicide in those prescribed lithium (although this finding was associated with significant heterogeneity). Surprisingly few trials included data on deliberate self-harm/attempted suicide. Therefore, results for these outcomes were combined with those for suicide. There was a lowered risk of fatal or nonfatal suicidal behaviour, this occurring in just 6 out of 670 patients in the lithium-treated groups compared with 18 out of 781 in the groups of patients receiving comparator drugs (Peto odds ratio 0.21, 95 % CI 0.08–0.50), although this result was also associated with significant heterogeneity. Finally, deaths from all causes were examined. There was a much-reduced overall risk of death from any cause in patients treated with lithium than in those in the comparison groups (Peto OR 0.42, 95 % CI 0.21–0.87), corresponding to 9/696 versus 22/788 events, respectively. Although this result was again associated with significant heterogeneity, the apparent beneficial effect of lithium on suicide risk was not offset by an increase in deaths from other causes. In post hoc analysis, there was no significant difference for the above outcomes between the results of trials that compared lithium with placebo and those that compared lithium with active drugs. Whilst the effects of lithium on suicidal behaviour were less in this meta-analysis of RCTs than reported in nonrandomized studies, lithium appeared to reduce the risk of suicide by approximately 60 %. The reduction for suicide and deliberate self-harm/attempted suicide combined was approximately 70 %.

Overall, the reduction in deaths from all causes was of the order of 60 %.

The results of such a meta-analysis offered more compelling evidence of an association between lithium treatment and reduced rates of suicide in patients with mood disorders. A methodological problem however remained in that the evidence of lithium's anti-suicidal effect in the RCTs discussed came from incidental findings, where suicide was not an outcome measure in the trial's design (Tondo and Baldessarini 2009). The opportunity to conduct placebo-controlled, randomized control trials to investigate the effects of lithium treatment on suicidality is, however, limited. Clearly, there are ethical implications for designing studies that would offer a placebo arm to participants experiencing suicidal ideation, particularly if one hopes to create a trial investigating suicide as a primary outcome measure. Investigating suicidal behaviours, even as a secondary outcome measure, has been impeded by the fact that suicidal ideations or actions are often exclusion criteria for large-scale clinical trials. Robust research evidence from placebo-controlled RCTs specifically designed to investigate the effects of lithium of suicidality has therefore been more challenging to accumulate than in other treatment areas. For instance, despite the high incidence of suicide and self-harming behaviour among adolescents, there are to date no randomized control trials to investigate the influence of lithium, or in fact any pharmaceutical agents on suicide, in younger populations (Ougrin et al. 2015). There are, however, now a number of RCTs investigating suicide in adults that have overcome these methodological and ethical barriers by comparing lithium to other mood stabilizers, antidepressants and antipsychotic treatments and therefore ensuring that all arms of the study offer treatment to participants. Researchers have called for more trials using this method and suggest that using suicide as a primary outcome measure is ethically feasible if all arms offered to the participants consist of some treatment—for example, antidepressant alone versus antidepressant plus lithium (Guzzetta et al. 2007).

The first of these placebo-controlled RCTs to specifically investigate lithium's relationship with suicidal behaviour was completed by Lauterbach and colleagues in 2008. This study explored the use of lithium as an adjunctive therapy, over a 1-year treatment period, versus placebo, to reduce suicidal behaviour in patients with depressive disorders ($n=170$). The findings of a survival analysis stated that there was no significant difference in suicidal acts, or attempts between the groups, with seven reported attempts in each group (hazard ratio (HR)=0.517; 95 % CI 0.18–1.43). Post hoc analysis, however, revealed that there was a significant difference in incidence rates of completed suicides ($p=0.049$), with three completed suicides in the placebo group, compared to no completed suicides in the lithium-treated group. There were a high number of dropouts, and the total number of suicidal events was just 17, such that these results should be interpreted with caution, but the study does appear to support a role for lithium in protecting those with affective disorders against suicidal acts with a fatal outcome. Khan and colleagues (2011) similarly conducted a randomized controlled trial comparing suicidality in depressed patients, assigning them to 4 weeks of treatment with citalopram and lithium or citalopram and placebo. In this study, a therapeutic dose of lithium (>0.5 mEq/L) was associated with prevention of suicidal thoughts and behaviour.

Oquendo and colleagues (2011) also completed an RCT using this comparator model, with a 2.5-year follow-up trial of patients with bipolar disorder and past history of suicide attempts ($n=98$). Patients were randomly assigned to treatment with lithium or valproate, in a double-blind trial. Over the 2.5-year follow-up, 45 suicide events were reported in 35 participants, including 18 suicide attempts (six suicide attempts in a lithium-treatment group and eight suicide attempts in a valproate group). An intent-to-treat analysis suggested there was no significant difference between treatment groups in terms of suicide event or time to attempt.

An additional methodological difficulty in this field is the fact that suicidal behaviour is statistically rare, making it difficult to achieve power in many of the studies, due to limited sample numbers and short follow-up periods. Many of the papers published in this area report findings comparing only one incidence of suicide in the participant group, compared to no incidences in another. This can create challenges, as small differences in count can have a large impact on the estimated effect size. It has been suggested that this low event rate may be reflective of the fact that suicidal patients are often excluded from participating in trials in the first place (Simon et al. 2006). Although Oquendo and colleagues' (2011) study did not show any significant effects of either treatment on suicidal behaviour, the use of patients with a history of suicide attempts did lead to an increased number of suicide events during the follow-up period than had been seen in previous studies. The authors describe this choice of participants with a higher baseline rate of the outcome of interest as a practical method for overcoming the challenges of studying such a low-frequency event as suicide.

In 2013, Cipriani and colleagues updated their previous 2005 meta-analysis, to allow for the inclusion of this new, more rigorously collected, data on lithium's anti-suicidal effects. This updated systematic review and meta-analysis incorporated all of the RCTs to date, totalling 48 published trials between 1968 and 2013 and analysing data from 6,674 randomized patients (Cipriani et al. 2013). This allowed for the synthesis of 16 more RCTs than were reported in the 2005 review, eight of which contributed new data. The authors concluded that the available randomized evidence suggests that lithium significantly reduces the risk of suicide and overall deaths from any cause, when compared to placebo as a long-term treatment for both unipolar depression and bipolar disorder (OR 0.13, 95 % CI 0.03–0.66). This indication that lithium reduces the risk of suicide and total deaths in people with both unipolar and bipolar depressive disorder was a new contribution to the literature. The effect appeared to be consistently and substantially evidenced in the literature reviewed in the paper. Lithium was not however found to be more effective at reducing suicide, deliberate self-harm or all-cause mortality than other pharmacological treatments. It was reported that lithium was significantly more effective than carbamazepine in reducing deliberate self-harming behaviour (OR 0.14, 95 % CI 0.02–0.83). Conducting a sensitivity analysis, to include only the studies of participants with bipolar disorder ($n = 19$), these results did not differ.

Such a review helps to overcome the issues associated with many of the methodological difficulties associated with research in this field and offers clear evidence of a consistently reported association between lithium treatment and reduced suicidality in people with bipolar disorder. More research into this phenomena is however needed, particularly RCTs specifically designed to investigate suicide and suicide attempts. Whilst the number of studies to review remains low, limitations to small samples are reported even in systematic reviews. The challenge of publication biases limiting such a review is also relevant. Due to a lack of reported suicides in a study being considered to be a favourable outcome, it could be deemed unlikely that publication bias may influence the findings of a review on suicide. However, because of the small sample of studies available, and the small number of events reported within these studies, even one or two unpublished trials demonstrating negative or neutral results may be influential on effect sizes in a review of studies with such low event rates (Cipriani et al. 2005, 2013; Egger and Smith 1998). The possible effect of publication bias on such meta-analyses must therefore be considered.

13.3 Mechanisms of Action

Currently, in spite of the drawbacks of lithium therapy, it appears that it may be the treatment of choice for people with bipolar disorder who are at risk of suicide (Malhi 2015) and may also have role in protecting those with depressive disorders against fatal suicidal acts (Guzzetta et al. 2007). The suicide-preventative action of lithium may be related to its mood-stabilizing properties. There is good evidence that lithium reduces the risk of relapse in patients with mood disorders (Geddes et al. 2004). It is not unreasonable to assume that, given most suicidal attempts occur in periods of mood instability and not in euthymia, if mood can be improved and euthymia prolonged, a reduction in suicide attempts will be seen.

Interestingly, however, accumulating evidence suggests that lithium has an independent suicide-preventative property. This notion has stemmed from research where reductions of suicide risk have been observed following lithium treatment even among patients whose mood symptoms have responded inadequately to lithium (Muller-Oerlinghausen et al. 1992). Some researchers have consequently suggested that lithium may have specific effects against suicide that are independent of its mood-stabilizing actions. Ahrens and Muller-Oerlinghausen (2001), for example, in a study of high-risk patients (mood disorder patients with a history of at least one suicide attempt), reported that suicide attempts were reduced in all participants, regardless of their improvement in affective symptoms. In the report of the findings, a reduction in suicide rates was evident in those with an excellent response to lithium, those with a moderate response and also those observed to have a poor response to lithium treatment in terms of affective symptomatology. They concluded from these findings that the suicide-preventative properties of lithium may be distinct from those that work to stabilize mood.

Such conclusions are also supported by evidence from comparison studies between lithium and other treatments for depression. Despite being superior in its action of preventing suicide to other antidepressant treatments (Kovacsics et al. 2009; Toffol et al. 2015), lithium is reported to have less potent effects on mood in acute phase treatment than other treatments when used in monotherapy (Geddes et al. 2004). This contrast in effect may be taken to suggest that lithium's treatment of affective symptoms is unlikely to be the mechanism through which suicide is prevented during treatment.

These indications of a separate action have led to suggestions of a possible use of lithium as an additional treatment to reduce suicidal behaviour, where other pharmacological treatments may be more effective for mood stabilization. In these cases, there is some evidence that supplementing antidepressant therapy with lithium may reduce the risk of suicide more than treating patients with antidepressants alone. Khan et al. (2011), for example, completed a study assigning half the patients ($n=40$) to citalopram and lithium treatment and the other half ($n=40$) to citalopram and placebo using a randomized, double-blind, parallel group design. The results of Khan and colleagues' trial indicated that citalopram may have some effect on suicidal thoughts and behaviour, but by augmenting citalopram therapy with lithium at a therapeutic dose, this effect could be increased (Khan et al. 2011).

The recognition that these anti-suicidal effects might not depend on mood stabilization sparked off an intensive analysis of the suicide-preventing effects of lithium. Together with colleagues from Canada and Europe, Mogens Schou founded the International Group for the Study of Lithium-Treated Patients. This group aimed to assemble a cohort large enough to analyse suicide-related excess mortality in patients with affective disorders on and off lithium. Analyses indicated that lithium reduced mortality to that seen in the general population (Müller-Oerlinghausen et al. 1994, 1996). Other groups reported similar findings (Coppen et al. 1990; Wolf et al. 1996). One such study came from a report of a 38-year, longitudinal cohort study of 406 participants with affective disorders in Zurich, published by Angst and colleagues (2005). This study similarly reported mortality rates in lithium-treated participants that were indifferent to those of the general population. These findings together suggest that lithium cannot only have a suicide-preventative effect but also potentially some ability to reduce rates of mortality from other causes. Given the significantly high rates of suicide and all-cause mortality in patients with affective disorders, this is an encouraging finding. The indications of an independent anti-suicidal action also have clinical implications for those withdrawing lithium treatment due to insufficient mood-stabilizing effects, as, despite observing a lack of improvement in affective symptoms, lithium may be providing a protective effect against suicidal behaviour. This suggests a possible clinical case for maintaining lithium treatment in patients at risk of suicide who have had a lack of response to lithium treatment, rather than switching to new treatment entirely (Rihmer and Gonda 2013).

Despite the influx of research surrounding a reduction in the suicide rate following lithium treatment, the causal mechanism by which this reduction occurs is still unclear. Several authors have indicated that a reduction in impulsivity and aggression may be key (Müller-Oerlinghausen and Lewitzka 2010; Mühlbauer and Müller-Oerlinghausen 1985). Aggression and impulsivity are associated with suicidal behaviour (Baldessarini et al. 2006) and are common traits of patients with bipolar disorder. Both aggression and impulsivity are known to have heritable components, and in numerous studies, both retrospective and prospective, of people who have attempted, or completed, suicide, biological family members have been shown to have increased levels of aggression and impulsivity (Kovacsics et al. 2009).

The anti-aggressive properties of lithium are also well reported (Müller-Oerlinghausen and Lewitzka 2010). Initially most of these reports came from case studies, but more rigorous controlled studies have now been conducted and report consistent reductions in aggressive behaviour following lithium treatment. Similarly, lithium treatment has been shown to be associated with reductions in impulsivity. Although to date the evidence found for this effect is not as strong, it has been suggested that this may be down to a lack of research rather than a lack of effect (Kovacsics et al. 2009). Additionally to studies of human behaviour, there is some evidence of a reduction in aggressive and impulsive behaviours in animal models. The effect of lithium on aggressive behaviour has been extensively studied in rodents (Kovacsics et al. 2009). Whilst recently, the effect of lithium on impulsivity has also been replicated in animal models. Ohmura and colleagues (2012), for example, completed a study investigating the underlying mechanisms of mood stabilizers, through exploring the effect of mood-stabilizing drugs on impulsive-like actions in rats. They concluded that lithium may decrease suicidal behaviour through its suppression of impulsive behaviour. Although it is impossible to fully investigate suicidal behaviour in animal models, such research is an important indication of possible mechanisms behind the suicide-preventative effect of lithium.

The hypothesis that lithium's anti-suicidal effect is a result of its action in reducing aggressive and impulsive behaviours may also offer an explanation for the lack of anti-suicidal effects of other treatments for mood disorders. The lack of consistency in reports of the efficacy of antidepressant therapy in reducing suicidal behaviour is interesting, given that most suicide attempts occur in depressive states (Tondo et al. 1998). It has been suggested that these findings may be due to the lack of effect of antidepressant treatment of symptoms associated with suicide, such as agitation, restlessness, irritability and anger. Antidepressant treatment does not focus on the treatment of such symptoms (Guzzetta et al. 2007) and in some cases may worsen them (Baldessarini et al. 2005). In comparison, lithium treatment is associated with a reduction of these symptoms and may therefore reduce suicidal behaviour through this mechanism (Tondo and Baldessarini 2009). This hypothesis may also offer an explanation for the high incidence of suicide in patients with bipolar disorder, as patients are more likely to suffer with symptoms of agitated depression than are patients diagnosed with major depressive disorder (Tondo and Baldessarini 2009).

Several other possible causal mechanisms by which lithium confers its anti-suicidal properties are also reported in the literature. Kalkman (2011), for example, has recently published evidence of a role of glutamine synthetase in lithium's effects on suicide. This paper discusses evidence that brain glutamine synthetase function has been shown to be suppressed in patient groups in which suicide rates are highest, for example, mood disorder, epilepsy and diabetes. Moreover, reduced glutamine synthetase activity has been reported in cases of completed suicide in both depressed and nondepressed individuals. In an animal experiment, Kalkman demonstrated that lithium, a glycogen synthase kinase 3 (GSK3) inhibitor, increased the expression of glutamine synthetase and brain glutamine levels following 7 days of treatment. It is therefore proposed that the reduction of suicides as a result of lithium treatment is associated with increased glutamine synthetase expression (Kalkman 2011).

Very recently, it has also been suggested that the role of lithium in improving decision-making in patients with bipolar disorder may be linked to lithium's protective effects against suicide. Following evidence that impaired decision-making was associated with an increased risk of suicidal acts, Adida and colleagues (2015) conducted a study to investigate the relationship between lithium and decision-making. In this trial, a decision-making task was completed by euthymic outpatients with a diagnosis of bipolar disorder. Participants were split into three groups: those treated with lithium (both in monotherapy and when combined with an anticonvulsant or antipsychotic), those without lithium (treated with an anticonvulsant, antipsychotic and combination treatment) and matched healthy controls. The results demonstrated that patients treated with lithium ($p=0.007$) and healthy controls ($p=0.001$) were significantly more likely to select cards from the "safe decks" than patients who were not treated with lithium (Adida et al. 2015). The authors thus propose a potential association between lithium and reduced suicidality as a result of improved decision-making in patients.

Another suggested mechanism by which lithium may have anti-suicidal effects is through increased access to clinical care as a result of its monitoring requirements (Tondo et al. 2006). It is suggested that the level of clinical monitoring needed to maintain lithium treatment may enhance opportunities to recognize and treat early warning signs such as dysphoria and agitation, as well as helping to identify and address suicidal ideation. Clozapine, a drug used to treat schizophrenia, is the only other drug known to have significant anti-suicidal effects, and it shares the opportunity for increased clinical care through a need for physical monitoring of side effects, like lithium (Guzzetta et al. 2007). A large-scale study comparing clozapine to olanzapine has, however, demonstrated a consistent finding of anti-suicidal properties of clozapine whilst controlling for contact-time with clinicians (Meltzer et al. 2003). This has led researchers to doubt that increased clinical care plays a pivotal role in the suicide-preventative effects of lithium (Tondo and Baldessarini 2009). The impact of potential benefits of additional clinical impact on patient samples must nevertheless not be disregarded. It is, however, thought more likely that the similar positive effect of clozapine and lithium on suicidal behaviour is down to their pronounced effects on serotonin. Dysfunction in the serotonergic system has been associated with suicidal behaviour (Lee and Kim 2011), and it is suggested that the change in serotonergic functioning as a result of treatment with these drugs may reduce the diathesis for suicidal acts (Mann 2002).

Conversely to the abundance of evidence of lithium's suicide-preventative mechanism, it is worth noting that there is some evidence of a temporary increase in suicide risk at the point of discontinuation from long-term lithium treatment (Baldessarini et al. 1999). This risk has been reported by several authors and has been demonstrated to be especially high if lithium is not titrated down slowly before discontinuation, resulting in the risk of suicidal behaviour increasing by up to 20-fold (Tondo et al. 1998). This unexpected observation has led to increasing interest into the causal action by which lithium's anti-suicidal mechanism operates. Such findings also have direct clinical implications for those considering withdrawing treatment due to lack of efficacy or side effects. Discontinuing lithium must always be completed with caution, particularly if suicidality is deemed to be a risk (Young 2014).

Despite this abundance of research, the mechanism by which lithium confers this protective effect remains unclear. Further research into the causal pathways of lithium's preventative effect on suicide is needed to bring more light to this area. Insight into the underlying neurobiological mechanisms by which lithium prevents suicide would allow for the development of novel treatments, which may harness these protective effects, whilst limiting the undesirable side-effect profile of standard lithium therapy (Kovacsics et al. 2009). This may help to both improve the quality of life for patients with mood disorders and decrease the incidence of suicide in these groups. Moreover, developments in this field may help researchers to understand the neurobiological basis for suicide itself and work to prevent incidences across the population (Kovacsics et al. 2009).

13.4 Summary

The anti-suicidal actions of lithium have been widely reported over the past 40 years in studies completed all over the world. Many studies consistently support the impression that risks of suicide and of life-threatening attempts are far lower during treatment with lithium than without treatment. The vast majority of these reports have come from studies designed to assess the therapeutic efficacy of treatments and have incidentally found evidence of a preventative effect on suicide, rather than measuring suicide as a specific outcome measure in the study's original design. This is because of the ethical and methodological challenges of studying suicidal behaviour. Despite this pitfall, the homogeneity of results suggesting an anti-suicidal effect of lithium treatment, when combined in meta-analysis such as that conducted by Cipriani and colleagues (2005, 2013), allows for convincing evidence to be demonstrated, despite the methodological challenges in collecting data in this field. Further research is needed to investigate the causal mechanisms by which lithium reduces rates of suicide. Moreover, determining the threshold of treatment required (e.g. the length of treatment or recommended dose) to achieve a preventative effect would be beneficial for prescribers.

Notwithstanding the evidence of lithium's suicide-preventative effects and corresponding recommendations in national and international guidelines for the acute and maintenance therapy of affective disorders (Pfennig and Bauer 2013; Grunze et al. 2013; Malhi et al. 2015; NICE 2014), the use of lithium is still underrepresented, and there has been an observed decline in the prescription of lithium in favour of new drugs (Young 2014). It has been suggested that the decline in use may potentially be related to its lack of publicity in comparison to other commercially marketed drugs (Young 2014). Other authors have reported that this decline in use is perhaps because of the difficulties in prescribing and monitoring associated with lithium treatment. Lithium is associated with a series of adverse effects, for instance, renal function, hypothyroidism and hypercalcaemia (McKnight et al. 2012). Routine monitoring of serum levels is therefore required for effective treatment (Shine et al. 2015). These challenges may discourage some clinicians and patients from choosing lithium treatment. Many of the significant side effects are however rare, treatable and comparable to those side effects experienced by patients taking other psychotropic medications (Shine et al. 2015). Moreover, we now know that some patients are more prone to experiencing adverse events, such as renal problems, than are others: for example, women under 60 years old and those with frequently high lithium serum levels (Shine et al. 2015). It is therefore possible to weigh up the potential risk of experiencing such side effects against the prospective benefits of treatment. Additionally, clinicians can take protective measures to reduce the risk of their patients experiencing such side effects, by, for instance, closely monitoring serum levels to avoid sustained periods where lithium levels are high (Malhi 2015). Such new research allows for the negative consequences of lithium treatment to be minimized, so that potential benefits can be utilized. Lithium's long-term mood-stabilizing properties, together with its protective effect against suicide, mean that, for many, it remains the best treatment choice for bipolar disorder (Malhi 2015).

References

Adida M, Jollant F, Clark L, Guillaume S, Goodwin GM, Azorin J-M, Courtet P (2015) Lithium might be associated with better decision-making performance in euthymic bipolar patients. Eur Neuropsychopharmacol 25:788–797

Ahrens B, Müller-Oerlinghausen B (2001) Does lithium exert an independent antisuicidal effect? Pharmacopsychiatry 34:132–136

Angst J, Angst F, Gerber-Werder R, Gamma A (2005) Suicide in 406 mood-disorder patients with and without long-term medication: a 40 to 44 years' follow-up. Arch Suicide Res 9:279–300

Bakst S, Braun T, Hirshberg R, Zucker I, Shohat T (2014) Characteristics of suicide completers with a psychiatric diagnosis before death: a post-mortem study of 98 cases. Psychiatry Res 220:556–563

Baldessarini RJ, Tondo L (2009) Suicidal risks during treatment of bipolar disorder patients with lithium versus anticonvulsants. Pharmacopsychiatry 42:72–75

Baldessarini RJ, Tondo L, Hennen J (1999) Effects of lithium treatment and its discontinuation on suicidal behavior in bipolar manic-depressive disorders. J Clin Psychiatry 60:77–84

Baldessarini RJ, Pompili M, Tondo L, Tsapakis E, Soldani F, Faedda GL, Hennen J (2005) Antidepressants and suicidal behaviour: are we hurting or helping? Clin Neuropsychiatry 2:73–75

Baldessarini RJ, Tondo L, Davis P et al (2006) Decreased risk of suicides and attempts during long-term lithium treatment: a meta-analytic review. Bipolar Disord 8:625–639

Barraclough BM (1972) Suicide prevention, recurrent affective disorder and lithium. Br J Psychiatry 121:391–392

Barraclough B, Bunch J, Nelson B, Sainsbury P (1974) A hundred cases of suicide: clinical aspects. Br J Psychiatry 125:355–373

Bowden CL, Calabrese JR, Sachs G, Yatham LN, Asghar SA, Hompland M, Montgomery P, Earl N, Smoot TM, DeVeaugh-Geiss J (2003) A placebo-controlled 18-month trial of lamotrigine and lithium maintenance treatment in recently manic or hypomanic patients with bipolar I disorder. Arch Gen Psychiatry 60:392–400

Calabrese JR, Bowden CL, Sachs G, Yatham LN, Behnke K, Mehtonen OP, Montgomery P, Ascher J, Paska W, Earl N, DeVeaugh-Geiss J (2003) A placebo-controlled 18-month trial of lamotrigine and lithium maintenance treatment in recently depressed patients with bipolar I disorder. J Clin Psychiatry 64:1013–1024

Cavanagh JT, Carson AJ, Sharpe M, Lawrie SM (2003) Psychological autopsy studies of suicide: a systematic review. Psychol Med 33:395–405

Cipriani A, Pretty H, Hawton K, Geddes JR (2005) Lithium in the prevention of suicidal behaviour and all-cause mortality in patients with mood disorders: a systematic review of randomized trials. Am J Psychiatry 162:1805–1819

Cipriani A, Hawton K, Stockton S, Geddes JR (2013) Lithium in the prevention of suicide in mood disorders: updated systematic review and meta-analysis. BMJ 346:3646

Collins JC, McFarland BH (2008) Divalproex, lithium and suicide among Medicaid patients with bipolar disorder. J Affect Disord 107:23–28

Coppen A, Standish-Barry H, Bailey J, Houston G, Silcocks P, Hermon C (1990) Long-term lithium and mortality. Lancet 335:1347

Department of Health (2002) National suicide prevention strategy for England. Department of Health, London

Egger M, Smith GD (1998) Bias in location and selection of studies. BMJ 316:61–66

Geddes J, Burgess S, Hawton K, Jamison K, Goodwin GM (2004) Long-term lithium therapy for bipolar disorder: systematic review and meta-analysis of randomized controlled trials. Am J Psychiatry 161:217–222

Geddes JR, Goodwin GM, Rendell J, Azorin JM, Cipriani A, Ostacher MJ, Morriss R, Alder N, Juszczak E (2010) BALANCE investigators and collaborators. Lithium plus valproate combination therapy versus monotherapy for relapse prevention in bipolar I disorder (BALANCE): a randomized open-label trial. Lancet 375:385–395

Goodwin FK, Fireman B, Simon GE, Hunkeler EM, Lee J, Revicki D (2003) Suicide risk in bipolar disorder during treatment with lithium and divalproex. JAMA 290:1467–1473

Greil W, Ludwig-Mayerhofer W, Erazo N, Engel RR, Czernik A, Giedke H, Müller-Oerlinghausen B, Osterheider M, Rudolf GA, Sauer H, Tegeler J, Wetterling T (1996) Comparative efficacy of lithium and amitriptyline in the maintenance treatment of recurrent unipolar depression: a randomised study. J Affect Disord 40:179–190

Greil W, Ludwig-Mayerhofer W, Erazo N, Schöchlin C, Schmidt S, Engel RR, Czernik A, Giedke H, Müller-Oerlinghausen B, Osterheider M, Rudolf GA, Sauer H, Tegeler J, Wetterling T (1997a) Lithium versus carbamazepine in the maintenance treatment of bipolar disorders—a randomised study. J Affect Disord 43:151–161

Greil W, Ludwig-Mayerhofer W, Erazo N, Engel RR, Czernik A, Giedke H, Müller-Oerlinghausen B, Osterheider M, Rudolf GA, Sauer H, Tegeler J, Wetterling J (1997b) Lithium vs carbamazepine in the maintenance treatment of schizoaffective disorder: a randomised study. Eur Arch Psychiatry Clin Neurosci 247:42–50

Grunze H, Vieta E, Goodwin GM, Bowden C, Licht RW, Moller HJ, Kasper S, on behalf of the WFSBP Task Force on Treatment Guidelines for Bipolar Disorder (2013) The World Federation of Societies of Biological Psychiatry (WFSBP) guidelines for the biological treatment of bipolar disorders: update 2012 on the long-term treatment of bipolar disorder. World J Biol Psychiatry 14:154–219

Guzzetta F, Tondo L, Centorrino F, Baldessarini RJ (2007) Lithium treatment reduces suicide risk in recurrent major depressive disorder. J Clin Psychiatry 68:380–383

Harris EC, Barraclough B (1997) Suicide as an outcome for mental disorders. A meta-analysis. Br J Psychiatry 170:205–228

Isometsä E, Sund R, Pirkola S (2014) Post-discharge suicides of inpatients with bipolar disorder in Finland. Bipolar Disord 16:867–874

Kalkman HO (2011) Circumstantial evidence for a role of glutamine-synthetase in suicide. Med Hypotheses 76:905–907

Kessler RC, Berglund P, Borges G, Nock M, Wang PS (2005) Trends in suicide ideation, plans, gestures, and attempts in the United States, 1990–1992 to 2001–2003. JAMA 293(20): 2487–2495

Khan A, Khan SR, Hobus J, Faucett J, Mehra V, Giller EL, Rudolph RL (2011) Differential pattern of response in mood symptoms and suicide risk measures in severely ill depressed patients assigned to citalopram with placebo or citalopram combined with lithium: role of lithium levels. J Psychiatr Res 45:1489–1496

Kovacsics CE, Gottesman II, Gould TD (2009) Lithium's antisuicidal efficacy: elucidation of neurobiological targets using endophenotype strategies. Annu Rev Pharmacol Toxicol 49:175–198

Latalova K, Kamaradova D, Prasko J (2014) Suicide in bipolar disorder: a review. Psychiatr Danub 26:108–114

Lauterbach E, Felber W, Müller-Oerlinghausen B, Ahrens B, Bronisch T, Meyer T, Kilb B, Lewitzka U, Hawellek B, Quante A, Richter K, Broocks A, Hohagen F (2008) Adjunctive lithium treatment in the prevention of suicidal behaviour in depressive disorders: a randomised, placebo-controlled, 1-year trial. Acta Psychiatr Scand 118:469–479

Lee B-H, Kim Y-K (2011) Potential peripheral biological predictors of suicidal behavior in major depressive disorder. Prog Neuropsychopharmacol Biol Psychiatry 35:842–847

Malhi GS (2015) Lithium therapy in bipolar disorder: a balancing act? Lancet 386:415–416, (published online 20 May 2015) [http://dx.doi.org/10.1016/S0140-6736(14)62123-1]

Malhi GS, Bassett D, Boyce P, Bryant R, Fitzgerald PB, Fritz K, Hopwood M, Lyndon B, Mulder R, Murray G, Porter R (2015) Singh AB Royal Australian and New Zealand College of Psychiatrists clinical practice guidelines for mood disorders. Aust N Z J Psychiatry 49: 1087–1206

Mann JJ (2002) A current perspective of suicide and attempted suicide. Ann Intern Med 136:302–311

McKnight RF, Adida M, Budge K, Stockton S, Goodwin GM, Geddes JR (2012) Lithium toxicity profile: a systematic review and meta-analysis. Lancet 379:721–728

Meltzer HY, Alphs L, Green AI, Altamura AC, Anand R, Bertoldi A, Bourgeois M, Chouinard G, Islam MZ, Kane J, Krishnan R, Lindenmayer J-P, Potkin S (2003) Clozapine treatment for suicidality in schizophrenia: International Suicide Prevention Trial (InterSePT). Arch Gen Psychiatry 2003(60):82–91

Mühlbauer HD, Müller-Oerlinghausen B (1985) Fenfluramine stimulation of serum cortisol in patients with major affective disorders and healthy controls: further evidence for a central serotonergic action of lithium in man. J Neural Transm 61:81–94

Müller-Oerlinghausen B, Lewitzka U (2010) Lithium reduces pathological aggression and suicidality: a mini-review. Neuropsychobiology 62:43–49

Müller-Oerlinghausen B, Müser-Causemann B, Volk J (1992) Suicides and parasuicides in a high-risk patient group on and off lithium long-term medication. J Affect Disord 25:261–269

Müller-Oerlinghausen B, Wolf T, Ahrens B, Schou M, Grof E, Grof P et al (1994) Mortality during initial and during later lithium treatment. A collaborative study by the International Group for the Study of Lithium-treated patients. Acta Psychiatr Scand 90:295–297

Müller-Oerlinghausen B, Wolf T, Ahrens B, Glaenz T, Schou M, Grof E et al (1996) Mortality of patients who dropped out from regular lithium prophylaxis: a collaborative study by the International Group for the Study of Lithium-treated patients (IGSLI). Acta Psychiatr Scand 94:344–347

National Institute for Health and Clinical Excellence (2014) Bipolar disorder: the assessment and managementof bipolar disorder in adults, children and young people in primary and secondary care http://www.nice.org.uk/

Nordentoft M, Mortensen PB, Pedersen CB (2011) Absolute risk of suicide after first hospital contact in mental disorder. Arch Gen Psychiatry 68:1058–1064

Ohmura Y, Tsutsui-Kimura I, Kumamoto H, Minami M, Izumi T, Yamaguchi T, Yoshida T, Yoshioka M (2012) Lithium, but not valproic acid or carbamazepine, suppresses impulsive-like action in rats. Psychopharmacology 219:421–432

Oquendo MA, Galfalvy HC, Currier D, Grunebaum MF, Sher L, Sullivan GM, Burke AK, HarkavyFriedman J, Sublette ME, Parsey RV, Mann JJ (2011) Treatment of suicide attempters with bipolar disorder: a randomized clinical trial comparing lithium and valproate in the prevention of suicidal behaviour. Am J Psychiatry 168:1050–1056

Ougrin D, Tranah T, Stahl D, Moran P, Asarnow JR (2015) Therapeutic interventions for suicide attempts and self-harm in adolescents: systematic review and meta-analysis. J Am Acad Child Adolesc Psychiatry 54:97–107

Pfennig A, Bauer M (2013) S3guidelines on bipolar disorders are contemporary and important instruments for clinical practice. Nervenarzt 84:874–875

Prien RF, Klett CJ, Caffey EM Jr (1973a) Lithium carbonate and imipramine in prevention of affective episodes. A comparison in recurrent affective illness. Arch Gen Psychiatry 29:420–425

Prien RF, Caffey EM Jr, Klett CJ (1973b) Prophylactic efficacy of lithium carbonate in manic-depressive illness. Report of the Veterans Administration and National Institute of Mental Health collaborative study group. Arch Gen Psychiatry 28:337–341

Rihmer Z, Gonda X (2013) Pharmacological prevention of suicide in patients with major mood disorders. Neurosci Biobehav Rev 37:2398–2403

Saunders KE, Hawton K (2009) The role of psychopharmacology in suicide prevention. Epidemiol Psichiatr Soc 18:172–178

Schaffer A, Isometsa ET, Tondo L, Moreno DH, Turecki G, Reis C, Cassidy F, Sinyor M, Azorin J-M, Kessing LV, Ha K, Goldstein T, Weizman A, Beautrais A, Chou Y-H, Diazgranados N, Levitt AJ, Zarate CA Jr, Rihmer Z, Yatham LN (2015) International society for bipolar disorders task force on suicide: meta-analyses and meta-regression of correlates of suicide attempts and suicide deaths in bipolar disorder. Bipolar Disord 17:1–16

Shine B, McKnight RF, Leaver L, Geddes JR (2015) Long-term effects of lithium on renal, thyroid, and parathyroid function: a retrospective analysis of laboratory data. Lancet 386:1–7, (published online May 21.) [http://dx.doi.org/S0140-6736(14)61842-0.]

Simon GE, Savarino J, Operskalski B, Wang PS (2006) Suicide risk during antidepressant treatment. Am J Psychiatry 163:41–47

Toffol E, Hätönen T, Tanskanen A, Lönnqvist J, Wahlbeck K, Joffe G, Tiihonen J, Haukka J, Partonen T (2015) Lithium is associated with decrease in all-cause and suicide mortality in high-risk bipolar patients: a nationwide registry-based prospective cohort study. J Affect Disord 183:159–165

Tondo L, Baldessarini RJ (2007) Suicidal risks among 2826 Sardinian major affective disorder patients. Acta Psychiatr Scand 116:419–428

Tondo L, Baldessarini RJ (2009) Long-term lithium treatment in the prevention of suicidal behavior in bipolar disorder patients. Epidemiol Psichiatr Soc 18:179–183

Tondo L, Baldessarini RJ, Hennen J, Floris G, Silvetti F, Tohen M (1998) Lithium treatment and risk of suicidal behavior in bipolar disorder patients. J Clin Psychiatry 59:405–414

Tondo L, Isacsson G, Baldessarini RJ (2003) Suicide in bipolar disorder: risk and prevention. CNS Drugs 17:491–511

Tondo L, Albert MJ, Baldessarini RJ (2006) Suicide rates in relations to health care access in the United States: an ecological study. J Clin Psychiatry 67:517–523

US Department of Health Human Services (2001) National strategy for suicide prevention: goals and objectives for action. US Department of Health and Human Services, Public Health Service, Washington, DC

WHO (2014) Preventing suicide: a global imperative. http://www.who.int/mental_health/suicide-prevention/world_report_2014/en/

Wolf T, Müller-Oerlinghausen B, Ahrens B, Grof P, Schou M, Felber W, Grof E, Lenz G, Nilsson A, Simhandl C, Thau K, Vestergaard P, Wolf R (1996) How to interpret findings on mortality of long-term lithium treated manic-depressive patients?! Critique of different methodological approaches. J Affect Disord 39:127–132

Young A (2014) Lithium and suicide. Lancet Psychiatry 6:483–484

Lithium Discontinuation and Reintroduction

14

Nicolas Schaad and Jean-Michel Aubry

Abstract

Stopping medication in patients who are lithium responders is associated with a high risk of early recurrence of bipolar illness, especially mania, in the following months. A shorter time to recurrence has been shown after rapid versus slow discontinuation. A loss of the preventive effect of lithium on suicidal risk has also been shown after lithium discontinuation. Possible refractoriness to treatment was described more than 20 years ago following lithium discontinuation; however, available data do not provide convincing evidence that lithium is less effective when it is reintroduced after treatment discontinuation.

N. Schaad
Pharmacie Interhospitalière de la Côte, Chemin du Crêt 2, 1110 Morges, Switzerland

Department of Neuroscience, Faculty of Medicine, University of Geneva, CH-1211 Geneva, Switzerland

J.-M. Aubry, MD (✉)
Division of Pyschiatric Specialties, Mood Disorders Unit, Department of Mental Health and Psychiatry, University Hospitals of Geneva, and Faculty of Medicine, University of Geneva, Rue de Lausanne 20bis, CH- 1201 Geneva, Switzerland
e-mail: jean-michel.aubry@hcuge.ch

G.S. Malhi et al. (eds.), *The Science and Practice of Lithium Therapy*, DOI 10.1007/978-3-319-45923-3_14

Key Points

- Discontinuation of lithium is associated with an increased risk of mood episode recurrence in the following months. Time to recurrence is shorter with rapid, versus slow, discontinuation.
- No data are available on the effectiveness of lithium reintroduction after it has been stopped during pregnancy and/or breastfeeding. The same is true regarding suicide prevention.
- Altogether, available data do not suggest that lithium is less effective when it is restarted after discontinuation, compared to uninterrupted treatment.
- Studies addressing this issue are still limited.

14.1 Introduction

There is general agreement that lithium does not lose its efficacy in patients who are good responders during long-term maintenance treatment. Thus, a tolerance effect has not been shown for long-term lithium treatment (Kleindienst et al. 1999). However, possible refractoriness following lithium discontinuation in nine previously stable patients (Post et al. 1992) was described more than 20 years ago. No similar change in efficacy seems to be described with other mood-stabilizing agent, which merits discussion of the data on lithium discontinuation and reintroduction. This subject has been reviewed by Post (2012). Interestingly, most clinical practice guidelines do not address either the question of maintenance-treatment interruption in bipolar symptom-free patients or the issue of refractoriness.

14.2 Lithium as Maintenance Treatment: Discontinuation and Course of the Disease

Lithium is still the only mood-stabilizing agent whose efficacy is demonstrated in the very long term. As its use can be associated with several side effects, such as weight gain and renal and thyroid dysfunction (Grandjean and Aubry 2009), questions pertaining to the interruption of treatment are of primary importance for the good care of patients. Back in 1970, Baastrup and colleagues described the resurgence of bipolar symptoms upon discontinuation of lithium therapy in 12 % of 50 patients when switched to a placebo, whereas none presented a relapse when lithium was continued. Patients were treated on average for 4 years (12–84 months) with lithium, and resurgence of symptoms occurred within the 5 months following medication cessation. These initial results were later confirmed by Suppes and colleagues (1991), who collected data from 179 bipolar I patients included in nine studies and evaluated the risk of new episodes upon lithium discontinuation in a blind manner. Sixty-three percent of patients switched to placebo presented a relapse the following months, whereas only 10 % presented these symptoms when lithium was continued. Including the data arising from non-blinded studies, the authors conclude that the time to recurrence of mania was 5.2 times earlier than for depression (2.7 vs 14 months).

Baldessarini and colleagues (1999b) collected data from 28 reports pertaining to the efficacy of lithium, collected in different clinical setting (controlled and not controlled) upon discontinuation. He found that the recurrence rate of symptoms averaged 1.5 ± 2.40 % per month during treatment with lithium versus 26.0 ± 34.0 % per month without lithium. This 17-fold risk difference after discontinuation of lithium was highly significant ($p < 0.001$). After discontinuation, he found that recurrence rates were three times higher in bipolar than in unipolar patients, whereas the number of monthly episodes was identical upon lithium.

Biel and colleagues (2007) followed 213 patients who were stable on lithium monotherapy for 30 months. Lithium was slowly discontinued over a period of 10 weeks in 54 patients, and they were then followed for 6 years. The risk of relapse was three times higher during the first year in the discontinuation group than in the lithium-free cohort. Patients who continued lithium had, on average, a survival time to recurrence of 7.3 years (95 % confidence interval (CI) 5.7–9.7), and for the group of patients who stopped lithium, the survival time was 1.3 years (95 % CI 0.3–2.3). Interestingly, when subgroups were analysed in the lithium continuation cohort, those who were maintained on relatively low lithium levels (<0.80 mM/L) during the 2 years of clinical stability were less likely (relative risk, RR 0.21) to relapse than those maintained at high lithium levels (1.00–1.26 mM/L). The group of patients whose lithium levels were intermediate (0.80–0.99 mM/L) had an RR of 0.54. Even if the discontinuation group did not benefit from a placebo treatment, these results confirm the long-term effectiveness of lithium.

Therefore, there is convincing evidence that stopping lithium medication, even after several years of treatment, is associated with a high risk of early recurrence of mood episodes, especially mania, in the following months.

14.3 Lithium Discontinuation: Rapid Versus Slow Dose Tapering

The reoccurrence of bipolar symptoms following the end of lithium therapy could result, at least partially, from a withdrawal effect, since the mechanism of lithium effectiveness appears to involve biochemical adaptive processes in the central nervous system (Malhi et al. 2012; Schloesser et al. 2012; Alda 2015).

In a study with 64 bipolar I and II patients, Faedda and colleagues (1993) compared rapid (1–14 days) versus more gradual discontinuation (>15 days). Time to 50 % risk of a first recurrence of depression or mania was more than fivefold shorter when lithium was stopped rapidly compared to gradual discontinuation. These results were further confirmed in 78 Sardinian bipolar I and II patients (Baldessarini et al. 1997), as well as in a secondary analysis of data pooled from both studies plus 19 new added cases (Baldessarini et al. 1996). Of note, and as stated by the authors, patients discontinuing lithium while starting an episode of hypomania or depression were not included in these studies. Lithium discontinuation was not part of a study protocol. The most common reason for stopping lithium (55 % of patients) was the patient's wish to discontinue treatment after several years of stability; pregnancy or significant adverse effects were the reasons for stopping lithium in 21 % of patients (Baldessarini et al. 1999b). The remaining 24 % were patients refusing ongoing treatment with lithium after years of affective stability, but who were open to trying other mood stabilizers in the future if required.

After increasing the total number to 227 bipolar I and II patients, Baldessarini and colleagues (1998, 1999b) carried out a pooled analysis on the effects of rapid (1–14 days) versus gradual (15–30 days) discontinuation. Before discontinuation, patients were treated on average for 4.2 ± 3.3 years. Average lithium level was of 0.60 ± 0.13 mM/L. Time to 50 % risk of a first recurrence was three times shorter after rapid versus gradual discontinuation (6 versus 24 months). Without alternative maintenance treatment, the chances of affective stability for up to 3 years after lithium discontinuation were nearly three times greater after gradual versus rapid discontinuation in these patients (Baldessarini et al. 1999b). The consequences of the interruption of treatment were mostly observed during the first few months after discontinuation, and no clinical predictor allows us to identify which patients will relapse.

In line with these observations, it has been reported that abrupt reduction of serum lithium concentration by about 50 % may also have a clinical impact (Gelenberg et al. 1989; Waters et al. 1982; Rosenbaum et al. 1992).

14.4 Lithium Discontinuation/Reintroduction and Suicide Rate

Several long-term studies and meta-analyses including large samples of bipolar subjects have shown a preventive effect of lithium on suicide risk (Cipriani et al. 2005; Lewitzka et al. 2012; Baldessarini et al. 2006; Goodwin et al. 2003).

The effects of lithium discontinuation/reintroduction on completed and attempted suicide were evaluated in a sample of 300 bipolar I and II patients (Baldessarini et al. 1999a, b; Tondo et al. 1998). Before lithium prescription, the rate was 2.3/100 patient-years. It diminished to 0.36 during lithium maintenance, corresponding to 6.5-fold decrease. After lithium discontinuation, the rate rose to 4.9/100 patient-years, corresponding to a nearly 14-fold increase. The risk of completed suicide rose 13-fold (Baldessarini et al. 1999a; Tondo et al. 1998). Interestingly, suicide risk rose 20-fold during the 12 months after discontinuation and then decreased at a rate approaching pre-lithium treatment. To our knowledge, there are no available data on lithium's effect on suicide attempts/fatalities after reintroduction.

14.5 Lithium Discontinuation and Pregnancy

For women taking lithium as maintenance treatment, anticipation or occurrence of pregnancy is a reason to discontinue the drug. Several studies suggest that pregnancy does not have a protective effect on mood instability and that about 50 % of women with bipolar disorder are symptomatic during pregnancy and carry a risk of recurrence up to 70 %, particularly in the post-partum period (Viguera et al. 2011, 2007).

In a prospective observational study, Viguera and colleagues showed that discontinuation of lithium or other mood stabilizer during pregnancy, particularly abruptly, is associated with a high risk of mood episode recurrence (Viguera et al. 2007).

In a recent retrospective case series, Deiana and colleagues (2014) compared five women off lithium with seven women on lithium during pregnancy. Of note, all the off-lithium pregnancies were unplanned, and none of them was taking concomitant medication, except one, who was on T4. Lithium was discontinued 24–48 h before caesarean delivery, as suggested by the ISBD guideline (Ng et al. 2009) and was reintroduced when patients were medically stable. Four out of five patients who were off lithium presented an episode during pregnancy and during the post-partum period, whereas two occurred (one possibly due to low lithium dosage) in the post-partum in the seven women on lithium and none during pregnancy.

Even though the number was small, these data suggest a higher rate of recurrence among women who discontinued lithium and a greater risk in the post-partum period (Deiana et al. 2014). Although lithium is frequently reintroduced in previous responders after treatment cessation during pregnancy, there are no published data on the effectiveness of its reintroduction in women who stopped treatment during pregnancy and/or breastfeeding.

14.6 Lithium Reintroduction in Patients Who Were Lithium Good Responders

The phenomenon of lithium refractoriness, that is, a loss of effectiveness when lithium is reintroduced, was proposed by Post and colleagues and subsequently by other authors (Appleby et al. 2006; Bauer 1994; Maj et al. 1995; Oostervink et al. 2000; Post et al. 1992). Other studies have not confirmed these observations (Coryell et al. 1998; Tondo et al. 1997; Baldessarini et al. 1999b).

Two studies included in a meta-analysis suggest that lithium is less effective after discontinuation and reintroduction, and three found no change in effectiveness (de Vries et al. 2013), which allows the authors to conclude that their results do not rule out the possibility of refractoriness in selected subgroups of patients. However, available data do not provide convincing evidence that lithium is less effective when treatment is restarted after discontinuation (de Vries et al. 2013). Since the mechanism of lithium action remains largely unknown, it is difficult to suggest a pharmacological explanation for this possible observation.

14.7 Summary

It is now clearly established that lithium discontinuation is associated with an increased risk of mood episode recurrence in the following months. Moreover, abrupt or rapid discontinuation has been shown to decrease time to recurrence.

Discontinuation of lithium during pregnancy has also been shown to be associated with a high risk of recurrence. Similarly, suicide risk increases markedly during the 12 months after discontinuation.

The phenomenon of lithium refractoriness, or loss of effectiveness, when lithium is reintroduced, described by some authors, is not confirmed by available data. Therefore, reintroduction of lithium treatment for patients who were good responders until lithium was discontinued is justified.

References

Alda M (2015) Lithium in the treatment of bipolar disorder: pharmacology and pharmacogenetics. Mol Psychiatry 20(6):661–670. doi:10.1038/mp.2015.4

Appleby B, Wise T, Isaac A (2006) A case of refractoriness to lithium therapy following its discontinuation in a previously responsive patient. Harv Rev Psychiatry 14(6):330–332. doi:10.1080/10673220601070039

Baastrup PC, Poulsen JC, Schou M, Thomsen K, Amdisen A (1970) Prophylactic lithium: double blind discontinuation in manic-depressive and recurrent-depressive disorders. Lancet 2(7668):326–330, doi:S0140-6736(70)92870-9

Baldessarini RJ, Tondo L (1998) Recurrence risk in bipolar manic-depressive disorders after discontinuing lithium maintenance treatment: an overview. Clin Drug Investig 15(4):337–351. doi:10.2165/00044011-199815040-00010

Baldessarini RJ, Tondo L, Faedda GL, Suppes TR, Floris G, Rudas N (1996) Effects of the rate of discontinuing lithium maintenance treatment in bipolar disorders. J Clin Psychiatry 57(10):441–448

Baldessarini RJ, Tondo L, Floris G, Rudas N (1997) Reduced morbidity after gradual discontinuation of lithium treatment for bipolar I and II disorders: a replication study. Am J Psychiatry 154(4):551–553

Baldessarini RJ, Tondo L, Hennen J (1999a) Effects of lithium treatment and its discontinuation on suicidal behavior in bipolar manic-depressive disorders. J Clin Psychiatry 60(Suppl 2):77–84, discussion 111–116

Baldessarini RJ, Tondo L, Viguera AC (1999b) Discontinuing lithium maintenance treatment in bipolar disorders: risks and implications. Bipolar Disord 1(1):17–24

Baldessarini RJ, Tondo L, Davis P, Pompili M, Goodwin FK, Hennen J (2006) Decreased risk of suicides and attempts during long-term lithium treatment: a meta-analytic review. Bipolar Disord 8(5 Pt 2):625–639. doi:10.1111/j.1399-5618.2006.00344.x

Bauer M (1994) Refractoriness induced by lithium discontinuation despite adequate serum lithium levels. Am J Psychiatry 151(10):1522

Biel MG, Peselow E, Mulcare L, Case BG, Fieve R (2007) Continuation versus discontinuation of lithium in recurrent bipolar illness: a naturalistic study. Bipolar Disord 9(5):435–442. doi:10.1111/j.1399-5618.2007.00389.x

Cipriani A, Pretty H, Hawton K, Geddes JR (2005) Lithium in the prevention of suicidal behavior and all-cause mortality in patients with mood disorders: a systematic review of randomized trials. Am J Psychiatry 162(10):1805–1819. doi:10.1176/appi.ajp.162.10.1805

Coryell W, Solomon D, Leon AC, Akiskal HS, Keller MB, Scheftner WA et al (1998) Lithium discontinuation and subsequent effectiveness. Am J Psychiatry 155(7):895–898

de Vries C, van Bergen A, Regeer EJ, Benthem E, Kupka RW, Boks MP (2013) The effectiveness of restarted lithium treatment after discontinuation: reviewing the evidence for discontinuation-induced refractoriness. Bipolar Disord 15(6):645–649. doi:10.1111/bdi.12105

Deiana V, Chillotti C, Manchia M, Carta P, Bocchetta A, Ardau R et al (2014) Continuation versus discontinuation of lithium during pregnancy: a retrospective case series. J Clin Psychopharmacol 34(3):407–410. doi:10.1097/JCP.0000000000000059

Faedda GL, Tondo L, Baldessarini RJ, Suppes T, Tohen M (1993) Outcome after rapid vs gradual discontinuation of lithium treatment in bipolar disorders. Arch Gen Psychiatry 50(6):448–455

Gelenberg AJ, Kane JM, Keller MB, Lavori P, Rosenbaum JF, Cole K, Lavelle J (1989) Comparison of standard and low serum levels of lithium for maintenance treatment of bipolar disorder. N Engl J Med 321(22):1489–1493. doi:10.1056/NEJM198911303212201

Goodwin FK, Fireman B, Simon GE, Hunkeler EM, Lee J, Revicki D (2003) Suicide risk in bipolar disorder during treatment with lithium and divalproex. JAMA 290(11):1467–1473. doi:10.1001/jama.290.11.1467

Grandjean EM, Aubry JM (2009) Lithium: updated human knowledge using an evidence-based approach: part III: clinical safety. CNS Drugs 23(5):397–418. doi:10.2165/00023210-200923050-00004

Kleindienst N, Greil W, Ruger B, Moller HJ (1999) The prophylactic efficacy of lithium – transient or persistent? Eur Arch Psychiatry Clin Neurosci 249(3):144–149

Lewitzka U, Scheffczyk R, Ritter D, Doucette S, Bauer M, Bschor T (2012) No correlation between lithium serum levels and psychopathological features during the euthymic interval of patients with recurrent affective disorder. Pharmacopsychiatry 45(1):1–6. doi:10.1055/s-0031-1286343

Maj M, Pirozzi R, Magliano L (1995) Nonresponse to reinstituted lithium prophylaxis in previously responsive bipolar patients: prevalence and predictors. Am J Psychiatry 152(12):1810–1811

Malhi GS, Tanious M, Das P, Berk M (2012) The science and practice of lithium therapy. Aust N Z J Psychiatry 46(3):192–211. doi:10.1177/0004867412437346

Ng F, Mammen OK, Wilting I, Sachs GS, Ferrier IN, Cassidy F et al (2009) The International Society for Bipolar Disorders (ISBD) consensus guidelines for the safety monitoring of bipolar disorder treatments. Bipolar Disord 11(6):559–595. doi:10.1111/j.1399-5618.2009.00737.x

Oostervink F, Nolen WA, Hoenderboom AC, Kupka RW (2000) Risk of inducing resistance upon stopping and restarting lithium after long-term usage. Ned Tijdschr Geneeskd 144(9):401–404

Post RM, Leverich GS, Altshuler L, Mikalauskas K (1992) Lithium-discontinuation-induced refractoriness: preliminary observations. Am J Psychiatry 149(12):1727–1729

Post RM (2012) Acquired lithium resistance revisited: Discontinuation-induced refractoriness versus tolerance. J Affect Disord 140(1):6–13. doi:10.1016/j.jad.2011.09.021

Rosenbaum JF, Sachs GS, Lafer B, Kane JM, Keller MB, Gelenberg AJ (1992) High rates of relapse in bipolar patients abruptly changed from standard to low plasma lithium levels in a double-blind, controlled trial. In: Proceedings of the Annual Meeting of the New Clinical Drug Evalutations Unit (NCDEU). 1992. National Institute of Mental Health, Washington, DC

Schloesser RJ, Martinowich K, Manji HK (2012) Mood-stabilizing drugs: mechanisms of action. Trends Neurosci 35(1):36–46. doi:10.1016/j.tins.2011.11.009

Suppes T, Baldessarini RJ, Faedda GL, Tohen M (1991) Risk of recurrence following discontinuation of lithium treatment in bipolar disorder. Arch Gen Psychiatry 48(12):1082–1088

Tondo L, Baldessarini RJ, Floris G, Rudas N (1997) Effectiveness of restarting lithium treatment after its discontinuation in bipolar I and bipolar II disorders. Am J Psychiatry 154(4):548–550

Tondo L, Baldessarini RJ, Hennen J, Floris G, Silvetti F, Tohen M (1998) Lithium treatment and risk of suicidal behavior in bipolar disorder patients. J Clin Psychiatry 59(8):405–414

Viguera AC, Whitfield T, Baldessarini RJ, Newport DJ, Stowe Z, Reminick A et al (2007) Risk of recurrence in women with bipolar disorder during pregnancy: prospective study of mood stabilizer discontinuation. Am J Psychiatry 164(12):1817–1824. doi:10.1176/appi.ajp.2007.06101639, quiz 1923

Viguera AC, Tondo L, Koukopoulos AE, Reginaldi D, Lepri B, Baldessarini RJ (2011) Episodes of mood disorders in 2,252 pregnancies and postpartum periods. Am J Psychiatry 168(11):1179–1185. doi:10.1176/appi.ajp.2011.11010148

Waters B, Lapierre Y, Gagnon A, Cahudhry R, Tremblay A, Sarantidis D et al (1982) Determination of the optimal concentration of lithium for the prophylaxis of manic-depressive disorder. Biol Psychiatry 17(11):1323–1329

Short- and Midterm Side Effects of Lithium Therapy

15

Rebecca F. McKnight, John R. Geddes, and Guy M. Goodwin

Abstract

Lithium is a widely used and effective treatment for mood disorders. In the short term, patients may experience transitory gastrointestinal (GI) side effects or mild tremor, whilst a proportion of patients develop a reversible nephrogenic diabetes insipidus. Endocrine disorders—goitre, hypothyroidism, hyperthyroidism, hyperparathyroidism and weight gain—may occur in the long term. Exacerbation or development of skin and hair conditions due to lithium therapy has been reported in the literature, but good quality evidence is still needed to verify this association. Regular monitoring whilst taking lithium is essential to identify patients at risk of developing adverse effects and to provide safe and effective treatment.

Key Points

- Lithium is an effective treatment for mood disorders but one that is often associated with side effects of varying degree and duration.
- Short-term side effects include nausea, diarrhoea, increased thirst, dysgeusia and mild tremor.
- Long-term side effects include neuroendocrine changes such as hypo- and hyperthyroidism and hyperparathyroidism. These need to be monitored and treated appropriately.
- In some patients lithium is also associated with significant weight gain, and this is important because it can decrease treatment adherence.
- To avoid side effects and minimise their consequences, plasma lithium levels should be carefully monitored on a regular basis.

R.F. McKnight, BA(Oxon), MA(Oxon), BMBch (✉) • G.M. Goodwin, DPhil
Department of Psychiatry, University of Oxford, Warneford Hospital, Headington, Oxford OX3 7JX, UK
e-mail: rebecca.mcknight@psych.ox.ac.uk; guy.goodwin@psych.ox.ac.uk

J.R. Geddes, MD
University Department of Psychiatry, Oxford University, Warneford Hospital, Oxford OX3 7JX, UK
e-mail: john.geddes@psych.ox.ac.uk

G.S. Malhi et al. (eds.), *The Science and Practice of Lithium Therapy*,
DOI 10.1007/978-3-319-45923-3_15

15.1 Introduction

Lithium is a widely available and effective treatment for mood disorders. Its use is probably restricted by the fact that it has a low therapeutic index and is dangerous in overdose and by the perception that it causes a number of potential serious adverse effects at therapeutic doses. In consequence, and given the requirement for regular blood monitoring of lithium levels, alternative drugs with less evidence of efficacy have increasingly been used in the long-term treatment of bipolar disorder (Netto 2014).

Lithium was introduced before modern pharmacovigilance, and although its adverse effects have been reported since the 1960s, only recently have large-scale pharmacoepidemiological studies been conducted to accurately estimate their prevalence. Its range of probable actions on the coupling of receptors to second messenger systems in cells means that it has wide-ranging effects on multiple organ systems, including the kidneys, thyroid, parathyroid and weight/metabolism. Long-term renal effects are described in Chapter 16. The endocrine systems, in particular, will be considered in this chapter. As endocrine disorders tend themselves to have multisystem effects, recognition and treatment in patients taking lithium are clinically important. The increased risks of physical problems for patients taking lithium are very likely real, but there are two difficulties with current estimates of the magnitude of the risks simply attributable to lithium. The first is that lithium is a more medicalised treatment than most other drugs used in psychiatry, with regular blood monitoring. This increases the probability that other abnormalities will be detected. Secondly, patients with mood disorders appear to be at increased risk of other physical disorders, so any association between lithium and such disorders may be, in part, confounded by indication.

Other important adverse effects of lithium include teratogenicity (Chapter 18) and the symptoms and management of acute toxicity (Chapter 17).

The side effects of lithium therapy are best considered as being short and longer term in nature. Short-term conditions tend to be mild and/or short-lived, whilst those that occur later have the potential to cause significant morbidity and require complex management.

15.2 Short-Term Effects of Lithium Therapy

Over the 60 years that it has been prescribed, the short-term mild side effects of lithium have been extensively characterised. The most common minor side effects—usually occurring within hours to days of administration—are fine tremor, mild gastrointestinal upset (especially diarrhoea), ankle oedema and increased thirst and urination (Vestergaard et al. 1980; Dols et al. 2013). Many patients also complain of fatigue, a general slowing (or cognitive blunting) of thought processes and poor concentration. Although the mechanism by which lithium acts as a mood stabiliser is poorly understood, throughout the body it is handled in a similar way to sodium. Disruption to usual sodium levels or quantity can affect lithium levels and related symptoms adversely, and vice versa. Many of the side effects of lithium are predictable if the physiology of sodium is considered.

15.2.1 Gastrointestinal Side Effects

Gastrointestinal (GI) side effects affect about half of patients who take lithium, usually mildly. The usual complaints are of diarrhoea, abdominal pain or nausea (Vestergaard et al. 1980). The symptoms start within a few days of commencing lithium and tend to improve within 1–3 weeks. There is a dose-response relationship between the severity of GI symptoms and plasma lithium: keeping to the recommended 0.5–0.8 mmol/L helps to minimise symptoms (Persson 1977). It is important to reassure patients that these troublesome symptoms are likely to improve if they persevere: GI complaints at the start of treatment are a common reason for failure of a trial of lithium.

15.2.2 Fine Tremor

Fine tremor (8–12 Hz) is a frequent finding in patients taking lithium, with approximately half of treated individuals reporting it when specifically asked (Lydiard and Gelenberg 1982; Vestergaard et al. 1980). The tremor rarely impedes function but can be a nuisance in everyday life. Those with pre-existing essential tremor or with a family history of movement disorders are at greatest risk. The fine tremor relating to therapeutic lithium levels should be distinguished from the coarse tremor found in intoxication. The former is benign, whereas the latter requires urgent treatment. Similarly to essential tremor, the lithium-induced tremor tends to be worsened by anxiety or caffeine and dampened down by alcohol or other central nervous system (CNS) sedatives. Many patients get used to the tremor, but if it does need treating, there are two main options: lowering plasma lithium level or starting a beta-blocker such as propranolol. The latter is best given in two to three divided doses daily and is usually effective (Taylor and Kapur 2012).

15.2.3 Lithium-Induced Nephrogenic Diabetes Insipidus

Increased thirst and urination (polydipsia and polyuria) are reported by the majority of patients on chronic lithium therapy (Duncavage et al. 1983). They are caused by lithium inducing a nephrogenic diabetes insipidus which reduces the maximum urinary concentrating ability by approximately 15 % (McKnight et al. 2012). Lithium enters the collecting ducts of the kidney through an epithelial sodium channel and acts to reduce the permeability of the membrane to water. Lithium decreases the activation of aquaporin-2 channels, which usually merge with the apical membrane to allow water reabsorption. It does this through three mechanisms: prevention of aquaporin-2 channel activation by argentine vasopressin (AVP); inhibition of the protein kinase A that activates aquaporin-2 by phosphorylation; and, with chronic lithium treatment, downregulation of the aquaporin-2 gene (Bedford et al. 2008; Kalita et al. 2013). The result of these actions is that less water is reabsorbed into the circulation, decreasing urinary concentration and increasing urinary volume. Whilst a lack of AVP can cause a life-threatening diabetes insipidus, there is no evidence that this occurs with lithium treatment—it seems to be a benign form of the condition. There is evidence that lithium-induced nephrogenic diabetes insipidus is reversible on stopping lithium, but there does seem to be significant individual variability (Blount et al. 2010).

Clinically, many patients can adapt to the symptoms relatively quickly, but occasionally treatment is requested. Polyuria tends to be worse when lithium is given in divided doses, so a single daily dose may be more practical. Lowering the dose of lithium to the minimum level needed for mood stabilisation is also beneficial. If these approaches fail, amiloride (an epithelial sodium channel blocker) may be helpful. It reduces lithium uptake into the collecting duct, thereby reducing its ability to block aquaporin function (Bedford et al. 2008; Lydiard and Gelenberg 1982).

15.3 Medium- to Long-Term Effects of Lithium Therapy

15.3.1 Thyroid Function

Clinicians have been aware of the effects of lithium upon the thyroid gland since it first became widely used in the mid twentieth century: goitre was one of the earliest described side effects (Schou et al. 1968). Like several other medications (e.g. amiodarone), lithium can cause a wide spectrum of thyroid abnormalities, which may or may not be clinically relevant. The pathophysiology within the thyroid gland is complex and incompletely understood, but a brief account is included below.

15.3.1.1 Lithium and the Physiology of the Thyroid Gland

Lithium is readily taken up into the thyroid gland and accumulates there at high concentrations (Berens et al. 1970; Lazarus 1998). In vitro and in vivo studies have both shown that lithium then alters the uptake of iodine into the thyroid; this can be an increase or decrease, but usually the latter. A decrease in iodine uptake is thought to be partly due to lithium-iodine competition for transport across the cell membrane and partly due to downregulation of the iodide transporter gene (Kibirige et al. 2013). Lithium also inhibits synthesis and release of both thyroxine (T4) and thyronine (T3). It reduces the iodination of the thyroid hormone precursors by inhibiting an iodinase, a crucial step in hormone production. Lithium then interferes with hormone release via two mechanisms: altering tubule polymerisation, which limits transportation to the cell membrane, and inhibition of thyroid-stimulating hormone (TSH) activation of C-AMP at the level of the membrane (Lazarus 1998).

These effects of lithium do not appear to be dose dependent or related to the length of treatment with lithium. Uptake into the thyroid gland occurs from the first administration, but not all patients develop thyroid disorders. How the cellular level effects of lithium lead to the clinical manifestations are still uncertain, especially for cases of hyperthyroidism.

15.3.1.2 Goitre

Schou and colleagues first described goitre in patients treated with lithium in 1968 (Schou et al. 1968). Since then it has become well established that many patients—including those who remain euthyroid throughout treatment—have an increased thyroid volume during lithium therapy. A goitre is often not noticed by the patient but may present as a diffuse non-tender neck swelling. When lithium is first taken, its rapid inhibition of thyroid hormone synthesis and release stimulates the release of large amounts of TSH. This leads to thyroid cell proliferation and gland enlargement (Burrow et al. 1971). A goitre takes some time to develop but can occur within weeks of starting therapy.

A large number of studies have reported the epidemiology of lithium-induced goitre, with highly variable results. This is partly explained by the variable baseline of goitre across the world, which relates to naturally different levels of dietary iodine intake. Schou and colleagues initially reported a prevalence of 3.6 %, which is no greater than that of the general population (2–4 %). However, this early study was based on clinical examination alone, and since the introduction of thyroid ultrasonography the detection of goitre has become more precise. Bocchetta and colleagues reported a cross-sectional study of 150 patients who had been taking lithium for 2–180 months (Bocchetta et al. 1991). Using clinical examination, they report a 51 % prevalence of goitre. There was no association with age, gender or family history of thyroid disease, but those with a raised TSH (signifying some level of hypothyroidism) almost invariably had a goitre. Similarly, other studies using ultrasound have reported a prevalence of goitre among lithium-treated patients of 50–60 % (Bauer et al. 2007; Ozsoy et al. 2010; Perrild et al. 1990). If the patient remains euthyroid and the goitre is not so large that it is causing mass effects, no treatment is needed; hypothyroidism and hyperthyroidism should be treated as below.

15.3.1.3 Hypothyroidism

Hypothyroidism (diagnosed by a raised TSH and low T4/T3) is a common condition affecting approximately 2 % of women and 0.5 % of men in Western Europe (Boelaert 2005). The rates are slightly higher in geographical locations deplete of iodine and slightly lower in countries which iodinate staple foods (e.g. salt, flour). Even more common is subclinical hypothyroidism—an isolated raised TSH with normal levels of T4/T3—affecting an additional 6–8 % of women and 3 % of men. So hypothyroidism occurs at a high base rate, but there is now high-quality evidence that lithium significantly increases the risk in patients who take it long term.

Bocchetta and colleagues undertook a prospective longitudinal study of 150 patients on lithium therapy in which they recorded their thyroid function biannually (Bocchetta et al. 2007). At 10-year follow-up of 118 patients, 17 % had required thyroxine replacement therapy for clinical or subclinical hypothyroidism. Additionally, 10 % of males and 27 % of females within the cohort were found to be positive for thyroid autoantibodies (antithyroid peroxidise or antithyroglobulin).

This compares to the general population rates (although epidemiological studies report variable results) of 2–5 % and 10–15 % for males and females, respectively (Bjoro et al. 2000). They concluded that lithium increases the risk of hypothyroidism and increases the likelihood of having positive thyroid autoantibodies.

A multitude of other studies—mostly small samples followed for 1–2 years—have reported a significantly increased risk of hypothyroidism in patients on lithium therapy. A large systematic review assessed 77 publications in this area and pooled the results of eight prospective cohort studies (McKnight et al. 2012). These studies compared the prevalence of subclinical or clinical hypothyroidism in patients given lithium ($n = 1{,}402$) for a mean of 70 months with the prevalence in controls ($n = 1{,}032$). A random-effects meta-analysis showed an almost sixfold increase in hypothyroidism in patients given lithium compared to controls (odds ratio 5·78 (2·00–16·67); $p = 0\cdot001$). In the same review, meta-analysis of case-control studies (cases = 645, controls = 377) showed an increase in TSH concentrations in patients taking lithium compared to controls (weighted mean difference (WMD) 4.00 iU/mL, 95 % CI 3.90–4.10, $p < 0.0001$).

A large recent study has compared the thyroid function in patients taking lithium ($n = 2{,}040$, as recorded by at least two lithium levels) to that of controls ($n = 198{,}156$) over a 25-year period (Shine et al. 2015). A survival analysis shows a significantly greater chance of developing hypothyroidism in patients who have taken lithium compared to controls (RR 7.06, 95 % CI4.7–10.6, $p = 0.00$). This figure is similar in magnitude to that found in our meta-analysis (McKnight et al. 2012).

In the general population, females—especially those aged 20–50—are at considerably higher risk of developing hypothyroidism than males. The factors that predispose patients taking lithium therapy to develop hypothyroidism have been investigated in several studies. The large longitudinal survival analysis above found young females to be at greatest risk of developing a raised TSH (Shine et al. 2015). Bocchetta and colleagues also found that females, older patients and those with positive thyroid autoantibodies were at increased risk within the lithium-treated cohort (Bocchetta et al. 2007). A recent cross-sectional study of 68 patients found that females were at increased risk of hypothyroidism, but did not find a positive correlation with the plasma concentration of lithium or the duration of lithium therapy (Ahmadi-Abhari et al. 2003). Similarly, Ozpoyraz and colleagues' cross-sectional study of older adults found a greatest risk in females, but also in those with a positive family history of thyroid disease. Again no positive correlation with length of lithium therapy was found (Ozpoyraz et al. 2002). The dose and length of therapy data in many trials have been unreported or highly variable: more data are needed to determine whether length of time on lithium or higher plasma levels do predispose to hypothyroidism. There is some evidence that cigarette smoking may increase an individual's risk of developing thyroid disease whilst on lithium (Bocchetta et al. 1996).

It is uncertain whether subclinical hypothyroidism should be treated if the patient is asymptomatic (Pham 2008). Those that are treated (irrespective of whether or not they are on lithium therapy) seem to benefit from greater energy and better mood

(Khandelwal and Tandon 2012). Treatment also prevents progression to overt hypothyroidism and subsequent complications. There is some evidence to suggest that, in mood disorders, treatment of subclinical hypothyroidism can help to stabilise or improve mood (Haggerty and Prange 1995), but there is no specific evidence relating to the lithium-treated population. It may well be beneficial to treat these patients, but in those with a bipolar illness, there is always the risk of triggering a hypomanic/manic episode.

Given the high prevalence of lithium-related hypothyroidism, assessing thyroid function at baseline and annually throughout lithium therapy is highly recommended. More frequent assessment—e.g. every 6 months—may be beneficial in women and those with positive thyroid autoantibodies or a family history of thyroid disease (Kibirige et al. 2013). Treatment of clinical or subclinical hypothyroidism should be with usual thyroxine replacements. In bipolar patients with rapid cycling or treatment-resistant depression, addition of thyroxine to their mood stabilisers can be beneficial even if they are biochemically euthyroid (Chakrabarti 2011).

15.3.1.4 Hyperthyroidism

The evidence linking lithium to hypothyroidism is of high quality, but until recently there has been much less available evidence surrounding hyperthyroidism. At least 30 case reports have been published since the 1970s, but there have been no larger scale epidemiological studies (Rosser 1976). Several case series have not found an increased risk from that of the general population (Bocchetta and Loviselli 2006), and a large prospective cohort found only one case in 150 patients after 15 years follow-up (Bocchetta et al. 2007). McKnight and colleagues' systematic review identified four case-control studies which reported the prevalence of hyperthyroidism in patients taking lithium ($n = 178$) compared to controls ($n = 181$); they found a non-significant increase in rates of hyperthyroidism in lithium-treated patients (odds ratio 1.45, 95% CI 0.23–9.35, $p = 0.69$) (McKnight et al. 2012). The largest sample size currently available is the unpublished work by Shine and colleagues referred to above (Shine et al. 2015). They compared a sample of 181,904 controls who had not previously had hypothyroidism to 1,680 patients taking lithium and found an almost 30% increased risk of hyperthyroidism (as defined by suppressed TSH) (relative risk 1.28, 95% CI 1.07–1.54, $p = 0.01$). Although the meta-analysis results did not reach significance, they produced a very similar risk estimate to Shine and colleagues. There was also evidence that females were at higher risk of developing a suppressed TSH than males.

When lithium-related hyperthyroidism does present, it tends to be a relatively short-lived painless thyroiditis. This may be related to a direct toxic effect of lithium upon the thyroid in some patients (Miller and Daniels 2001)—although this is poorly characterised—or could be related to autoantibody status. As described above, more patients on lithium therapy are positive for thyroid autoantibodies than in the general population.

Assessment of thyroid function should be as described for hypothyroidism. Treatment of lithium-induced thyrotoxicosis is along standard protocols, usually with antithyroid drugs such as carbimazole. Treatment is often only needed for a short period and a conservative approach may be appropriate. Some patients will spontaneously remit, but many go on to develop hypothyroidism later. Patients who develop Graves' disease or a more aggressive disorder may benefit from radioiodine treatment or thyroidectomy.

One confounding factor is that—as in multinodular goitre disease—a single patient may move between being euthyroid, hypothyroid and hyperthyroid whilst on lithium. This means that cross-sectional studies do not always capture an accurate picture and prospective studies with regular assessment of thyroid function are needed to fully understand the epidemiology.

One remaining question is whether or not thyroid disorders remit once lithium therapy is stopped. Souza and colleagues reported a trial in which thyroid and adrenal function was monitored in patients who had been on lithium for at least 1 year but discontinued at entry into the study. They found a significant increase in T4 and decrease in TSH in the cohort at 1 month after lithium discontinuation (Souza et al. 1991). At baseline this cohort had a tendency towards hypothyroidism, so this certainly suggests that lithium withdrawal may lead to a return to euthymia. However, more robust data are required.

15.3.2 Parathyroid Function

The parathyroid glands are located on the posterior side of the thyroid gland and a number between four and six in humans. They produce parathyroid hormone (PTH), which is released in response to low plasma calcium and acts to increase calcium uptake from the intestines and release from bones. Hypercalcaemia (caused by an excess of PTH—hyperparathyroidism) can cause a multitude of symptoms, commonly summarised as 'moans, groans, stones and bones'. Depression, abdominal pain, renal stones, osteoporosis, fatigue, constipation and memory loss are all common symptoms.

By 2012, 60 studies relating to the effect of lithium upon the parathyroid gland had been published (McKnight et al. 2012). These included 113 case reports (some as case series) and various observational studies. No randomised controlled trials including lithium treatment have yet reported parathyroid data.

The case reports linking lithium with parathyroid disorders have been homogenous in content: all 113 reported hyperparathyroidism, with 44 cases of coexistent parathyroid adenoma. Approximately half of the cases had associated thyroid disorders, the most common being a multinodular goitre with hypothyroidism. The latter is unsurprising, as the endocrine conditions are often comorbid.

One cross-sectional study compared the calcium and PTH levels of 204 patients who had been taking lithium for at least 6 years (Bendz 2003). They reported a twofold increase in the rates of hypercalcaemia and hyperparathyroidism in cases compared to controls.

McKnight and colleagues pooled the data from 14 observational studies that reported calcium and PTH levels in patients taking lithium ($n=739$) compared to controls ($n=699$) and found a 10% increase in both calcium and PTH in lithium-treated patients (calcium 0·09 mmol/L (95% CI 0·02–0·17); $p=0·009$, PTH 7·32 pg/mL (95% CI 3·42–11·23); $p<0·0001$) (McKnight et al. 2012). This increase is enough to tip a large number of lithium-treated patients over into clinical hyperparathyroidism. Shine and colleagues found a 1.5-fold increased risk of hyperparathyroidism (signified by a high plasma calcium) in patients taking lithium ($n=1,420$) compared to controls ($n=210,227$) (Shine et al. 2015) (relative risk 1.47, 95% CI 1.30–1.66, $p=0.00$).

Given these relatively large-scale analyses, it is highly likely that lithium does cause hyperparathyroidism in some patients, but the prevalence and risk factors for this association are yet to be fully characterised. In contrast to thyroid disorders, there does not seem to be an increased risk in females compared to males (Shine et al. 2015).

Treatment of hyperparathyroidism has traditionally been parathyroidectomy. This is a tricky surgical procedure, as the parathyroids are difficult to locate and hard to remove completely, especially without removing or damaging the thyroid gland and surrounding structures. Recent surgical advances include the use of intra-operative parathyroid hormone monitoring, which helps the surgeon to know when the parathyroids have been appropriately removed and have improved the success of the procedures (Hundley et al. 2005). Hundley and colleagues have shown this to be a successful treatment in lithium-induced hyperparathyroidism.

Similarly to thyroid disorders, it is not clear cut as to when a patient needs aggressive treatment of lithium-induced hyperparathyroidism. Patients who are symptomatic (especially with serious physical sequelae) clearly need a parathyroidectomy, but those with asymptomatic mildly raised calcium may not. As raised calcium is linked to depression, some patients with treatment-resistant mood disorders may benefit from having their calcium normalised. Until recently, clinical guidelines have not included any guidance on when and if to monitor calcium levels in patients on lithium therapy. Given the increasing recognition of the elevated incidence of parathyroid disorders on lithium treatment, calcium should be measured at baseline and at least annually whilst lithium therapy continues.

15.3.3 Weight and Metabolism

Patients with mental health disorders are more likely to be overweight than the general population, and this is especially true in bipolar disorder. A third to a half of patients with bipolar disorder are classed as obese (BMI >30), and 10–30% meet the criteria for type 2 diabetes mellitus (Kim et al. 2009). Being overweight carries a high risk of physical health complications (hypertension, diabetes, ischaemic heart disease, etc.) but can also negatively affect mental health. Obese patients tend to have longer and more severe depressive episodes, and women, in particular, may struggle with their self-esteem and body image (Vanina et al. 2002). Unfortunately, psychotropic drugs do not help patients, as many, including lithium, have a tendency to increase body weight. In clinical trials, patients are classified overweight according to World Health Organisation BMI criteria, and a significant weight gain across the course of a trial is arbitrarily defined to be >7 % of body weight. Weight gain is one of the most common complaints about lithium in the outpatient department and a frequent reason for non-compliance with therapy.

There is good evidence that lithium increases body weight, although—similarly to antipsychotic medications—the mechanism of this is not fully understood. A variety of mechanisms have been proposed and it may well be a combination of these:

- Lithium appears to increase appetite, perhaps through direct stimulation of the hypothalamus (Vanina et al. 2002).
- Increase in energy consumption, especially thirst-quenching high-calorie fluids (Elmslie et al. 2001; Vieweg et al. 1988).
- Accumulation of peripheral oedema (especially in the early stages of treatment).
- Increased carbohydrate storage (Vanina et al. 2002).
- Reduced energy levels—this may lead to reduced exercise or increased food consumption to try and compensate.
- The effects of hypothyroidism (see references above).

There has been speculation that lithium may act to increase body weight through influencing the hormone leptin. Leptin is a peptide hormone released from adipocytes that acts to inhibit appetite and reduce weight when adipose tissue is plentiful. Unfortunately investigations into the association between lithium and leptin have produced inconsistent results. Atamaca and colleagues reported that leptin levels increased in a cohort of 15 patients treated with lithium for 8 weeks (Atmaca et al. 2002). Contrastingly, neither a small case-control study nor a prospective cohort found any change in leptin levels before and after lithium therapy, or between cases and controls (Himmerich et al. 2005; Ozbulut et al. 2007).

Various observational studies have confirmed that lithium tends to increase body weight—typically by 4–10 kg within a year of starting treatment (Vanina et al. 2002). However, the variation between individuals is enormous, and some studies have not reported a significant weight increase when outliers are removed.

In our meta-analysis, 14 randomised controlled trials (RCTs) comparing lithium with placebo or other drug treatment reported weight change data (McKnight et al. 2012). Clinically significant weight gain (>7 %) was more frequent in patients receiving lithium than in those receiving placebo (odds ratio 1.89, 95 % CI 1.27–2.82, $p=0.002$). When lithium was compared to olanzapine, patients taking lithium showed significantly less weight gain (odds ratio 0.32, 95 % CI 0.21–0.49, $p<0.001$). Unfortunately, there were insufficient data on objective change in weight in the RCTs for the results to be statistically pooled. The values for mean weight change in kilograms for the RCTs included above ranged from 0 to 6 kg across 12 months.

The risk factors for weight gain whilst taking lithium appear to be primarily related to gender, age and polypharmacy. Females seem to gain weight more easily—although this is not related to reproductive hormones (Baptista et al.

2000) —as do older patients and those taking other psychotropic medications. Individuals with subclinical or clinical hypothyroidism are at high risk for clinically relevant weight gain.

As lithium is the best evidence-based treatment for many patients with mood disorders, the prescribing clinician should work hard to avoid unnecessary weight gain. The patient should be informed of the risk and recommended to eat a healthy diet and take regular exercise. As weight gain is dose related, lithium levels should be kept to the lowest possible level to maintain a stable mood. Co-administration of other weight-increasing medications should be minimised. There is some evidence that the addition of the emerging mood stabiliser topiramate (which has a tendency to reduce weight) may benefit some patients struggling with their weight (Kotwal et al. 2006; Mahmoudi-Gharaei et al. 2012). Weight and BMI should be recorded before lithium therapy begins and at least every 6 months thereafter. Close communication with the patient's family doctor can be beneficial in helping patients engage in weight management programmes.

15.3.4 Dermatological Conditions

Skin disorders are a frequent adverse effect of drug therapy and are poorly reported in the literature. They are associated with various mood stabilisers, especially sodium valproate and lamotrigine. At the time of writing, 78 publications pertaining to the effects of lithium upon the skin are within the literature; 69 of which are case reports or series (a summary up to 2012 can be found in McKnight et al. 2012). These include 137 patients and report either new onset or an exacerbation of a wide range of dermatological conditions (Table 15.1). There is clearly no distinct lithium-related 'rash', and it appears to be variable as to whether or not there is full recovery after lithium withdrawal.

Table 15.1 Dermatological conditions reported in the literature as being caused or exacerbated by lithium therapy (search done January 2014)

Dermatological condition	Primarily new onset or exacerbations	Number of case reports	Tendency to remit with lithium withdrawal
Psoriasis (plaque or guttate)	Exacerbations and new onset cases	44	Variable
Maculopapular rashes	New onset	16	Yes
Acne	Exacerbations and new onset	17	Yes
Hidradenitis suppurativa	Exacerbation	2	No
Mucosal disorders (esp. stomatitis)	New onset	7	Yes
Folliculitis	New onset	30	Yes
Darier's disease	All exacerbations	6	Yes
Miscellaneous	Variable	15	Variable

Whilst there are plenty of individual reports, there are very few high-quality trial data supporting an association between lithium and skin disorders. Unfortunately the majority of large RCTs have not collected data on skin disorders and/or not reported it. A meta-analysis pooling the results of two large RCTs showed no significant difference in the prevalence of skin disorders between patients on lithium and those given placebo (odds ratio 1.28, 95 % CI 0.49–3.36, $p=0.62$) (McKnight et al. 2012). However, the data collected were patient reports only on any 'skin rash', so may well be incomplete and heterogeneous in nature.

Psoriasis is probably the most well-established link with lithium—many textbooks list it as a 'common trigger' for the condition. Brauchli and colleagues have reported a very large case-control study to look further into this association (Brauchli et al. 2009). They used logistic regression to establish the odds of having used lithium in 36,702 patients with psoriasis and an equal number of matched controls. There was a small increase in the risk of psoriasis in those who had been exposed to lithium (odds ratio 1.68, 95 % CI 1.18–2.39, $p<0.01$).

Skin conditions are very common and some may very well be exacerbated or triggered by lithium therapy. Whilst the evidence is not strong, if patients do present with skin problems that they attribute to their lithium, common sense should be taken towards management. Normal dermatological treatment from their GP—or a dermatologist—is indicated. If lithium is successful in maintaining a stable mood, then the patient should be encouraged to continue with therapy where possible. There is a suggestion that psoriasis may be more difficult to treat in those taking lithium, but this has yet to be substantiated (Jafferany 2008).

15.3.5 Hair Disorders

Hair disorders are another set of conditions that are frequently associated with psychotropic medications, including lithium. There is a known strong association between alopecia and sodium valproate. As with skin disorders, the evidence is very limited, mostly comprising case reports. Electronic searching (by the author RM, January 2014) found a total of 25 publications reporting an adverse effect on lithium upon hair, 15 of which were case reports. These are summarised in Table 15.2. Many of the patients described in these reports were taking several psychotropic medications: confounders may therefore be clouding the picture.

Table 15.2 Summary of case reports published on the association between lithium and hair disorders (Jan 2014)

Hair condition	Primarily new onset or exacerbations	Number of case reports	Associated with skin disorders
Diffuse scalp hair loss	New onset	34	3× psoriasis, 1× acne, 1× eczema
Alopecia areata	New onset	2	No
Alopecia totalis	New onset	4	No
Miscellaneous	New onset	8	No

One RCT comparing lithium ($n=91$) to placebo ($n=94$) for 12 months reported hair loss in 8 % of patients in the lithium group compared with 6 % in the placebo group; this was a non-significant difference (Bowden et al. 2000). Similarly, another RCT also reported no significant difference in hair loss between lithium-treated patients (3 %) and placebo (0 %) (Calabrese et al. 2005). Several cross-sectional studies on groups of patients taking lithium reported 'hair thinning' during the 1970s and 1980s (see summary in McKnight et al. (2012)). However, in all these cases, the patients self-reported a problem with their hair—often spontaneously—and there were no diagnostic criteria or guidance as to what and how it should be recorded. To reliably characterise the association between lithium and skin/hair disorders, further evidence is needed. Trials should gather routine data as part of adverse events reporting.

15.4 Clinical Recommendations

It is clear from the above that, whilst lithium is an effective drug, it brings with it a high risk of dose-related adverse effects. To ensure continued safe prescribing, it is recommended that patients should have baseline tests before commencing therapy and at regular intervals thereafter. Based on the evidence outlined above (and the recent UK clinical guidance (NICE 2014)), Table 15.3 outlines recommendations for clinical monitoring.

15.5 Summary

Lithium is a classic exemplar of a psychotropic drug: it is highly efficacious at stabilising mood in both unipolar and bipolar mood disorders but has a multitude of short- and long-term side effects. There is good evidence for its association with

Table 15.3 Clinical monitoring of patients on lithium therapy

Baseline tests	Baseline function	Long-term monitoring (maximum time intervals)
Lithium levels		3 months
Renal function (eGFR)	Yes	6 months
Thyroid function (TSH, T4, T3)	Yes	6 months
Thyroid autoantibodies	If available	If available or if thyroid function is abnormal
Calcium	Yes	12 months
Weight and BMI	Yes	6 months
ECG	Yes	If becomes toxic or develops cardiac symptoms
Counselling about teratogenic risk	Yes	Yearly for women of reproductive age

thyroid disorders, hyperparathyroidism, weight gain and various short-term conditions. Further evidence is needed to fully understand the risk factors and epidemiology of these adverse effects and to substantiate claims of a link with hair and skin disorders.

Given that lithium is an established effective treatment, future clinical trials are likely to compare it to newer therapeutics, probably over a relatively short time period of several months. Routine baseline and follow-up data on endocrine, metabolic, skin and hair parameters would provide high-quality data and fill the current knowledge gaps, whilst biobanks and other databases may add other physiological and genetic data to clinical observations. Large prospective cohorts—perhaps those set up for other purposes, e.g., new mood-monitoring technology—could help to refine our epidemiological knowledge of this aspect of lithium's side effects (Bopp et al. 2010).

Managing patients on lithium will always be a matter of balancing risks; improving our knowledge of the adverse effects of lithium will allow clinicians to provide the highest quality of care for mood disorders safely and effectively. Elucidating its mechanism of action and developing less toxic analogues should remain a priority.

References

Ahmadi-Abhari SA, Ghaeli P, Fahimi F, Esfahanian F, Farsam H, Dehpour AR, Jahanzad I, Hatmi ZN, Dashti S (2003) Risk factors of thyroid abnormalities in bipolar patients receiving lithium: a case control study. BMC Psychiatry 3:4

Atmaca M, Kuloglu M, Tezcan E, Ustundag B (2002) Weight gain and serum leptin levels in patients on lithium treatment. Neuropsychobiology 46(2):67–69

Baptista T, Lacruz A, De Mendoza S, Guillen MM, Burguera JL, De Burguera M, Hernandez L (2000) Endocrine effects of lithium carbonate in healthy premenopausal women: relationship with body weight regulation. Prog Neuropsychopharmacol Biol Psychiatry 24(1):1–16. doi:10.1016/s0278-5846(99)00085-8

Bauer M, Blumentritt H, Finke R, Schlattmann P, Adli M, Baethge C, Bschor T, Mueller-Oerlinghausen B, Berghoefer A (2007) Using ultrasonography to determine thyroid size and prevalence of goiter in lithium-treated patients with affective disorders. J Affect Disord 104(1–3):45–51. doi:10.1016/j.jad.2007.01.033

Bedford JJ, Weggery S, Ellis G, McDonald FJ, Joyce PR, Leader JP, Walker RJ (2008) Lithium-induced nephrogenic diabetes insipidus: renal effects of amiloride. Clin J Am Soc Nephrol 3(5):1324–1331. doi:10.2215/cjn.01640408

Bendz H (2003) Renal and parathyroid function in psychiatric patients on lithium treatment. Nord J Psychiatry 57(1):68–69

Berens SC, Wolff J, Murphy DL (1970) Lithium concentration by thyroid. Endocrinology 87(5):1085

Bjoro THJ, Krüger O, Midthjell K, Hunstad K, Schreiner T, Sandnes L, Brochmann H (2000) Prevalence of thyroid disease, thyroid dysfunction and thyroid peroxidase antibodies in a large, unselected population. The Health Study of Nord-Trondelag (HUNT). Eur J Endocrinol 143(5):639–647

Blount MA, Sim JH, Zhou R, Martin CF, Lu W, Sands JM, Klein JD (2010) Expression of transporters involved in urine concentration recovers differently after cessation of lithium treatment. Am J Physiol Ren Physiol 298(3):F601–F608. doi:10.1152/ajprenal.00424.2009

Bocchetta A, Loviselli A (2006) Lithium treatment and thyroid abnormalities. Clin Pract Epidemiol Ment Health 2:23

Bocchetta A, Bernardi F, Pedditzi M, Loviselli A, Velluzzi F, Martino E, DelZompo M (1991) Thyroid abnormalities during lithium treatment. Acta Psychiatr Scand 83(3):193–198

Bocchetta A, Cherchi A, Loviselli A, Mossa P, Velluzzi F, Derai R, DelZompo M (1996) Six-year follow-up of thyroid function during lithium treatment. Acta Psychiatr Scand 94(1):45–48

Bocchetta A, Cocco F, Velluzzi F, Del Zompo M, Mariotti S, Loviselli A (2007) Fifteen-year follow-up of thyroid function in lithium patients. J Endocrinol Invest 30(5):363–366

Boelaert KFJ (2005) Thyroid hormone in health and disease. J Endocrinol 187(1):1–15

Bopp JM, Miklowitz DJ, Goodwin GM, Stevens W, Rendell JM, Geddes JR (2010) The longitudinal course of bipolar disorder as revealed through weekly text messaging: a feasibility study. Bipolar Disord 12(3):327–334. doi:10.1111/j.1399-5618.2010.00807.x

Bowden CL, Calabrese JR, McElroy SL, Gyulai L, Wassef A, Petty F, Pope HG, Chou JCY, Keck PE, Rhodes LJ, Swann AC, Hirschfeld RMA, Wozniak PJ, Divalproex Maintenance SG (2000) A randomized, placebo-controlled 12-month trial of divalproex and lithium in treatment of outpatients with bipolar I disorder. Arch Gen Psychiatry 57(5):481–489

Brauchli YB, Jick SS, Curtin F, Meier CR (2009) Lithium, antipsychotics, and risk of psoriasis. J Clin Psychopharmacol 29(2):134–140. doi:10.1097/JCP.0b013e31819a4b7c

Burrow GN, Burke WR, Himmelho JM, Spencer RP, Hershman JM (1971) Effect of lithium on thyroid function. J Clin Endocrinol Metab 32(5):647

Calabrese JR, Shelton MD, Rapport DJ, Youngstrom EA, Jackson K, Bilali S, Ganocy SJ, Findling RL (2005) A 20-month, double-blind, maintenance trial of lithium versus divalproex in rapid-cycling bipolar disorder. Am J Psychiatr 162(11):2152–2161

Chakrabarti S (2011) Thyroid functions and bipolar affective disorder. J Thyroid Res 2011:306367. doi:10.4061/2011/306367

Dols A, Sienaert P, van Gerven H, Schouws S, Stevens A, Kupka R, Stek ML (2013) The prevalence and management of side effects of lithium and anticonvulsants as mood stabilizers in bipolar disorder from a clinical perspective: a review. Int Clin Psychopharmacol 28(6): 287–296

Duncavage MB, Nasr SJ, Altman EG (1983) Subjective side-effects of lithium-carbonate – a longitudinal-study. J Clin Psychopharmacol 3(2):100–102

Elmslie JL, Mann JI, Silverstone JT, Williams SM, Romans SE (2001) Determinants of overweight and obesity in patients with bipolar disorder. J Clin Psychiatry 62(6):486–491

Haggerty JJ, Prange AJ (1995) Borderline hypothyroidism and depression. Annu Rev Med 46:37–46. doi:10.1146/annurev.med.46.1.37

Himmerich H, Koethe D, Schuld A, Yassouridis A, Pollmacher T (2005) Plasma levels of leptin and endogenous immune modulators during treatment with carbamazepine or lithium. Psychopharmacology (Berl) 179(2):447–451. doi:10.1007/s00213-004-2038-9

Hundley JC, Woodrum DT, Saunders BD, Doherty GM, Gauger PG (2005) Revisiting lithium-associated hyperparathyroidism in the era of intraoperative parathyroid hormone monitoring. Surgery 138(6):1027–1031. doi:10.1016/.j.surg.2005.09.028

Jafferany M (2008) Lithium and psoriasis: what primary care and family physicians should know. Prim Care Companion J Clin Psychiatry 10(6):435–439

Kalita P, Bahn A, Bedford J, Leader J, Walker R (2013) Long-term modulation of lithium-induced nephrogenic diabetes insipidus by amiloride. Nephrology 18:37–37

Khandelwal D, Tandon N (2012) Overt and subclinical hypothyroidism: who to treat and how. Drugs 72(1):17–33. doi:10.2165/11598070-000000000-00000

Kibirige D, Luzinda K, Ssekitoleko R (2013) Spectrum of lithium induced thyroid abnormalities: a current perspective. Thyroid Res 6(1):3–3. doi:10.1186/1756-6614-6-3

Kim B, Kim S, McIntyre RS, Park HJ, Kim SY, Joo YH (2009) Correlates of metabolic abnormalities in bipolar I disorder at initiation of acute phase treatment. Psychiatry Invest 6(2):78–84

Kotwal R, Guerdjikova A, McElroy SL, Keck PE Jr (2006) Lithium augmentation of topiramate for bipolar disorder with comorbid binge eating disorder and obesity. Hum Psychopharmacol Clin Exp 21(7):425–431. doi:10.1002/hup.783

Lazarus JH (1998) The effects of lithium therapy on thyroid and thyrotropin-releasing hormone. Thyroid 8(10):909–913. doi:10.1089/thy.1998.8.909

Lydiard RB, Gelenberg AJ (1982) Hazards and adverse-effects of lithium. Annu Rev Med 33:327–344. doi:10.1146/annurev.me.33.020182.001551

Mahmoudi-Gharaei J, Shahrivar Z, Faghihi T, Mohammadi MR, Tehrani-Doost M, Alaghband-Rad J, Ghaeli P (2012) Topiramate versus valproate sodium as adjunctive therapies to a combination of lithium and risperidone for adolescents with bipolar I disorder: effects on weight and serum lipid profiles. Iran J Psychiatry 7(1):1–10

McKnight RF, Adida M, Budge K, Stockton S, Goodwin GM, Geddes JR (2012) Lithium toxicity profile: a systematic review and meta-analysis. Lancet 379(9817):721–728. doi:10.1016/s0140-6736(11)61516-x

Miller KK, Daniels GH (2001) Association between lithium use and thyrotoxicosis caused by silent thyroiditis. Clin Endocrinol (Oxf) 55(4):501–508

Netto IPR (2014) Lithium prescribing trends: review. Int J Healthc Biomed Res 2(2):95–103

NICE (2014) Bipolar disorder: Assessment and management. CG185. https://www.nice.org.uk/guidance/cg185

Ozbulut O, Guler O, Kurt E, Alatas G, Serteser M, Gecici O (2007) Effects of lithium treatment on ghrelin, leptin and cholesterol levels in patients with euthymic bipolar disorder. Neurol Psychiatry Brain Res 14(3):127–130

Ozpoyraz N, Tamam L, Kulan E, Sert M (2002) Thyroid functions in lithium-treated patients. Eur Neuropsychopharmacol 12:S435–S435

Ozsoy S, Mavili E, Aydin M, Turan T, Esel E (2010) Ultrasonically determined thyroid volume and thyroid functions in lithium-naive and lithium-treated patients with bipolar disorder: a cross-sectional and longitudinal study. Hum Psychopharmacol Clin Exp 25(2):174–178

Perrild H, Hegedus L, Baastrup PC, Kayser L, Kastberg S (1990) Thyroid-function and ultrasonically determined thyroid size in patients receiving long-term lithium treatment. Am J Psychiatr 147(11):1518–1521

Persson G (1977) Lithium side effects in relation to dose and to levels and gradients of lithium in plasma. Acta Psychiatr Scand 55(3):208–213. doi:10.1111/j.1600-0447.1977.tb00163.x

Pham CBSA (2008) Should we treat subclinical hypothyroidism? BMJ 337:a834

Rosser R (1976) Thyrotoxicosis and lithium. Br J Psychiatry 128:61–66

Schou M, Amdisen A, Jensen SE, Olsen T (1968) Occurrence of goitre during lithium treatment. Br Med J 3(5620):710

Shine B, Mcknight R, Leaver L, Geddes J (2015) Long-term effects of lithium on renal, thyroid, and parathyroid function: a retrospective analysis of laboratory data. The Lancet 386(9992):461–468

Souza FGM, Mander AJ, Foggo M, Dick H, Shearing CH, Goodwin GM (1991) The effects of lithium discontinuation and the non-effect of oral inositol upon thyroid-hormones and cortisol in patients with bipolar affective-disorder. J Affect Disord 22(3):165–170. doi:10.1016/0165-0327(91)90050-3

Taylor DPC, Kapur S (2012) Prescribing guidelines in psychiatry, 11th edn. Wiley-Blackwell, London

Vanina Y, Podolskaya A, Sedky K, Shahab H, Siddiqui A, Munshi F, Lippmann S (2002) Body weight changes associated with psychopharmacology. Psychiatr Serv 53(7):842–847. doi:10.1176/appi.ps.53.7.842

Vestergaard P, Amdisen A, Schou M (1980) Clinically significant side effects of lithium treatment a survey of 237 patients in long-term treatment. Acta Psychiatr Scand 62(3):193–200. doi:10.1111/j.1600-0447.1980.tb00607.x

Vieweg WVR, Godleski LS, Hundley PL, Yank GR (1988) Lithium, polyuria and abnormal diurnalweight-gaininpsychosis.ActaPsychiatrScand78(4):510–514.doi:10.1111/j.1600-0447.1988.tb06375.x

Lithium and Renal Failure

16

Aude Servais, Marc Masson, and Jean-Pierre Grünfeld

Abstract

Lithium has a narrow therapeutic index and can consequently result in considerable toxicity if levels are not monitored regularly. Among lithium's potential side effects are two types of renal toxicity: decreased renal concentrating ability and chronic renal failure. Lithium-induced polyuria is frequent, estimated to affect up to 40 % of patients, and usually develops early. It may be irreversible, especially if the treatment has been prescribed for more than 15 years. Chronic renal failure is observed in patients treated for more than 10–20 years. Its prevalence is estimated at 12 % after 19 years of treatment. Some patients (0.5 %) may reach end-stage renal disease.

A. Servais (✉)
Department of Nephrology, Assistance publique – Hôpitaux de Paris, Necker – Enfants Malades Hospital, Paris, France

Université Paris Descartes, Paris, France

Department of Nephrology, Necker Hospital, 149, rue de Sèvres, 75015 Paris, France
e-mail: aude.servais@nck.aphp.fr

M. Masson
Clinique du Château, Garches, France
e-mail: marc.masson@clinique-garches.com

J.-P. Grünfeld
Department of Nephrology, Assistance publique – Hôpitaux de Paris, Necker – Enfants Malades Hospital, Paris, France

Université Paris Descartes, Paris, France
e-mail: jean-pierre.grunfeld@nck.aphp.fr

G.S. Malhi et al. (eds.), *The Science and Practice of Lithium Therapy*,
DOI 10.1007/978-3-319-45923-3_16

Key Points

- Nephrogenic diabetes insipidus is observed in 40–50 % of patients.
- This nephrogenic diabetes insipidus is due to a negative regulation and a decrease of trafficking towards the membrane of aquaporin-2 channels (AQP2) in the principal cells of the collecting duct.
- Long-term lithium administration may induce chronic tubulointerstitial nephropathy leading to renal failure, even in the absence of episodes of lithium intoxication.
- The mean annual loss of eGFR is slow, but lithium-induced nephropathy can progress to end-stage renal disease over several decades.
- When lithium therapy is stopped, the probability of renal improvement is higher when the estimated glomerular filtration rate is above 40 mL/min at lithium discontinuation.

16.1 Introduction

Since its introduction into the field of psychiatry more than half a century ago, lithium has become established as a valuable and effective agent in the treatment of acute mania and in the prophylaxis of bipolar and unipolar affective disorders (Geddes et al. 2004; Cipriani et al. 2005; Geddes and Miklowitz. 2013). However, lithium has a narrow therapeutic index and can result in considerable toxicity. The potential side effects of lithium have always been an issue, and among these the possible impact on renal function of long-term lithium treatment has given rise to considerable concern (Markowitz et al. 2000). Lithium is also known to affect renal concentrating ability, and lithium-induced polyuria is frequent and estimated to affect up to 40 % of patients (Stone 1999). In addition, lithium is known to have effects on calcium homeostasis. Indeed, long-term treatment may be associated with persistent hyperparathyroidism and hypercalcaemia (Grandjean and Aubry 2009).

16.2 Physiology

Lithium is a monovalent cation that acts as sodium in the kidneys (Fig. 16.1). It is freely filtered through the glomeruli. Seventy-five percent of the filtered load is then reabsorbed in the proximal tubule. In the case of sodium, an additional 20 % is reabsorbed in the large ascending loop of Henle in the proximal tubule. A small fraction is reabsorbed in the collecting duct, through the apical sodium epithelial channel (ENaC) (Trepiccione and Christensen 2010).

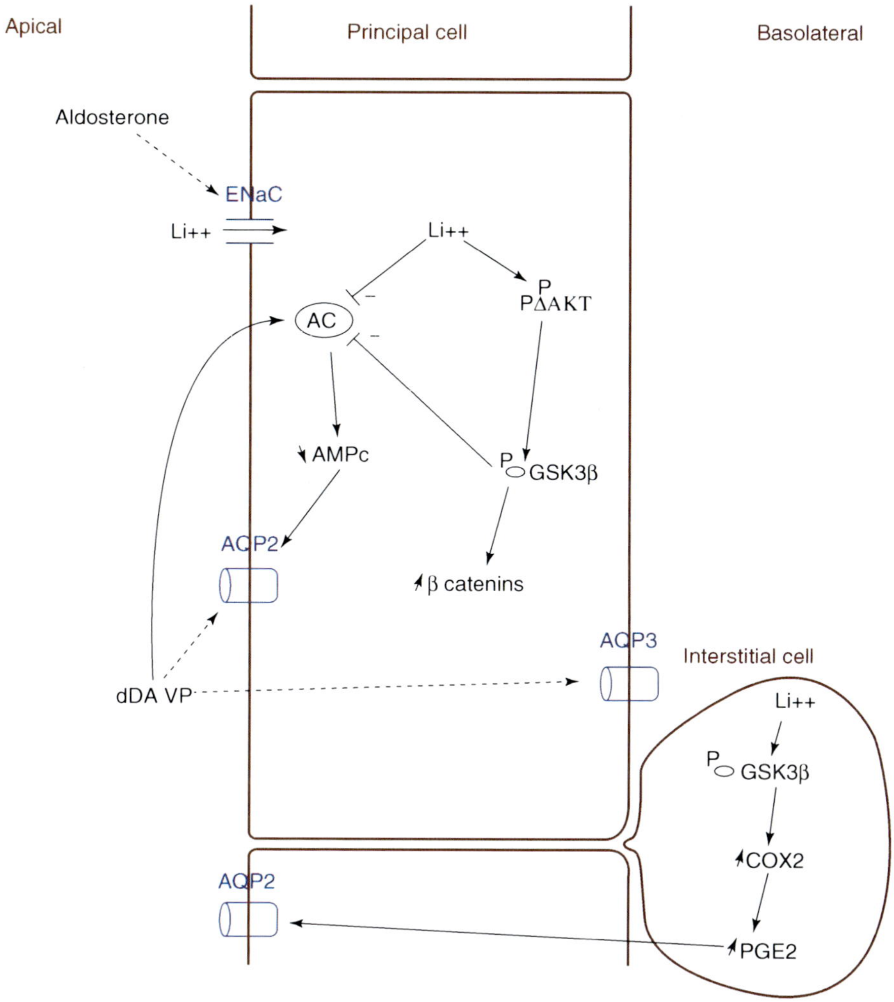

Fig. 16.1 Mechanism of nephrogenic diabetes insipidus. Lithium produces a negative regulation and a decrease of trafficking towards the membrane of aquaporin-2 channels (AQP2) in the principal cells of the collecting duct. Lithium enters into the principal cells through ENaC channel and is not able to exit the cell. Then, in the cell, the lithium inhibits adenylate cyclase activity and cyclic adenosine monophosphate 3-5′(AMPc) production. The regulation of adenylate cyclase also depends on glycogen synthase kinase 3β (GSK3β). Other key enzymes that are regulated by lithium include the phosphorylation of Akt, which leads to the inhibition of GSK3β via a mechanism of phosphorylation. In medullary interstitial cells, lithium inhibits GSK3β and increases the expression of COX-2 and PGE-2 which also regulates the AQP2

16.3 Nephrogenic Diabetes Insipidus

Nephrogenic diabetes insipidus is observed in 40–50 % of patients prescribed lithium. This reduced ability to concentrate the urine is correlated to the duration of lithium exposure (Bedford et al. 2008). However, in healthy volunteers, the ability to concentrate the urine was already reduced after just 1 month of treatment (Walker et al. 2005). Nephrogenic diabetes insipidus is characterised by a polyuro-polydipsic syndrome, which may be abundant—up to 10 l per day. Nephrogenic diabetes insipidus may occur less frequently with once-a-day dosing (Ljubicic et al. 2008). In patients, it usually develops early, after 8 weeks of treatment (Boton et al. 1987). When treatment is stopped, symptoms diminish slowly over 8 weeks (Bucht and Wahlin 1980). However, it can be irreversible, especially if lithium has been prescribed for more than 15 years (Bendz et al. 1996a).

Lithium-induced nephrogenic diabetes insipidus is caused by a negative regulation and a decrease of trafficking towards the membrane of aquaporin-2 channels (AQP2) in the principal cells of the collecting duct. AQP3 channels are also negatively regulated by lithium (Fig. 16.1) (Kwon et al. 2000). The ENaC channel is also dysregulated. Indeed, lithium enters into the principal cells of the collecting duct through the ENaC channel and is not able to exit out of the cell. Once in the cell, the lithium inhibits adenylate cyclase activity and cyclic adenosine monophosphate 3–5′(AMPc) production. Regulation of the adenylate cyclase also depends on glycogen synthase kinase 3β (GSK3β) (Kuure et al. 2007). Other key enzymes are regulated by lithium: the phosphorylation of Akt leads to the inhibition of GSK3β via a mechanism of phosphorylation (Nielsen et al. 2008). The inhibition of GSK3β is associated with an accumulation of β-catenin in the principal cells. This protein has a role in cellular adhesion and plays a role in the transcriptional coregulation of the genes of cellular growth. This can induce the high proliferation rate of principal cells observed with lithium (Christensen et al. 2006).

Medullary interstitial cells are also involved in lithium-induced nephrogenic diabetes insipidus. In these cells, lithium inhibits GSK3β and increases the expression of COX-2 and PGE-2, which also regulates the AQP2 (Rao et al. 2005).

In patients, Bedford and colleagues have shown a decrease of urinary excretion of AMPc and AQP2 that was correlated with the duration of lithium exposure (Bedford et al. 2008). The use of amiloride has been proposed for cases of severe nephrogenic diabetes insipidus: it inhibits the ENaC channel and thus the entry of lithium into the principal cells of the collecting duct (Batlle et al. 1985). In a cross-over study, Bedford and colleagues treated patients for 6 weeks and observed an increase in urinary osmolality (Bedford et al. 2008). In that study, lithemia remained stable, but kalemia and creatininemia should be monitored carefully. However, polyuro-polydipsic syndrome is usually well tolerated and does not require any treatment.

16.4 Lithium-Induced Nephropathy

Even when serum lithium concentrations are maintained within the therapeutic range, the glomerular filtration rate (GFR) falls slightly in about 20 % of patients (Grandjean and Aubry 2009). In the early 1990s, it was reported that there was no evidence that patients taking lithium were at risk of progressive renal failure (Groleau 1994). Longitudinal studies showed a progressive decline in GFR over time that was considered not to exceed that seen with normal ageing (Gitlin 1999). However, most authors now disagree with that view (Grandjean and Aubry 2009). Nephrologists increasingly report patients with renal insufficiency and no other medical history than lithium therapy (Presne et al. 2003). In a laboratory database study, a very high percentage of lithium-treated outpatients had an eGFR below 60 mL/min/L.73 m^2, ranging from 39 % in the 20–39 year age group to 85 % in patients aged over 70 years (Bassilios et al. 2008). This was much higher than in patients from the general population. Recently, Minay and colleagues also compared eGFR in patients treated with lithium to a control group and showed that eGFR was below 60 mL/min/L.73 m in 17 % of patients, which was significantly more than in controls (13 %) (Minay et al. 2013). Therefore, the suggestion that only a few patients receiving long-term lithium are at increased risk of progressive renal insufficiency should be viewed with caution. It is now unequivocally established that long-term lithium administration may induce chronic tubulointerstitial nephropathy leading to renal failure, even in the absence of episodes of lithium intoxication (Hestbech et al. 1977; Markowitz et al. 2000).

Several studies have demonstrated the impact that time spent on lithium treatment has on the prevalence of lithium-induced nephropathy (Bendz 1983; Bendz et al. 1994, 2010; Presne et al. 2003; Bassilios et al. 2008). The maximal number of new cases of renal failure was observed after 16–20 years of treatment in a cohort study including 114 patients treated with lithium (Lepkifker et al. 2004). Presne and colleagues also showed a correlation between the estimated glomerular filtration rate and the duration of treatment with lithium (Presne et al. 2003). Since bipolar disorder often begins at a relatively young age, renal failure may develop by 40 years of age. Two multicentre studies have investigated the prevalence of lithium-induced nephropathy (Bendz 1983; Bendz et al. 1994) and showed that, after a mean treatment duration of 6.5 years, only 4 % of patients had elevated serum creatinine levels, whereas after 19 years of exposure, this proportion increased to 12 %. Recently, the same group established that the prevalence of chronic kidney disease (defined as plasma creatinine over 150 μmol/L) in the general lithium-treated population was about 1.2 % (excluding patients on renal replacement therapy) (Bendz et al. 2010). Time on lithium was the only identified risk factor in that study. Similarly, in the cohort described by Lepkifer and colleagues, mean duration of lithium treatment was 16.8 years and 21 % of patients had an abnormal creatininemia, defined by a creatininemia above 133 μmol/L (Lepkifker et al. 2004).

A recent meta-analysis of case-control studies compared the eGFR of lithium-treated patients and controls and showed a reduction in eGFR of 5 mL/min per year in lithium-treated patients. However, the very short follow-up periods of these case-control studies does not fully allow for thorough assessment of lithium nephrotoxicity, which is a much longer term concern (McKnight et al. 2012).

The mean annual loss of eGFR is slow, but lithium-induced nephropathy can progress to end-stage renal disease over several decades (Markowitz et al. 2000; Presne et al. 2003; Bendz et al. 2010). The prevalence of end-stage renal disease was sixfold higher in lithium-treated patients than in the general population in the study by Bendz and colleagues: 18 of 3,369 patients (0.5 %) reached end-stage renal disease (Bendz et al. 2010). Patients were on dialysis at a mean age of 46 years, after a mean duration of treatment of 23 years, and more than half of them had stopped lithium treatment for more than 10 years. Among French dialysis patients, the prevalence of lithium nephropathy was estimated at 0.22 % (Presne et al. 2003). In Australian and New Zealand registries, the prevalence of end-stage renal disease due to lithium nephropathy was estimated at between 0.2 % and 0.7 % in 2000–2003. In the French cohort, the only risk factor of end-stage renal disease was initial eGFR. The risk was similar between patients who had stopped lithium and those who had not. A delay of 20 years was observed between the beginning of treatment and dialysis (Presne et al. 2003). However, patients who received more than 750 mg/day were three times more likely to experience an annual eGFR decline above 2 mL/min than those who received a lower dose. Similarly, eGFR inversely correlated with the mean serum lithium concentration (Presne et al. 2003). Furthermore, in those patients who underwent renal biopsy in that study, the degree of interstitial fibrosis was related to the cumulative dose of lithium salt.

16.5 Clinical Presentation and Diagnoses

Lithium-induced nephropathy is an asymptomatic tubulointerstitial nephropathy, without hypertension or proteinuria. There is often an associated defect in the ability to concentrate urine. The main risk factor is the length of lithium exposure. Some comorbidities, such as diabetes, may also play a role in the progression of renal failure. Diagnosis is based on the association of long-term lithium treatment (more than 10 years) and a chronic tubulointerstitial nephropathy, without any other aetiology. Tubular cysts may be observed.

The presence of tubular cysts on imaging, magnetic resonance imaging or ultrasound, in normal-size kidneys, is highly characteristic of lithium toxicity (Alexander et al. 2008) and observed in 33–62 % of cases (Markowitz et al. 2000; Farres et al. 2003). Renal calcifications may be observed in cases of hyperparathyroidism. A renal ultrasound should be performed in patients with renal failure or those who have been treated with lithium for more than 10 years. Even if renal function is normal, the presence of characteristic lesions should lead to a reduction of lithium dose if possible. A renal biopsy may be indicated if the diagnosis is uncertain and will show a chronic tubulointerstitial nephropathy pattern. Interstitial fibrosis may be present early, even 5 years after initiation of treatment (Presne et al. 2003). There are usually no glomerular lesions. Clinical evolution is usually slow with a mean decrease of eGFR of 2.2 mL/min/year.

16.6 Hypercalcaemia

Long-term lithium treatment is also associated with hypercalcaemia and hyperparathyroidism. When examining the end point of isolated hypercalcaemia, the reported prevalence among those treated with lithium varies greatly, from 10 % up to 42 % (Bendz et al. 1996b; Presne et al. 2003; Grandjean and Aubry 2009). However, the prevalence of hyperparathyroidism in chronic lithium users (more than 10 years) has been estimated at 10–15 % in retrospective case series (Hundley et al. 2005). The prevalence of hyperparathyroidism is 7.5 times higher in patients with lithium for more than 15 years than it is in the general population (Bendz et al. 1996b). 'False' hypercalcaemia due to plasma volume depletion resulting from nephrogenic diabetes insipidus should be excluded in such individuals (Grunfeld and Rossier 2009). These data support a recommendation to routinely follow a serum calcium level in those patients treated with lithium therapy (Livingstone and Rampes 2006). Some cases of lithium-associated hyperparathyroidism are associated with nephrocalcinosis, nephrolithiasis and/or osteoporosis (Dwight et al. 2002; Awad et al. 2003; Hundley et al. 2005; Carchman et al. 2008).

A proposed mechanism is an alteration in the calcium-sensing receptor, resulting in a shift to the right of the calcium–parathyroid hormone (PTH) response curve (Livingstone and Rampes 2006). This then reduces suppression of serum PTH due to an increased threshold for calcium levels (Brown 1981; Riccardi and Gamba 1999). However, it is not entirely understood whether lithium causes hyperparathyroidism directly or somehow potentiates hyperparathyroidism in patients with early parathyroid dysregulation. One possible effect of lithium on parathyroid tissue may be promotion of the growth of pre-existing abnormal tissue (Saxe and Gibson 1991; Saxe et al. 1995).

Complete resolution of hyperparathyroidism may thus require total or partial parathyroidectomy (Bendz et al. 1996b; Awad et al. 2003). Ablation of a single parathyroid adenoma usually leads to normocalcaemia, even in patients who continue to receive lithium (Awad et al. 2003; Hundley et al. 2005; Grunfeld and Rossier 2009). In contrast, surgical treatment of multiglandular disease is technically difficult and hazardous (Nemeth et al. 1998). A conservative approach to the therapy of hypercalcaemia may also be appropriate. Cinacalcet hydrochloride is an allosteric activator of the calcium-sensing receptor present in chief cells of the parathyroid glands (Quarles et al. 2003; Peacock et al. 2005; Wuthrich et al. 2007). This calcimimetic action lowers the threshold for activation of the calcium-sensing receptor by extracellular calcium and decreased PTH secretion. Activation of the calcium-sensing receptor by calcimimetics could thus specifically antagonise the effects of lithium on the parathyroid glands (Sloand and Shelly 2006; Gregoor and de Jong 2007; Szalat et al. 2009).

16.7 Increased Risk of Solid Renal Tumours in Lithium-Treated Patients

Some acquired cystic kidney disease and other toxic tubulointerstitial nephropathies have been associated with an increased occurrence of renal and urothelial carcinomas, and some cases of renal tumours have been described in lithium-treated patients (Markowitz et al. 2000; Kjaersgaard et al. 2012). We recently studied a cohort of 170 lithium-treated patients and identified 14 patients (8.2 %) who developed renal solid tumours, including seven cases of malignant and seven of benign tumours (Zaidan et al. 2014a). The mean duration of lithium exposure at diagnosis was 21.4 ± 10.3 years. The renal cancers included three clear cell renal cell carcinomas (RCC), two papillary RCC, one hybrid tumour and one rare tumour (clear cell carcinoma with leiomyomatous stroma). The benign tumours included four oncocytomas, one mixed epithelial and stromal tumour and two angiomyolipomas. The percentage of renal tumours, particularly cancers and oncocytomas, was significantly higher in lithium-treated patients compared to 340 sex-, age- and eGFR-matched lithium-free patients. The standardised incidence ratio of renal cancer was also significantly higher in lithium-treated patients compared to the general population: 7.51 (95 % CI 1.51–21.95) and 13.69 (95 % CI 3.68–35.06) in men and women, respectively. These results indicated an increased risk of renal tumours in lithium-treated patients (Zaidan et al. 2014a). Collectively, we reported an association and not a causal effect (Licht et al. 2014; Zaidan et al. 2014b). Our work did not question the benefit of lithium in the management of bipolar disorders but should alert nephrologists and psychiatrists to the possibility of late tumour development, which should encourage regular renal screening of patients until further studies are conducted.

16.8 Pathophysiology

Accumulation of lithium in cells of the distal nephron and the early collecting duct via ENaC could account for the chronic nephrotoxic effects. Increased inhibitory phosphorylation of GSK3β leads to the activation of Wnt/β-catenin signalling, a critical pathway for progression of cystic kidney diseases (Sinha et al. 2005; Nielsen et al. 2008; Kjaersgaard et al. 2012). Altogether, lithium seems to interfere with multiple critical signalling pathways that regulate tubular cell proliferation, differentiation and apoptosis, any of which may explain the onset of tubular dilatations, cysts, subsequent renal damage and eventually tumours (Sinha et al. 2005; Nielsen et al. 2008; Kjaersgaard et al. 2012).

16.9 Deciding Whether to Stop Lithium Therapy

When renal failure is diagnosed, the question of interrupting lithium treatment has to be discussed (Fig. 16.2). A reduction in lithium dose may first be considered. However, the psychiatric risk also has to be taken into account. Despite the availability of other mood stabilisers, lithium is the only compound that has clearly

Lithium and Estimated GFR (MDRD or CKD-EPI)

>60 ml/min	40-60 ml/min	25-40 ml/min	<25 ml/min
EUC, calcium and PTH annually	Discuss therapeutic alternatives especially in young patients	eGFR may continue to decline even if lithium is ceased	Cessation of lithium may not prevent progressive renal decline
Or at the time of any inter-current dehydration event	Renal ultrasound and nephrology review	Nephrology review + ultrasound if not already done	Requires close nephrology follow-up
	EUC, calcium, PTH every 6 months	EUC, calcium, PTH and FBC every 3-4 months	EUC, calcium, PTH and FBC every 1-3 months
	Target lithium level <0.4-0.6 mmol/L	Target lithium level <0.4-0.6 mmol/L	Target lithium level <0.4-0.6 mmol/L
	Mood follow-up Ensure adequate fluid intake Avoid other nephrotoxic medications	Mood follow-up Ensure adequate fluid intake Avoid other nephrotoxic medications	Mood follow-up Review fluid intake to avoid fluid overloading Avoid other nephrotoxic medications

Fig. 16.2 Discussion of interruption of lithium and recommended follow-up in the case of renal failure

demonstrated antisuicide effects in the maintenance treatment of major affective disorders (Tondo et al. 1997; Cipriani et al. 2005). Moreover, the risk of early recurrence of bipolar illness appears very high following lithium discontinuation (Suppes et al. 1991; Biel et al. 2007). In practice, some patients, whose illness is well controlled by lithium therapy, refuse to consider interruption and substitution. Therefore, the decision to interrupt lithium and to substitute another drug should involve the patient, the psychiatrist and the nephrologist.

From a nephrologist's point of view, it has been shown that the probability of renal improvement is higher when eGFR is above 40 mL/min at lithium discontinuation than when it is lower (Presne et al. 2003). If eGFR is above 40 mL/min, one may expect a mild renal improvement or a stabilisation of renal function. If eGFR is below 40 mL/min, the rate of decrease in eGFR may improve, but renal function will probably worsen progressively. If eGFR is below 25 mL/min, most patients will have a similar decrease of renal function whether or not they stop lithium treatment. Furthermore, in other series, patients progressed to end-stage renal disease over the course of several years, despite having discontinued lithium treatment (Markowitz et al. 2000; Presne et al. 2003; Lepkifker et al. 2004; Bendz et al. 2010). Thus, there is probably a point of no return, where renal fibrosis continues to progress despite suppression of the triggering toxic insult.

Strategies to reduce the risk of lithium toxicity should include careful lithium dosing with regular monitoring of serum levels. Low doses might have a lower impact on renal function (Aprahamian et al. 2014). In the case of renal failure, the goal would be to obtain the minimal efficient dose with lithemia between 0.4 and 0.6 mmol/L and once-a-day dosing (Fig. 16.2). It is also recommended that intraerythrocytic concentrations are controlled between 0.10 and 0.15 mEq/L with lithium 400 mg or between 0.05 and 0.10 mEq/L with lithium 250 mg.

16.10 Summary

Lithium is a very important and widely used therapeutic agent for bipolar disorders, and, as the number of patients on long-term treatment is increasing, more patients may be at risk of nephropathy. Nephrogenic diabetes insipidus is a frequent complication but is not a risk factor of nephropathy. The duration of lithium therapy is the major determinant of nephrotoxicity, suggesting that regular monitoring of renal function in these patients is mandatory. The decision to discontinue lithium in a patient with declining kidney function should be carefully considered, particularly as renal function does not always improve after lithium withdrawal.

Acknowledgement We acknowledge the assistance of Dr Bruce Cooper (Head, Department of Renal Medicine, Royal North Shore Hospital) with Figure 16.2.

References

Alexander MP, Farag YM, Mittal BV et al (2008) Lithium toxicity: a double-edged sword. Kidney Int 73(2):233–237

Aprahamian I, Santos FS, dos Santos B et al (2014) Long-term, low-dose lithium treatment does not impair renal function in the elderly: a 2-year randomized, placebo-controlled trial followed by single-blind extension. J Clin Psychiatry 75(7):e672–e678

Awad SS, Miskulin J, Thompson N (2003) Parathyroid adenomas versus four-gland hyperplasia as the cause of primary hyperparathyroidism in patients with prolonged lithium therapy. World J Surg 27(4):486–488

Bassilios N, Martel P, Godard V et al (2008) Monitoring of glomerular filtration rate in lithium-treated outpatients – an ambulatory laboratory database surveillance. Nephrol Dial Transplant 23(2):562–565

Batlle DC, von Riotte AB, Gaviria M et al (1985) Amelioration of polyuria by amiloride in patients receiving long-term lithium therapy. N Engl J Med 312(7):408–414

Bedford JJ, Weggery S, Ellis G et al (2008) Lithium-induced nephrogenic diabetes insipidus: renal effects of amiloride. Clin J Am Soc Nephrol 3(5):1324–1331

Bendz H (1983) Kidney function in lithium-treated patients. A literature survey. Acta Psychiatr Scand 68(5):303–324

Bendz H, Aurell M, Balldin J et al (1994) Kidney damage in long-term lithium patients: a cross-sectional study of patients with 15 years or more on lithium. Nephrol Dial Transplant 9(9):1250–1254

Bendz H, Sjodin I, Aurell M (1996a) Renal function on and off lithium in patients treated with lithium for 15 years or more. A controlled, prospective lithium-withdrawal study. Nephrol Dial Transplant 11(3):457–460

Bendz H, Sjodin I, Toss G et al (1996b) Hyperparathyroidism and long-term lithium therapy – a cross-sectional study and the effect of lithium withdrawal. J Intern Med 240(6):357–365

Bendz H, Schon S, Attman PO et al (2010) Renal failure occurs in chronic lithium treatment but is uncommon. Kidney Int 77(3):219–224

Biel MG, Peselow E, Mulcare L et al (2007) Continuation versus discontinuation of lithium in recurrent bipolar illness: a naturalistic study. Bipolar Disord 9(5):435–442

Boton R, Gaviria M, Batlle DC (1987) Prevalence, pathogenesis, and treatment of renal dysfunction associated with chronic lithium therapy. Am J Kidney Dis 10(5):329–345

Brown EM (1981) Lithium induces abnormal calcium-regulated PTH release in dispersed bovine parathyroid cells. J Clin Endocrinol Metab 52(5):1046–1048

Bucht G, Wahlin A (1980) Renal concentrating capacity in long-term lithium treatment and after withdrawal of lithium. Acta Med Scand 207(4):309–314

Carchman E, Ogilvie J, Holst J et al (2008) Appropriate surgical treatment of lithium-associated hyperparathyroidism. World J Surg 32(10):2195–2199

Christensen BM, Kim YH, Kwon TH et al (2006) Lithium treatment induces a marked proliferation of primarily principal cells in rat kidney inner medullary collecting duct. Am J Physiol Renal Physiol 291(1):F39–F48

Cipriani A, Pretty H, Hawton K et al (2005) Lithium in the prevention of suicidal behavior and all-cause mortality in patients with mood disorders: a systematic review of randomized trials. Am J Psychiatry 162(10):1805–1819

Dwight T, Kytola S, Teh BT et al (2002) Genetic analysis of lithium-associated parathyroid tumors. Eur J Endocrinol 146(5):619–627

Farres MT, Ronco P, Saadoun D et al (2003) Chronic lithium nephropathy: MR imaging for diagnosis. Radiology 229(2):570–574

Geddes JR, Miklowitz DJ (2013) Treatment of bipolar disorder. Lancet 381(9878):1672–1682

Geddes JR, Burgess S, Hawton K et al (2004) Long-term lithium therapy for bipolar disorder: systematic review and meta-analysis of randomized controlled trials. Am J Psychiatry 161(2):217–222

Gitlin M (1999) Lithium and the kidney: an updated review. Drug Saf 20(3):231–243

Grandjean EM, Aubry JM (2009) Lithium: updated human knowledge using an evidence-based approach: part III: clinical safety. CNS Drugs 23(5):397–418

Gregoor PS, de Jong GM (2007) Lithium hypercalcemia, hyperparathyroidism, and cinacalcet. Kidney Int 71(5):470

Groleau G (1994) Lithium toxicity. Emerg Med Clin North Am 12(2):511–531

Grunfeld JP, Rossier BC (2009) Lithium nephrotoxicity revisited. Nat Rev Nephrol 5(5):270–276

Hestbech J, Hansen HE, Amdisen A et al (1977) Chronic renal lesions following long-term treatment with lithium. Kidney Int 12(3):205–213

Hundley JC, Woodrum DT, Saunders BD et al (2005) Revisiting lithium-associated hyperparathyroidism in the era of intraoperative parathyroid hormone monitoring. Surgery 138(6):1027–1031; discussion 1031–1022

Kjaersgaard G, Madsen K, Marcussen N et al (2012) Tissue injury after lithium treatment in human and rat postnatal kidney involves glycogen synthase kinase-3beta-positive epithelium. Am J Physiol Renal Physiol 302(4):F455–F465

Kuure S, Popsueva A, Jakobson M et al (2007) Glycogen synthase kinase-3 inactivation and stabilization of beta-catenin induce nephron differentiation in isolated mouse and rat kidney mesenchymes. J Am Soc Nephrol 18(4):1130–1139

Kwon TH, Laursen UH, Marples D et al (2000) Altered expression of renal AQPs and Na(+) transporters in rats with lithium-induced NDI. Am J Physiol Renal Physiol 279(3): F552–F564

Lepkifker E, Sverdlik A, Iancu I et al (2004) Renal insufficiency in long-term lithium treatment. J Clin Psychiatry 65(6):850–856

Licht RW, Grabenhenrich LB, Nielsen RE et al (2014) Lithium and renal tumors: a critical comment to the report by Zaidan et al. Kidney Int 86(4):857

Livingstone C, Rampes H (2006) Lithium: a review of its metabolic adverse effects. J Psychopharmacol 20(3):347–355

Ljubicic D, Letica-Crepulja M, Vitezic D et al (2008) Lithium treatments: single and multiple daily dosing. Can J Psychiatry 53(5):323–331

Markowitz GS, Radhakrishnan J, Kambham N et al (2000) Lithium nephrotoxicity: a progressive combined glomerular and tubulointerstitial nephropathy. J Am Soc Nephrol 11(8): 1439–1448

McKnight RF, Adida M, Budge K et al (2012) Lithium toxicity profile: a systematic review and meta-analysis. Lancet 379(9817):721–728

Minay J, Paul R, McGarvey D et al (2013) Lithium usage and renal function testing in a large UK community population; a case-control study. Gen Hosp Psychiatry 35(6):631–635

Nemeth EF, Steffey ME, Hammerland LG et al (1998) Calcimimetics with potent and selective activity on the parathyroid calcium receptor. Proc Natl Acad Sci U S A 95(7):4040–4045

Nielsen J, Hoffert JD, Knepper MA et al (2008) Proteomic analysis of lithium-induced nephrogenic diabetes insipidus: mechanisms for aquaporin 2 down-regulation and cellular proliferation. Proc Natl Acad Sci U S A 105(9):3634–3639

Peacock M, Bilezikian JP, Klassen PS et al (2005) Cinacalcet hydrochloride maintains long-term normocalcemia in patients with primary hyperparathyroidism. J Clin Endocrinol Metab 90(1):135–141

Presne C, Fakhouri F, Noel LH et al (2003) Lithium-induced nephropathy: rate of progression and prognostic factors. Kidney Int 64(2):585–592

Quarles LD, Sherrard DJ, Adler S et al (2003) The calcimimetic AMG 073 as a potential treatment for secondary hyperparathyroidism of end-stage renal disease. J Am Soc Nephrol 14(3):575–583

Rao R, Zhang MZ, Zhao M et al (2005) Lithium treatment inhibits renal GSK-3 activity and promotes cyclooxygenase 2-dependent polyuria. Am J Physiol Renal Physiol 288(4):F642–F649

Riccardi D, Gamba G (1999) The many roles of the calcium-sensing receptor in health and disease. Arch Med Res 30(6):436–448

Saxe AW, Gibson G (1991) Lithium increases tritiated thymidine uptake by abnormal human parathyroid tissue. Surgery 110(6):1067–1076; discussion 1076–1067

Saxe A, Gibson G, Silveira E (1995) Effects of long-term lithium infusion on normal parathyroid tissue. Surgery 117(5):577–580

Sinha D, Wang Z, Ruchalski KL et al (2005) Lithium activates the Wnt and phosphatidylinositol 3-kinase Akt signaling pathways to promote cell survival in the absence of soluble survival factors. Am J Physiol Renal Physiol 288(4):F703–F713

Sloand JA, Shelly MA (2006) Normalization of lithium-induced hypercalcemia and hyperparathyroidism with cinacalcet hydrochloride. Am J Kidney Dis 48(5):832–837

Stone KA (1999) Lithium-induced nephrogenic diabetes insipidus. J Am Board Fam Pract 12(1):43–47

Suppes T, Baldessarini RJ, Faedda GL et al (1991) Risk of recurrence following discontinuation of lithium treatment in bipolar disorder. Arch Gen Psychiatry 48(12):1082–1088

Szalat A, Mazeh H, Freund HR (2009) Lithium-associated hyperparathyroidism: report of four cases and review of the literature. Eur J Endocrinol 160(2):317–323

Tondo L, Jamison KR, Baldessarini RJ (1997) Effect of lithium maintenance on suicidal behavior in major mood disorders. Ann N Y Acad Sci 836:339–351

Trepiccione F, Christensen BM (2010) Lithium-induced nephrogenic diabetes insipidus: new clinical and experimental findings. J Nephrol 23(Suppl 16):S43–S48

Walker RJ, Weggery S, Bedford JJ et al (2005) Lithium-induced reduction in urinary concentrating ability and urinary aquaporin 2 (AQP2) excretion in healthy volunteers. Kidney Int 67(1):291–294

Wuthrich RP, Martin D, Bilezikian JP (2007) The role of calcimimetics in the treatment of hyperparathyroidism. Eur J Clin Invest 37(12):915–922

Zaidan M, Stucker F, Stengel B et al (2014a) Increased risk of solid renal tumors in lithium-treated patients. Kidney Int 86(1):184–190

Zaidan M, Landais P, Grunfeld JP et al (2014b) The authors reply. Kidney Int 86(4):857–858

Lithium Toxicity: Clinical Presentations and Management

17

Souleiman El Balkhi and Bruno Mégarbane

Abstract

Lithium may be responsible for a large spectrum of central nervous system toxicity, ranging from mental confusion to life-threatening and prolonged myoclonic encephalopathy. Three presentations of lithium poisoning should be clearly distinguished—acute, acute-on-chronic and chronic poisonings—as presentation has a significant influence on the relationship between severity and toxicity-to-serum lithium concentration. Poisoning management is mainly supportive based on saline diuresis but should also include therapies to decrease lithium absorption from the gastrointestinal tract. Haemodialysis is the reference technique to enhance lithium elimination. Recently the EXTRIP workgroup has published guidelines to help physicians manage extracorporeal treatment for lithium poisoning.

Key Points

- Lithium poisoning may be responsible for life-threatening toxicity.
- Three patterns of lithium overdose are usually distinguished: acute, acute-on-chronic and chronic poisonings. Presentation determines toxicity and its management.

S. El Balkhi
INSERM U1144, Paris, France
e-mail: souleiman.elbalkhi@chu-limoges.fr

B. Mégarbane, MD, PhD (✉)
INSERM UMRS-1144, Paris, France

Paris-Diderot University, Paris, France

Department of Medical and Toxicological Critical Care, Lariboisière Hospital – Paris Diderot University, 2 Rue Ambroise Paré, Paris 75010, France
e-mail: bruno-megarbane@wanadoo.fr; bruno.megarbane@lrb.aphp.fr

G.S. Malhi et al. (eds.), *The Science and Practice of Lithium Therapy*,
DOI 10.1007/978-3-319-45923-3_17

- It is not possible to predict the severity of lithium poisoning from serum lithium concentration alone.
- The management of lithium poisoning involves supportive care, gastrointestinal decontamination and elimination enhancement.
- Haemodialysis is the technique of choice to enhance lithium elimination.
- Clinical decisions about when to use extracorporeal treatment should take into account: serum lithium concentration, kidney function, pattern of lithium toxicity, patient's clinical status and availability of extracorporeal treatments.

17.1 Introduction

Lithium poisoning remains relatively rare, but when it occurs, it is often life threatening. Over the past decade, more than 6,000 exposure cases were reported annually to the American Association of Poison Control Centers. In 2013, 6,610 single exposures were declared, including 528 cases with no toxicity, 768 with minor toxicity, 1,168 with moderate toxicity and 153 with major toxicity, resulting in 2,883 admissions in healthcare facilities and five fatalities (Mowry et al. 2014). From 2002 to 2012, about 55 deaths were attributed to lithium poisoning in the USA. Unfortunately, there is no similar data for Europe.

In the acute overdose setting, lithium is responsible for neurological, renal and cardiac compromise, thus frequently requiring admission and monitoring in an intensive care unit. Lithium toxicity may also occur at therapeutic doses, generally after a prolonged period of treatment. Nephrotoxicity is the major consequence in this chronic setting. This chapter will review all the presentations of acute and chronic lithium toxicity, with a focus on clinical characteristics, suggested mechanisms of toxicity and recommended management.

17.2 Lithium Toxicity

Lithium has been used for more than 60 years as a primary therapy for bipolar disorder (Geddes and Miklowitz 2013). Although its exact mechanism of action is still poorly understood, lithium is able to regulate multiple cellular targets to achieve dramatic mood-stabilizing effects. Its efficacy in treating acute manic phases and in preventing recurrent mood disorders has enabled a lowering of both suicide and mortality rates in bipolar disorder patients (Geddes and Miklowitz 2013). However, the therapeutic window between the effective pharmacological dose and toxicity is narrow. Side effects and even toxicity may be observed in patients exhibiting serum lithium concentrations within the therapeutic range (Malhi et al. 2009; Mégarbane et al. 2014).

It is now widely agreed that lithium toxicity is not only related to lithium concentrations in the different body fluids but also to the duration of exposure to lithium. Accordingly, lithium accumulation in body tissues and especially in the central nervous system (CNS), as recently demonstrated in experimental models (Hanak et al. 2015), has been hypothesized as being a major determinant of neurotoxicity (Timmer and Sands 1999). Mechanisms of lithium transport in and out of the cell membranes are still not fully understood; however, it has been suggested that substitution for sodium or potassium at one of the numerous transport proteins may allow lithium entrance into the cell, while more active transport that does not automatically equilibrate lithium concentrations across the membranes could be responsible for its efflux, with possible accumulation in the intracellular compartment.

In situations of overdose, three patterns of lithium toxicity are usually distinguished in relation to the ingested dose and to the duration of exposure to lithium (Fig. 17.1) (Jaeger et al. 1993; Waring et al. 2007; Ivkovic and Stern 2014), as described below.

17.2.1 Acute Toxicity

Acute toxicity refers to poisonings in lithium-naïve patients, in whom symptoms may be absent or minor, despite high serum lithium concentrations. This situation can be observed in self-overdosed patients recently started on lithium or those having accidentally ingested another patient's treatment. Gastrointestinal symptoms (nausea, vomiting and diarrhoea) and slight neurological symptoms (drowsiness, slurred speech, apathy and confusion) are the usual side effects despite elevated serum lithium concentrations.

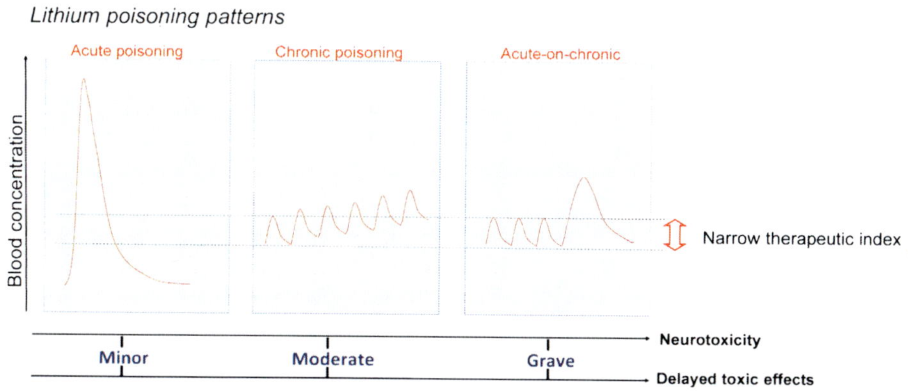

Fig. 17.1 Relationships between severity of clinical toxicity and serum lithium concentrations according to the three patterns of lithium poisoning in humans (acute, acute-on-chronic and chronic intoxications)

17.2.2 Acute-on-Chronic Toxicity

Acute-on-chronic toxicity occurs after acute lithium self-overdose in a previously lithium-treated patient. Severe symptoms are often observed. Concentrations of serum lithium are usually elevated and often higher than 3 mmol/L.

17.2.3 Chronic Toxicity

Chronic toxicity occurs insidiously due to lithium accumulation in a chronically lithium-treated patient. This kind of toxicity results from a rapid decrease in lithium renal clearance. Symptoms appear progressively when lithium concentrations are higher than 1.5 mmol/L, and the severity of presentation is usually correlated to the serum lithium concentration. Four major physiological factors, mirroring decrease in glomerular filtration rate (GRF) and long duration of treatment, have been associated with the risk of chronic poisoning: age over 50, thyroid dysfunction, impaired creatinine clearance and nephrogenic diabetes insipidus (NDI). A sudden salt restriction in the patient's diet, as well as instauration of nonsteroidal anti-inflammatory drugs (NSAID) or diuretics (loop diuretics and thiazides), angiotensin-converting enzyme inhibitors or cyclosporine, may reduce the GRF and potentially precipitate lithium toxicity. Pre-existing CNS disease and disruption of the blood-brain barrier may also worsen the risk of chronic poisoning. As potassium and calcium compete with lithium regarding transport at the cell membranes, lithium reuptake increases in the presence of hypokalaemia and hypocalcaemia, thus leading to an enhanced risk of lithium toxicity (Ivkovic and Stern 2014). An increased risk of chronic lithium toxicity has also been reported in the context of other illnesses, including hypertension, diabetes mellitus, chronic renal failure, heart failure, nephrotic syndrome, cystic fibrosis, epilepsy, anorexia and acute gastroenteritis, and following surgery and the rapid correction of hyponatraemia (Porto et al. 2009).

17.3 Features of Intoxication

Lithium overdose mainly results in the development of myoclonic encephalopathy (Timmer and Sands 1999; Waring et al. 2007; Ivkovic and Stern 2014; Okusa and Crystal 1994; Hansen and Amdisen 1978; Sheean 1991; Vodovar 2016). Acute toxicity is often associated with gastrointestinal symptoms. Severity is assessed using the Hansen and Amdisen score (Table 17.1) (Hansen and Amdisen 1978). Severe toxicity may result in a mortality rate of up to 15% and a 10% rate of neurological sequelae (Sheean 1991).

17.3.1 Neurotoxicity

Intrinsic lithium toxicity affects the CNS with a particular tropism to the cerebellum. At earlier stages of lithium intoxication, ataxia, coarse tremor, dyskinaesia, dysarthria, hyperreflexia and muscle weakness are observed (Timmer and Sands

Table 17.1 Hansen and Amdisen classification of lithium toxicity

Grade	Severity of poisoning	Signs and symptoms
Grade I	Mild intoxication	Nausea, vomiting, tremor, hyperreflexia, agitation, muscle weakness and ataxia
Grade II	Moderate intoxication	Stupor, muscular hypertonicity, rigidity and hypotension
Grade III	Severe intoxication	Altered mental status, convulsion, myoclonus and collapse

1999; Waring et al. 2007; Ivkovic and Stern 2014; Okusa and Crystal 1994; Hansen and Amdisen 1978; Sheean 1991). Typically, encephalopathy with altered cognitive and cerebellar functions occurs, along with seizures, pyramidal and extrapyramidal syndromes (Parkinsonism) and coma. Impairment in consciousness should be interpreted with caution, since it may hide a subclinical epilepticus state. Fever is frequent if neurotoxicity persists. Brainstem dysfunction is rare. Downbeat nystagmus, retrobulbar optic neuritis, persistent papilloedema, choreoathetoid movements, peripheral neuropathy and blindness have been reported on occasion. The majority of lithium-poisoned patients have abnormal electroencephalograms, typically including rhythmic slowing and polymorphic theta and delta waves of moderate-to-high voltage. Abnormalities can be continuous or paroxysmal, diffuse or focal, with epileptiform aspects sometimes mimicking status epilepticus. Patients with chronic or acute-on-chronic lithium toxicity are more likely to be symptomatic and to present higher score on the Hansen and Amdisen score and the poisoning severity score (Persson et al. 1998) than patients with acute toxicity (Waring et al. 2007; Ivkovic and Stern 2014). In severe presentations, myoclonic encephalopathy, usually delayed from the massive ingestion, may be prolonged up to 2 months but typically reverses in the absence of additional non-specific complications such as infection or brain anoxia. Lithium-related neurotoxicity is considered persistent when symptoms are present for more than 2 months after lithium elimination from the blood (Ivkovic and Stern 2014). Usually, most episodes of neurotoxicity are reversible on dose reduction, lithium cessation or elimination. Irreversible lithium-related neurotoxicity is uncommon (Porto et al. 2009; Niethammer and Ford 2007).

17.3.2 The SILENT

The term 'syndrome of irreversible lithium-effectuated neurotoxicity' (SILENT) was coined in 1987 to describe patients with 'long-lasting' neurologic symptoms induced by lithium toxicity that persisted for at least 2 months after discontinuation of the drug (Adityanjee 1987). In several case reports, SILENT was reported to persist for more than 1 year, in the absence of previous neurological impairment (Porto et al. 2009; Niethammer and Ford 2007). This syndrome includes persistent cerebellar dysfunction, extrapyramidal syndromes, brainstem dysfunction and dementia. Pathologic findings include extensive demyelination at several sites, especially in the cerebellum (Ivkovic and Stern 2014). Reports mentioned the loss of Purkinje cells, cerebellar atrophy and gliosis of the cerebellar cortex.

No risk factor has been evidenced for SILENT. No study has investigated relationships between the intoxication pattern and the likelihood of SILENT. SILENT may occur after acute, acute-on-chronic and chronic poisonings. SILENT has been observed over a broad range of serum lithium concentrations (0.1–8 mmol/L) and even reported in patients with concentrations in the therapeutic range.

17.3.3 Nephrotoxicity

Lithium-induced renal toxicity includes renal tubular acidosis, chronic tubulointerstitial nephritis (CTIN) and nephritic syndrome. The most common early complication in chronic lithium treatment is nephrogenic diabetes insipidus (NDI), with a prevalence of 20–70 % of patients (Porto et al. 2009; Alexander et al. 2008). Patients present with polyuria and polydipsia and an inability to concentrate their urine. Lithium interferes with the ability of antidiuretic hormone (ADH) to increase water permeability, and NDI occurs in relation to lithium accumulation in collecting tubule cells (Oliveira et al. 2010). Several mechanisms have been suggested, including decreased stimulation of adenylate cyclase, reduced density of ADH receptors and downregulation of aquaporin 2 in the apical plasma membrane of principal cells in the collecting ducts. NDI can be induced even if serum lithium concentrations are within the therapeutic range. The volume depletion induced by polyuria increases proximal reabsorption and consequently serum lithium. It is important to monitor this side effect closely, since patients with lithium-induced NDI have to maintain their oral fluid intake to compensate for their urinary losses, otherwise they are at risk of chronic poisoning (Timmer and Sands 1999).

Lithium-induced tubular toxicity is reversible in its earliest stages, but irreversible injury is possible (Alexander et al. 2008). CTIN is often irreversible and is preceded by the insidious development of renal insufficiency, with little or no proteinuria, and often after the onset of chronic NDI. However, whether lithium poses a serious risk for CTIN has been questioned. CTIN is characterized by the presence of tubular atrophy and interstitial fibrosis on renal biopsy. Chronic interstitial changes have even been described following acute intoxication or NDI (Alexander et al. 2008; Markowitz et al. 2000). Although CTIN represents a somewhat nonspecific pattern of disease, the presence of tubular cysts is a highly characteristic of lithium toxicity, having been reported in up to 40 % of cases. Glomerular toxicity of lithium therapy has also been described and lesions of focal segmental glomerulosclerosis reported (Markowitz et al. 2000). Finally, although earlier studies reported no (or only moderate) links between creatinine clearance, duration of lithium treatment and daily lithium dosage, the risk of developing chronic kidney disease in patients on long-term lithium treatment has now been well established by multiple clinical, histopathological and epidemiological studies (Alexander et al. 2008; Oliveira et al. 2010; Markowitz et al. 2000; Shine et al. 2015). Lithium therapy is associated with a progressive decline in creatinine clearance, resulting in an estimated interval of 20 years between the initiation of therapy and end-stage renal failure (Presne et al. 2003). Women younger than 60 and people with lithium concentrations higher than median are at the greatest risk (Shine et al. 2015). Thus, because lithium remains a treatment of choice for bipolar disorder, patients need baseline measurements of renal, thyroid and parathyroid function and regular long-term monitoring.

17.3.4 Cardiotoxicity

A wide range of cardiovascular complications, including arrhythmias, conduction disturbances and interstitial myocarditis, have been reported in lithium-poisoned patients (Timmer and Sands 1999; Waring et al. 2007; Ivkovic and Stern 2014; Okusa and Crystal 1994; Hansen and Amdisen 1978; Sheean 1991; Kayrak et al. 2010; Serinken et al. 2009). Variable electrocardiography (ECG) abnormalities have been observed, including QT prolongation, ST segment and T wave changes, sinus and junctional bradycardia and sinoatrial and atrioventricular blocks. Rarely, lithium poisoning has resulted in myocardial infarction (Puhr et al. 2008) or cardiac arrest from asystole. Brugada-pattern ECG changes have been reported, even at therapeutic levels, but in only a few case reports (Wright and Salehian 2010). The clinical consequences range from isolated ECG changes to cardiac syncope and sudden cardiac death. Removal of lithium has resulted in ECG normalization or conversion to type 2 or 3 Brugada ECG changes.

Lithium-related cardiac toxicity is attributed to various mechanisms, including competition with sodium, potassium, calcium and magnesium ions, which play important roles in cellular membrane physiology. Changes in ECG such as QT prolongation and non-specific abnormalities of ST segment and T wave are due to lithium-related effects on the myocardial cell membranes. Recent reports suggest that lithium dosage and concentrations should be considered as possible determinants of QTc prolongation (Mamiya et al. 2005). ECG changes are more likely to occur with chronic overdoses than with acute overdoses (Puhr et al. 2008). It is important to keep in mind that cardiovascular manifestations of lithium toxicity may be delayed by several days and tend to occur later than neurological effects (Waring et al. 2007). Interestingly, lithium-induced cardiotoxicity has been reported even in patients within therapeutic range (Singh et al. 2011).

17.3.5 Hormonal Toxicity

Lithium is reported to cause hypothyroidism, hyperparathyroidism, hypercalcaemia and weight gain (McKnight et al. 2012). It is taken up avidly by thyroid cells and blocks thyroid hormone release from thyroglobulin, which inhibits adenylate cyclase and prevents thyroid-stimulating hormone (TSH) from activating thyroid cells via TSH receptors. Lithium may also affect parathyroid hormone (PTH) synthesis, brought about by changes in calcium-sensing receptor sensitivity in the parathyroid gland causing a decrease in PTH secretion. In the long-term or overdose setting, hyperparathyroidism leads to hypothermia and hypercalcaemia with exacerbated psychiatric manifestations. Deleterious effects on mineral metabolism and urinary excretion due to hyperparathyroidism in intoxicated patients may also aggravate lithium-induced NDI (Timmer and Sands 1999; Oliveira et al. 2010).

17.4 The Role of the Laboratory

17.4.1 Diagnosing Lithium Poisoning

Despite all the potential confounding factors, the diagnosis of lithium poisoning relies on the measurement of serum lithium concentration as being equal to or above 1.5 mmol/L in conjunction with toxic symptoms. However, lithium intoxication may occur even when patients have serum lithium concentrations within the therapeutic range (Mégarbane et al. 2014). Thus, diagnosis of lithium poisoning should depend on the presence of clinical features of toxicity rather than laboratory investigations alone.

Analytically, the most used methods to measure lithium include flame emission spectrometry (FES), atomic absorption spectroscopy (AAS), automated colorimetry and ion-selective electrode (ISE) (El Balkhi et al. 2009). These techniques have limits of detection of around 0.05 mmol/L. More sensitive techniques, such as inductively coupled plasma optical emission spectroscopy (ICP-OES) and inductively coupled plasma mass spectrometry (ICP-MS), are less commonly used. Red blood cell (RBC) lithium can be measured directly after RBC isolation using density centrifugation then haemolysis, generally in de-ionized water. Proteins are then precipitated with trichloracetic acid, and lithium concentration is measured using FES, AAS or ISE. Internal standards (caesium or potassium) are used routinely to guarantee the quality of FES measurement. Importantly, one should be aware of the sampling conditions before diagnosing lithium poisoning, since lithium-heparinate can add up to 3 mmol/L of lithium, depending on the tube and the blood volumes.

Evaluation of lithium toxicity depends not only on the judicious interpretation of serum lithium concentrations but also on the determination of the poisoning pattern, the duration of exposure and the delay between the acute ingestion and presentation (Jaeger et al. 1993; Waring et al. 2007; Ivkovic and Stern 2014). However, distinction between the different patterns may be blurred and difficult to establish accurately, especially as physicians often rely on the observations of family members, who are subject to recall bias. Moreover, chronic lithium poisoning may present with non-specific clinical features, especially in older patients.

For years, RBC lithium concentration has been presented as a better reflection of brain lithium concentration than serum concentration. However, the utility of measuring RBC lithium concentration appears questionable (El Balkhi et al. 2009). Briefly, there is a large and complex inter- and intra-individual variability in RBC lithium distribution; methods used to measure RBC concentration vary from a simple calculation based on plasma concentration to direct measurement; and lithium transport from plasma varies between RBC, skeletal muscle and other tissues. Recently, it has been suggested that acute lithium exposure lowers serum tyrosine levels in the brain (McFarlane et al. 2011). However, whether chronic and acute-on-chronic poisonings exert different effects on serum tyrosine concentrations was not investigated. This would have offered an indirect biomarker to distinguish between the three patterns of lithium poisoning.

In practice, lithium toxicity is likely to correlate with local concentrations in brain and cardiac tissue, neither of which can be predicted easily from plasma or RBC concentrations alone. In addition, serum lithium concentrations have been reported to be similar between all patterns of lithium intoxications, and intoxicated patients may present to healthcare facilities several hours after serum peak. Thus, declining serum lithium concentrations alone may not offer reassurance and should be interpreted cautiously and in light of other clinical findings. Delayed neurotoxicity is commonly observed, and in the most severe poisonings, the peak of neurological impairment occurs from several hours to up to 2–3 days after peak serum lithium concentration has occurred. Serum lithium pharmacokinetics differ according to the poisoning pattern (Ferron et al. 1995). Lithium renal clearance in acute-on-chronic patients is approximately 28 mL/min, whereas it decreases to 12 mL/min in chronic patients. Lithium elimination in the serum and RBC is similar across all the patterns of poisoning, underscoring the limited usefulness of lithium measurement in RBC (El Balkhi et al. 2009). Nonetheless, poisoning management can be improved by serum concentration monitoring and calculation of the elimination half-lives of lithium.

17.4.2 Predicting Poisoning Prognosis

Lithium toxicity is weakly correlated with serum concentrations. Although mild toxicity usually appears at levels up to 2.5 mmol/L, severe toxicity at around 2.5–3.5 mmol/L and life-threatening effects at above 3.5 mmol/L (Waring et al. 2007; Ivkovic and Stern 2014; Waring 2007), there is still ongoing uncertainty about the significance of plasma lithium concentrations in poisoned patients. Patients with serum concentrations of up to 14 mmol/L have been reported without any toxicity. Chronic toxicity may also occur within the therapeutic range of serum lithium (Mégarbane et al. 2014). Thus, misevaluation of a patient's poisoning severity may result in delayed treatment and run the risk of neurological damage or death.

Based on 172 cases, including all three patterns of poisoning, the relationships between the presumed ingested lithium doses, serum lithium concentration and clinical severity of lithium toxicity have been investigated (Waring 2007). Neither the ingested dose nor the serum concentration predicted the likelihood of severe clinical features. Accordingly, serum lithium poorly reflects body impregnation by lithium, and measurements are usually performed when the patient is symptomatic, i.e., a long time after the serum peak. Experimental evidence suggests that peak lithium in the brain is delayed, with increasing concentrations up to 24 h after the serum peak. Lithium accumulation in the CNS is critical, especially in acute-on-chronic and chronic poisonings. These pharmacokinetic specificities are explained by the brain-blood barrier, which has relatively poor permeability for lithium, delaying not only its penetration but also its elimination from the brain, as recently demonstrated in rat models (Mégarbane et al. 2014). Since neurotoxicity is best predicted by the brain lithium burden, and since its measurement is impossible and not paralleled by serum lithium, prognosis and consequently treatment decisions in lithium poisoning are highly dependent on the pattern of lithium exposure (Vodovar 2016).

17.5 Management of Lithium Poisoning

Managing lithium poisoning is based on optimizing supportive care, gastrointestinal decontamination and elimination enhancement (Okusa and Crystal 1994; Hansen and Amdisen 1978; Waring 2006; Scharman 1997; Vodovar 2016).

17.5.1 Supportive Treatments

Treatment is based on adequate management of the patient's airways, ventilation and circulation, guided by the severity of the clinical presentation. Tracheal intubation should be considered in the presence of impaired consciousness or the occurrence of seizures due to the elevated risk of aspiration. Seizure activity should be screened daily using electroencephalography and treated with anticonvulsant drugs. Since lithium clearance is highly dependent on GFR, fluid resuscitation using isotonic saline is the cornerstone of therapy to normalize the patient's intravascular volume and to allow normal urine flow. Extracellular volume depletion, mainly related to digestive losses, results in the decrease of glomerular filtration and an increase in lithium tubular reabsorption, compromising lithium elimination (Scharman 1997). However, the risk of hypernatraemia is elevated due to saline infusion and lithium-induced NDI; thus in these situations, it is necessary to adapt the infused liquids to hypotonic ones. Standard treatments are required to treat pneumonia, sepsis, rhabdomyolysis and acute respiratory distress syndrome. Acute renal failure may require fluid repletion if related to hypovolaemia or haemodialysis, especially in the presence of elevated serum lithium concentrations.

17.5.2 Gastrointestinal Decontamination

Gastric lavage has limited effect on lithium removal unless it is performed before the plasma peak level is achieved, since lithium is rapidly absorbed in the gastrointestinal tract. International recommendations advise using gastric lavage if the poisoned patient is referred less than 2 h after lithium ingestion (Benson et al. 2013). Activated charcoal does not bind lithium and is therefore not useful. Whole bowel irrigation (WBI) using polyethylene glycol can remove unabsorbed lithium and appears useful, especially if sustained-release formulas are involved (Timmer and

Sands 1999; Okusa and Crystal 1994). An experimental study in ten healthy volunteers ingesting sustained-release 0.8 mg/kg lithium showed significant lowering in their lithium concentrations with WBI using 10 l of polyethylene glycol (PEG) solution administered 1 h after lithium ingestion over 5 h (Smith et al. 1991). A more recent observational study suggested that early WBI initiation (<12 h) was associated with less toxicity in lithium-poisoned patients (Bretaudeau Deguigne et al. 2013). However, several biases limited any definitive conclusion, since only gastrointestinal decontamination was studied and data regarding lithium elimination and initial neurologic status were poorly reported (Vodovar and Mégarbane 2013). Moreover, WBI feasibility is routinely limited in lithium-poisoned patients experiencing nausea and vomiting. International guidelines recommended WBI in sustained-release lithium poisoning in the absence of contraindications (i.e. significant neurologic impairment, unless the trachea is intubated, and bowel obstruction) (Thanacoody et al. 2015). WBI should be administered at a rate of 1,500–2,000 mL/h via a nasogastric tube in a seated or half-seated patient if not intubated or in an intubated patient until rectal effluent is clear.

Oral sodium polystyrene sulfonate was also reported to be efficacious to limit lithium bioavailability in chronically lithium-poisoned patients, with an additional contribution in increasing the renal clearance of lithium (Ghannoum et al. 2010). However, this treatment was shown to induce mild hypokalaemia, and its impact on the patient's neurologic condition and final outcome was not studied. The optimal dosage is unknown, but 1 g/kg seems reasonable. Consequently, since neither its effectiveness nor its toxicity profile is definitively known, sodium polystyrene sulfonate cannot be recommended on a routine basis for lithium poisoning.

17.5.3 Lithium Elimination Enhancement

Lithium is readily dialysable given its low molecular weight, small volume of distribution, the absence of protein binding, the absence of metabolism and exclusive renal clearance (Table 17.2). Thus, haemodialysis (HD) is the method of choice to enhance lithium elimination; however, its exact benefit to improve patient outcome has been questioned in the literature. According to the American Psychiatric Association treatment guidelines for lithium toxicity, HD is the only reliable method for rapidly lowering excess lithium levels and is more effective than peritoneal dialysis (Ivkovic and Stern 2014). Lithium is well cleared by haemodialysis (clearance: 80–120 mL/min) in comparison to its endogenous clearance (15–20 mL/min), with significant reduction in its half-life (3–6 h vs. 16–30 h) (Jaeger et al. 1993). High-flux membranes are additionally susceptible to enhance haemodialysis clearance of lithium. However, since intracellular lithium diffuses slowly from cellular compartments, rebound may occur in serum at the end of the haemodialysis session without necessarily being responsible for any worsening in the patient's clinical situation (Ivkovic and Stern 2014; Bosinski et al. 1998).

Due to the absence of any randomized controlled trial, the exact indications of haemodialysis remain debated, and its real impact on poisoning outcome is still not

Table 17.2 Lithium physicochemical and pharmacokinetic properties

Target therapeutic serum concentrations: 0.3–1.2 mmol/L (depending on the indications)
Conversion factor: 1 mmol/L = 1 mEq/L
Toxic dose (acute poisoning): >1 g elemental lithium
Oral bioavailability: 95–100 % (immediate release) and 60–90 % (modified release)
Serum peak: 30 min to 4 h after a dose (prolonged in delayed-release preparations)
Volume of distribution: 0.7–1.4 L/kg
Distribution to multiple tissues including brain, kidney and thyroid
Slow distribution to the CNS through the blood-brain barrier
Concentration sin the cerebrospinal fluid: ~40 % of the plasma concentration
Trace amounts in sweat, saliva and breast milk
Protein binding: negligible (free circulation as Li^+)
Metabolism: none
Route of elimination: renal
Filtration trough the glomerulus, reabsorption in the proximal tube (80 %), in the loop of Henle
Renal clearance: 13–56 mL/min
Elimination half-life: 12–27 h (longer in patients with long-term treatment; reduced in the elderly) and 12–50 h in poisonings

clear. If renal failure and severe neurological impairment are commonly accepted as haemodialysis criteria, the threshold of lithium concentration remains controversial, and the following have been proposed by different authorities: >4–7.5 mmol/L regardless of the clinical features and pattern of poisoning, >2.5–4 mmol/L in chronically exposed patients, >2.5 mmol/L with onset of severe toxicity and >2.5 mmol/L combined with renal failure. Others have not used lithium concentration thresholds but recommend instead determining the exact phase of lithium kinetics to decide when to start haemodialysis (Jaeger et al. 1993). Accordingly, increasing lithium concentrations during the first few hours corresponds to the persistence of absorption and thus supports gastrointestinal decontamination. A significant initial decrease in lithium concentrations corresponds to its distribution or preserved elimination, thus precluding any emergent indication for haemodialysis. In contrast, an increasing lithium half-life supports a delay in lithium elimination, suggesting initiating haemodialysis. These kinetic considerations may explain the discrepancies between the different thresholds of lithium concentrations proposed in the literature. Thus, Jaeger and colleagues propose the following criteria during the first 8–12 h after admission to initiate haemodialysis: severe lithium poisoning (grade III, particularly if acute-on-chronic or chronic poisoning), progressive clinical deterioration, increased serum half-life and renal impairment (Jaeger et al. 1993).

Recently the EXTRIP workgroup published guidelines to assist in the decision of whether to use renal replacement therapy to enhance lithium elimination (Table 17.3) (Decker et al. 2015). The workgroup reviewed all the studies and case reports regarding extracorporeal treatment of lithium-poisoned patients. It

Table 17.3 EXTRIP recommendations regarding extracorporeal treatments (ECTR) in lithium poisoning

General: ECTR is recommended in patients with severe lithium poisoning (1D)
Indications:
ECTR is recommended (1D)
If kidney function is impaired and the serum lithium concentration >4.0 mmol/L
In the presence of decreased level of consciousness, seizures or life-threatening dysrhythmias irrespective of serum lithium concentration
ECTR is suggested (2D)
If the serum lithium concentration >5.0 mmol/L
If confusion is present
If the expected time to obtain a serum lithium concentration >1.0 mEq/L with optimal management is >36 h
Cessation of ECTR: cessation is recommended (1D)
When the serum lithium concentration >1.0 mEq/L or clinical improvement is apparent
After a minimum of 6 h of ECTR if the serum lithium concentration is not readily available
After interruption of ECTR, serial serum lithium concentration measurements should be obtained over 12 h to determine use of subsequent ECTR sessions (1D)
Choice of ECTR:
Intermittent haemodialysis is the preferred ECTR (1D)
Continuous RRT is an acceptable alternative if intermittent haemodialysis is not available (1D)
After initial treatment, both continuous renal replacement therapies and intermittent haemodialysis are equally acceptable (1D)

Adapted from Decker et al. (2015) with permission from the American Society of Nephrology

concluded that lithium is dialysable (level of evidence = A) and supported the use of extracorporeal treatment in severe lithium poisoning (1D). However, the workgroup recommended that clinical decisions on when to use extracorporeal treatment should take into account serum lithium concentration, kidney function, pattern of lithium toxicity, patient's clinical status and availability of extracorporeal treatments. Extracorporeal treatment was 'recommended' if kidney function is impaired and serum lithium concentration exceeds 4.0 mmol/L or in the presence of a decreased level of consciousness and seizures or life-threatening dysrhythmias irrespective of the serum lithium concentration (1D). Extracorporeal treatment was 'advised' if serum lithium concentration is >5.0 mmol/L, significant confusion is present or the expected time to reduce the serum lithium concentration to <1.0 mmol/L is >36 h (2D). Extracorporeal treatment should be continued until clinical improvement is apparent or serum lithium concentration is <1.0 mEq/L (1D). Extracorporeal treatments should be continued for a minimum of 6 h if serum lithium concentration is not readily measurable (1D). Haemodialysis was chosen as the preferred extracorporeal treatment (1D), but continuous renal replacement therapies, including continuous veno-venous haemofiltration (CVVHF), continuous veno-venous haemodialysis (CVVHD) and continuous veno-venous haemodiafiltration (CVVHDF), was recognized as an acceptable alternative (1D). Higher clearance of lithium is constantly obtained with intermittent haemodialysis in comparison to continuous techniques (80–120 mL/min vs. 20–50 mL/min). However, lithium concentration rebound has been observed after haemodialysis (Bosinski et al. 1998), and some authors even suggest adding continuous haemofiltration after the initial intermittent haemodialysis (Meyer et al. 2001).

17.5.4 NDI Treatment

NDI is a possible consequence of lithium therapy and may occur during the institution of saline therapy when treating lithium poisoning. The key point in NDI is the loss of free water, thus requiring the administration of intravenous hypotonic fluids such as 2.5 % or 5 % dextrose, half-normal saline or the enteral administration of water. ADH is inefficient. However, reconstitution of the intracellular volume should be cautious to avoid additional metabolic complications. The following formulations can be used to manage fluids:

- Actual intracellular fluid (ICF) = [Normal ICF (estimated at 0.4 L/kg) × 2 × normal serum sodium concentration (140 mmol/L) × weight (kg)]/2 × measured serum sodium concentration
- Total free water deficit (FWD) = [(measured serum sodium concentration–140) × 0.66 × weight (kg)]/140

17.6 Conclusions

Lithium poisoning is rare but may be responsible for life-threatening presentations with neurological, renal and cardiovascular impairments. Nephrogenic diabetes insipidus is a potential complication of lithium at therapeutic doses and in poisoning. Toxicity is related to tissue and mainly CNS lithium concentrations and is not well correlated to serum lithium. Lithium is cleared by the kidneys, and factors causing a significant decrease in renal clearance may predispose to lithium toxicity in chronically treated patients. Haemodialysis enhances lithium clearance; however, its indications remain controversial and its exact benefit to improve patient outcome remains contentious.

17.7 Summary

The spectrum of lithium toxicity is large and includes gastrointestinal, central nervous system and cardiac and renal impairments. Neurotoxicity may be life threatening, resulting in delayed and prolonged myoclonic encephalopathy. Acute, acute-on-chronic and chronic lithium poisonings should be distinguished, since they result in different features and severity and have varying correlations with serum concentrations. Poisoning management should be supportive and based on saline diuresis, and therapies to decrease lithium absorption or to enhance lithium elimination should be considered. Haemodialysis is the technique of choice to increase lithium clearance; however its indications and exact clinical benefits are still a matter of debate.

References

Adityanjee (1987) The syndrome of irreversible lithium effectuated neurotoxicity. J Neurol Neurosurg Psychiatry 50:1246–1247

Alexander MP, Farag YM, Mittal BV et al (2008) Lithium toxicity: a double-edged sword. Kidney Int 73:233–237

Benson BE, Hoppu K, Troutman WG et al (2013) Position paper update: gastric lavage for gastrointestinal decontamination. Clin Toxicol (Phila) 51:140–146

Bosinski T, Bailie GR, Eisele G (1998) Massive and extended rebound of serum lithium concentrations following hemodialysis in two chronic overdose cases. Am J Emerg Med 16:98–100

Bretaudeau Deguigne M, Hamel JF, Boels D et al (2013) Lithium poisoning: the value of early digestive tract decontamination. Clin Toxicol (Phila) 51:243–248

Decker BS, Goldfarb DS, Dargan PI et al (2015) Extracorporeal treatment for lithium poisoning: systematic review and recommendations from the EXTRIP Workgroup. Clin J Am Soc Nephrol 10:875–887

El Balkhi S, Megarbane B, Poupon J et al (2009) Lithium poisoning: is determination of the red blood cell lithium concentration useful? Clin Toxicol (Phila) 47:8–13

Ferron G, Debray M, Buneaux F et al (1995) Pharmacokinetics of lithium in plasma and red blood cells in acute and chronic intoxicated patients. Int J Clin Pharmacol Ther 33:351–355

Geddes JR, Miklowitz DJ (2013) Treatment of bipolar disorder. Lancet 381:1672–1682

Ghannoum M, Lavergne V, Yue CS et al (2010) Successful treatment of lithium toxicity with sodium polystyrene sulfonate: a retrospective cohort study. Clin Toxicol (Phila) 48:34–41

Hanak AS, Chevillard L, El Balkhi S et al (2015) Study of blood and brain lithium pharmacokinetics in the rat according to three different modalities of poisoning. Toxicol Sci 143:185–195

Hansen HE, Amdisen A (1978) Lithium intoxication. Report of 23 cases and review of 100 cases from the literature. Q J Med 47:123–144

Ivkovic A, Stern TA (2014) Lithium-induced neurotoxicity: clinical presentations, pathophysiology, and treatment. Psychosomatics 55:296–302

Jaeger A, Sauder P, Kopferschmitt J et al (1993) When should dialysis be performed in lithium poisoning? A kinetic study in 14 cases of lithium poisoning. J Toxicol Clin Toxicol 31:429–447

Kayrak M, Ari H, Duman C et al (2010) Lithium intoxication causing ST segment elevation and wandering atrial rhythms in an elderly patient. Cardiol J 17:404–407

Malhi GS, Adams D, Berk M (2009) Is lithium in a class of its own? A brief profile of its clinical use. Aust N Z J Psychiatry 43:1096–1104

Mamiya K, Sadanaga T, Sekita A et al (2005) Lithium concentration correlates with QTc in patients with psychosis. J Electrocardiol 38:148–151

Markowitz GS, Radhakrishnan J, Kambham N et al (2000) Lithium nephrotoxicity: a progressive combined glomerular and tubulointerstitial nephropathy. J Am Soc Nephrol 11:1439–1448

McFarlane HG, Steele J, Vinion K et al (2011) Acute lithium administration selectively lowers tyrosine levels in serum and brain. Brain Res 1420:29–36

McKnight RF, Adida M, Budge K et al (2012) Lithium toxicity profile: a systematic review and meta-analysis. Lancet 379:721–728

Mégarbane B, Hanak AS, Chevillard L (2014) Lithium-related neurotoxicity despite serum concentrations in the therapeutic range: risk factors and diagnosis. Shanghai Arch Psychiatry 26:243–244

Meyer RJ, Flynn JT, Brophy PD, Smoyer WE, Kershaw DB, Custer JR, Bunchman TE (2001) Hemodialysis followed by continuous hemofiltration for treatment of lithium intoxication in children. Am J Kidney Dis 37:1044–1047

Mowry JB, Spyker DA, Cantilena LR Jr et al (2014) 2013 Annual Report of the American Association of Poison Control Centers' National Poison Data System (NPDS): 31st annual report. Clin Toxicol (Phila) 52:1032–1283

Niethammer M, Ford B (2007) Permanent lithium-induced cerebellar toxicity: three cases and review of literature. Mov Disord 22:570–573

Okusa MD, Crystal LJ (1994) Clinical manifestations and management of acute lithium intoxication. Am J Med 97:383–389

Oliveira JL, Silva Junior GB, Abreu KL et al (2010) Lithium nephrotoxicity. Rev Assoc Med Bras 56:600–606

Persson HE, Sjoberg GK, Haines JA et al (1998) Poisoning severity score. Grading of acute poisoning. J Toxicol Clin Toxicol 36:205–213

Porto FH, Leite MA, Fontenelle LF et al (2009) The Syndrome of Irreversible Lithium-Effectuated Neurotoxicity (SILENT): one-year follow-up of a single case. J Neurol Sci 277:172–173

Presne C, Fakhouri F, Noel LH et al (2003) Lithium-induced nephropathy: rate of progression and prognostic factors. Kidney Int 64:585–592

Puhr J, Hack J, Early J et al (2008) Lithium overdose with electrocardiogram changes suggesting ischemia. J Med Toxicol 4:170–172

Scharman EJ (1997) Methods used to decrease lithium absorption or enhance elimination. J Toxicol Clin Toxicol 35:601–608

Serinken M, Karcioglu O, Korkmaz A (2009) Rarely seen cardiotoxicity of lithium overdose: complete heart block. Int J Cardiol 132:276–278

Sheean GL (1991) Lithium neurotoxicity. Clin Exp Neurol 28:112–127

Shine B, McKnight RF, Leaver L et al (2015) Long-term effects of lithium on renal, thyroid, and parathyroid function: a retrospective analysis of laboratory data. Lancet 386:461–468

Singh LK, Praharaj SK, Munda SK et al (2011) Lithium-induced sinus node dysfunction at therapeutic serum levels. Natl Med J India 24:151–152

Smith SW, Ling LJ, Halstenson CE (1991) Whole-bowel irrigation as a treatment for acute lithium overdose. Ann Emerg Med 20:536–539

Thanacoody R, Caravati EM, Troutman B et al (2015) Position paper update: whole bowel irrigation for gastrointestinal decontamination of overdose patients. Clin Toxicol (Phila) 53:5–12

Timmer RT, Sands JM (1999) Lithium intoxication. J Am Soc Nephrol 10:666–674

Vodovar D, Mégarbane B (2013) Maintaining lithium elimination is at least as important as early gastrointestinal decontamination in acute-on-chronic poisonings. Clin Toxicol (Phila) 51:808

Vodovar D, El Balkhi S, Curis E, Deye N, Mégarbane B (2016) Lithium poisoning in the intensive care unit: predictive factors of severity and indications for extracorporeal toxin removal to improve outcome. Clin Toxicol (Pila) 54:615–623

Waring WS (2006) Management of lithium toxicity. Toxicol Rev 25:221–230

Waring WS (2007) Delayed cardiotoxicity in chronic lithium poisoning: discrepancy between serum lithium concentrations and clinical status. Basic Clin Pharmacol Toxicol 100:353–355

Waring WS, Laing WJ, Good AM et al (2007) Pattern of lithium exposure predicts poisoning severity: evaluation of referrals to a regional poisons unit. QJM 100:271–276

Wright D, Salehian O (2010) Brugada-type electrocardiographic changes induced by long-term lithium use. Circulation 122:e418–e419

Lithium and Pregnancy

18

Anne-Laure Sutter-Dallay and Stéphanie Brun

Abstract

During the perinatal period, women are at an increased risk of the onset or recurrence of most psychiatric disorders and this is especially so for bipolar disorders. Medical teams regularly question the prescription of lithium during pregnancy, although it remains a widely used mood stabiliser. Data from the current literature are reassuring, particularly regarding the risk of cardiac teratogenicity, which seems lower than that reported by earlier work. Thus it is possible to use lithium throughout the perinatal period, but only within a strictly monitored multidisciplinary framework that includes obstetricians, psychiatrists and paediatricians.

A.-L. Sutter-Dallay, MD, PhD (✉)
Department of Adult Psychiatry, Charles Perrens Hospital, Bordeaux, France

Bordeaux University, U657, 33000 Bordeaux, France

INSERM, U657, 33000 Bordeaux, France
e-mail: alsutter@ch-perrens.fr

S. Brun
Aliénor d'Aquitaine Maternity, University Hospital, Bordeaux, France

G.S. Malhi et al. (eds.), *The Science and Practice of Lithium Therapy*,
DOI 10.1007/978-3-319-45923-3_18

Key Points

- The likelihood of foetal and neonatal abnormalities with the prescription of lithium is modest.
- The risks associated with the treatment of maternal mood disorders with lithium during pregnancy and during the perinatal period need to be balanced against the risk of relapse if treatment is stopped. This decision requires careful deliberation.
- To ensure that the risks associated with lithium use are minimised it must be monitored closely, and its prescription should ideally be managed jointly within a multidisciplinary framework that includes obstetricians and paediatricians alongside psychiatrists.
- Risks such as cardiac abnormalities and neural tube defects associated with the prescription of lithium remain important considerations, but concerns regarding these potential complications appear to have been overestimated in earlier studies.

18.1 Introduction

18.1.1 Bipolar Disorders and Pregnancy

Interactions between bipolar disorders and the perinatal period are of great concern. In fact, 7 % of bipolar women present with their first episode in the context of the perinatal period (Viguera et al. 2011). In women with bipolar disorder prior to pregnancy, the risk of relapse reaches approximately 20 % during pregnancy (van der Lugt et al. 2012) and lies between 50 % and 70 % postpartum (Jones and Craddock 2005; Viguera et al. 2011). These perinatal episodes are predominantly mixed states or depressive episodes (Viguera et al. 2007). Conversely, bipolar disorder itself seems to have a potential influence on the course of pregnancy. A recent epidemiological study (Boden et al. 2012) shows that the risk of obstetric complications is higher in bipolar patients, treated or not, than in the general population.

Working with a preventive perspective during this period of life is crucial. We should provide relevant information to patients at the time of the diagnosis of a mood disorder and adopt multidisciplinary monitored and planned pregnancy, ideally starting prior to conception. Standard puerperium monitoring should involve examination for hypomanic symptoms during the early postpartum days that seem to characterise the clinical status of patients most likely to develop a postnatal episode (Heron et al. 2008). One effective and simple to-implement preventive approach is the promotion of sleep for new mothers. Indeed, changes in sleep architecture during the perinatal period, associated with painful physical events and the stress of birth, can lead to major sleep disorders that may favour the occurrence of mood episodes.

Finally, and beyond the frequency of these disorders, there is a real and vital risk to women during the perinatal period. An Australian report on maternal deaths (1994–2002) suggests that maternal mental illness is one of the principal causes of maternal perinatal death, with a large majority of violent deaths by suicide (Austin et al. 2007), suggesting a link with mood disorders. Equally, some studies show that many of those patients committing infanticide appear to present with episodes of bipolar disorder with psychotic features at the time of their regrettable actions (Kim et al. 2008).

Thus, patients should be informed of the potential influence of pregnancy on the course of their disorders, as well as of the risks associated with treatment or absence of treatment, through a 'risk-benefit' reflection. This discussion should involve the patient, if possible their partner, the psychiatrist, the obstetrician and the paediatricians. The goal should be to balance the risks of relapse in interrupting or modifying treatment versus the teratogenic and/or foetal risks linked to in utero exposure to psychotropic drugs. Indeed, maternal decompensation during pregnancy and/or the postpartum period can lead to risky behaviours but also to the worsening of the prognosis in the long term, because a perinatal episode is an additional episode. As regards to risks to the foetus, it is important to consider risks related to prenatal stress exposure, increased tobacco consumption and/or toxic exposure, as well as those related to exposure to higher (because curative) doses of psychotropic drugs. Finally, we must emphasise the potential risk of an inability of the mother to care for her child, and the risk of interaction disorders, requiring specialised care.

18.1.2 Metabolism, Teratogenicity and Pregnancy

A prescribed drug in a pregnant woman acts simultaneously on two individuals whose physiological and metabolic capabilities are very different. In the pregnant woman, fluid distribution is considerably changed, due to haemodynamic variations (increased heart rate and decreased blood pressure) and fluid augmentation (from a water extracellular volume of 14–20 l and for intracellular volume, from 28 to 30 l). Meanwhile, renal blood flow increases during pregnancy, leading to an increase in the renal elimination of drugs. In the foetus, blood pressure is low and is characterised by a very large cephalic vascularization, blood-brain barrier free, associated with a shunt of liver and lungs, limiting metabolic capabilities. The only elimination route is placental circulation, which allows a return to maternal circulation, since no urinary excretion exists. The amniotic fluid is urine being secreted and swallowed by the foetus. Finally, the term 'placental barrier' is not a reality, since any molecule weighing less than 600 Da crosses from mother to foetus.

The effect on the child depends on whether exposure occurs during the 'embryonic' (first trimester) or 'foetal' period (second and third trimester). The risk of birth defects is of particular concern during the first trimester of pregnancy, the period of organogenesis (although for certain organs, and especially the central nervous system, development continues throughout pregnancy and even after birth). Animal studies and neonatal monitoring of the children of treated mothers provide a useful assessment of those risks. However, we have less knowledge of the influence of psychotropic drugs on the course of pregnancy, on the foetus during the second and third trimesters and on the further development of the child. Research remains rare, albeit increasing, regarding psychotropic treatments, and is mainly retrospective and frequently carried out on small sample sizes, as ethical concerns currently make it impossible to develop prospective studies. In addition, no work takes into account the influence of maternal somatic disorders on foetal development and pregnancy issues, such as, for example, malnutrition, obesity, diabetes, gynaecological infections and much less the influence of environmental factors, such as an unhealthy lifestyle or domestic violence.

18.2 Lithium and Pregnancy

18.2.1 Placental Transfer

Placental transfer of lithium has been studied by Newport and colleagues (2005). This work combined the results of a prospective sample of ten women with 32 cases of neonatal dosages identified in the literature. Their results suggest that placental transfer of lithium is complete and that the balance between maternal and foetal circulation (average child/mother plasma lithium ratio = 1.05) is as between the different fluid compartments of an individual. Another recent study (van der Lugt et al. 2012) found neonatal blood lithium levels above 0.8 mEq/L in 2 of 30 examined children, 1 of whom presented signs of neonatal distress. This work does not provide the blood levels of the other children included in the sample and does not take into account co-prescriptions of other psychotropic drugs.

18.2.2 Embryonic Period Organogenesis

The first studies on lithium teratogenicity are animal studies dating from the end of nineteenth century (Giles and Bannigan 2006). In animals, with doses used for treatment in human beings, the studies do not show an increase in teratogenic risk. With very high doses, some studies reveal different types of malformations (central nervous system, skeletal, craniofacial, etc.), while others did not find any link between prenatal exposure to lithium and birth defects (Giles and Bannigan 2006). The first data on lithium's teratogenicity in humans come from the *Register of Lithium Babies*, which reported, retrospectively, the pregnancy outcomes of patients treated with lithium in Denmark, the United States and Canada. The first work reported 118 cases of Danish children whose mothers had taken lithium during the first trimester of pregnancy (Schou et al. 1973). Nine of them (7.6 %) presented with congenital malformations, six of which related to the cardiovascular system, including one Ebstein anomaly (severe malformation of the tricuspid valve). By adding data from the United States and Canada, the number of cases reported increased to 143. In this second sample, cardiovascular malformations accounted for 77 % of congenital defects, against 12.5 % in the general population. Ebstein's anomaly was largely overrepresented: 40 % of babies from the registry against 1.25 % of cardiac malformations in the general population. The final publication (Weinstein and Goldfield 1975) included 225 observations (at least three children were also exposed to other treatments), among which 11 % (n=25) of the children presented congenital malformations (against 2 % in the general population). Three quarters of these malformations were cardiovascular (n = 18) and 33 % (n=6) were Ebstein anomalies. The retrospective collection of data in this register represents a major bias, since it is likely that pathological cases have been more frequently reported than normal births. Subsequent retrospective studies have used more strict data collection methods and by and large did not reveal any statistically significant relationship between the use of lithium during pregnancy and the occurrence of cardiac malformations in the newborn, nor links between Ebstein's anomaly and prenatal exposure to lithium (Nora et al. 1974; Kozma 2005). In addition, a review of 59 cases of Ebstein diseases found no cases of children exposed to lithium in early pregnancy (Zalzstein et al. 1990). There are a few prospective studies (Jacobson et al. 1992; Bogen et al. 2012), some unpublished (Gentile 2012), that do not show a statistically significant link with birth defects, although some note the existence of sporadic cases of Ebstein's anomaly (Jacobson et al. 1992; Gentile 2012). Case-control studies have also failed to demonstrate a significant association between birth defects and in utero exposure to lithium (Zalzstein et al. 1990; Gentile 2012). Case reports in the literature indicate both heart defects, sometimes with Ebstein's anomaly, and other types of defects, including neural tube defects (Gentile 2012).

Finally, a very recent meta-analysis (McKnight et al. 2012) about the general toxicity of lithium concluded, inter alia, that the risk of congenital malformation after early in utero exposure to lithium is 'uncertain' and that the benefit-risk balance of a decrease or a discontinuation of treatment during pregnancy should be weighed for each clinical situation. If treatment with lithium is continued during the first trimester, it is recommended that foetal echocardiography is performed between 18 and 20 weeks of gestation (Nordon et al. 2007).

18.2.3 Foetal and Neonatal Periods

Data on late in utero exposure to lithium comes mainly from retrospective studies. A number of case studies report various perinatal complications: 'floppy infant syndrome' (hypotonia, poor sucking, tachypnoea, tachycardia, respiratory distress), macrosomia, polyhydramnios, hyperbilirubinemias, diabetes insipidus, hypothyroidism and hypoglycaemia. None of these studies has demonstrated statistically significant links between in utero exposure to lithium and any of these complications, which sometimes occur in situations of co-prescription, and remain largely inconsequential (Kozma 2005; Sutter-Dallay et al. 2010; van der Lugt et al. 2012).

18.2.4 Maternal Complications

An increase in the volume of distribution and renal excretion among pregnant women usually requires an increase of prescribed doses during pregnancy in order to maintain blood levels within the therapeutic range. Conversely, after birth, a decline of blood volume prompts the need to decrease dosages. Given these variations, it is recommended that serum assays are performed every 4 weeks up to 36 weeks of gestation, then weekly until birth and in the first 24 h after birth (NICE 2006). Thereafter, requirements will be adapted according to standard protocols, while maintaining special vigilance during the first 15 days postpartum. Some authors (Newport et al. 2005) propose achieving levels within the therapeutic window in the 24–48 h prior to delivery when scheduled or upon the start of delivery, with rapid reintroduction immediately afterwards. In pregnant and/or postpartum women, the complication that is most feared, more than overdose, is relapse. The decision to continue lithium in early pregnancy primarily depends on the severity and course of the disorder. The works of Viguera and colleagues (Viguera et al. 2007) emphasise the interest of maintaining a mood-stabilising treatment when necessary. In a cohort of bipolar patients compared to non-pregnant patients, those who discontinued lithium during pregnancy had approximately a twofold higher risk of postpartum relapse, and those relapses occurred four times faster and lasted five times longer.

18.3 Lithium and Lactation

In general, in women who breastfeed, the necessity of being 'available' continuously for the child leads to sleep deprivation. Breastfeeding may also represent a stress factor in itself, for example, with anxiety over the implementation of difficult breastfeeding and uncertainties about the amount of milk produced. In addition to the direct effect of medications on the newborn, these two dimensions must be taken into account when considering the risk-benefit balance of psychotropic drug prescription during breastfeeding.

Lithium is excreted in breast milk, and some data suggest that children receive between 30% and 40% of maternal doses (Nielsen and Damkier 2012). Renal metabolism associated with a high risk of rapid dehydration in newborns increases the risk of reaching toxic levels quickly. Given these findings, breastfeeding while treating with lithium remains controversial (Bogen et al. 2012). Although rare, serious complications have been reported (cyanosis, hypotonia) (Hale 2008). If the patient chooses to breastfeed anyway, it is preferable to administer treatment immediately after a feed so that the maternal plasma concentration is minimal at the next feed. Very regular monitoring of maternal plasma dosages, TSH and renal function is recommended (at 6 weeks of life and every 8–12 weeks thereafter), combined of course with clinical follow-up (Viguera et al. 2007).

Finally, remember that, for women who choose not to breastfeed, dopamine agonists such as bromocriptine, typically prescribed to inhibit lactation, should be banned because they can precipitate mood disorder relapse (Ansm 2013). The alternative is non-drug measures, which do not prevent lactation but minimise it (moderate water restriction, local anti-inflammatory creams, breast restraint).

18.4 Child Development

If intrauterine exposure to lithium does not seem to generate a significant excess risk of congenital malformations, the question of an effect on foetal brain development in the long term remains. Indeed, brain structures develop throughout pregnancy and may be particularly susceptible to the impact of drugs. An animal study reported long-term effects with 'anxious' persistent behaviour in the pups of treated dams (Youngs et al. 2006). The very few publications in humans are yet to identify any significant developmental peculiarities (Schou 1976; Jacobson et al. 1992; van der Lugt et al. 2012), except for a case of a minor neurological disorder. However, the extreme paucity of data does not allow any conclusions at present.

18.5 Summary

If a psychiatric decompensation requiring lithium prescription occurs perinatally, whether it is a first episode or a relapse, rapid drug therapy is crucial. For patients already being treated, it is advisable to maintain effective treatment. Regarding breastfeeding, the decision remains delicate and belongs to the parents, after the provision of specialised benefit-risk information, and with a specialised paediatric follow-up.

Finally, when possible, it is advisable to work with a specialised perinatal psychiatry team. If this type of approach is not possible, early multidisciplinary collaboration involving obstetric teams, the psychiatrist, the patient's partner and, when necessary, social services, must always be proposed so as to develop a specific pathway to care for each situation. Such collaborations allow an optimal monitoring of these 'at-risk' pregnancies (H.A.S. 2009), as well as perinatal accompanying of the patient and of the medical teams.

References

ANSM(2013)http://ansm.sante.fr/S-informer/Points-d-information-Points-d-information/Bromocriptine-Parlodel-R-et-Bromocriptine-Zentiva-R-le-rapport-benefice-risque-n-est-plus-favorable-dans-l-inhibition-de-la-lactation-Point-d-information

Austin MP, Kildea S, Sullivan E (2007) Maternal mortality and psychiatric morbidity in the perinatal period: challenges and opportunities for prevention in the Australian setting. Med J Aust 186:364–367

Boden R, Lundgren M, Brandt L, Reutfors J, Andersen M, Kieler H (2012) Risks of adverse pregnancy and birth outcomes in women treated or not treated with mood stabilisers for bipolar disorder: population based cohort study. BMJ 345:e7085

Bogen DL, Sit D, Genovese A, Wisner KL (2012) Three cases of lithium exposure and exclusive breastfeeding. Arch Womens Ment Health 15:69–72

Gentile S (2012) Lithium in pregnancy: the need to treat, the duty to ensure safety. Expert Opin Drug Saf 11:425–437

Giles JJ, Bannigan JG (2006) Teratogenic and developmental effects of lithium. Curr Pharm Des 12:1531–1541

H.A.S. (2009) Haute Autorité de Santé. Recommandations de bonnes pratiques: grossesses à risque: orientation des femmes enceintes entre les maternités en vue de l'accouchement

Hale TW (2008) Medication and mother' s milk, 13th edn. Hale Publishing, Amarillo

Heron J, McGuinness M, Blackmore ER, Craddock N, Jones I (2008) Early postpartum symptoms in puerperal psychosis. BJOG 115:348–353

Jacobson SJ, Jones K, Johnson K, Ceolin L, Kaur P, Sahn D et al (1992) Prospective multicentre study of pregnancy outcome after lithium exposure during first trimester. Lancet 339:530–533

Jones I, Craddock N (2005) Bipolar disorder and childbirth: the importance of recognising risk. Br J Psychiatry 186:453–454

Kendell RE, Chalmers JC, Platz C (1987) Epidemiology of puerperal psychoses. Br J Psychiatry 150:662–673

Kim JH, Choi SS, Ha K (2008) A closer look at depression in mothers who kill their children: is it unipolar or bipolar depression? J Clin Psychiatry 69:1625–1631

Kozma C (2005) Neonatal toxicity and transient neurodevelopmental deficits following prenatal exposure to lithium: another clinical report and a review of the literature. Am J Med Genet A 132:441–444

McKnight RF, Adida M, Budge K, Stockton S, Goodwin GM, Geddes JR (2012) Lithium toxicity profile: a systematic review and meta-analysis. Lancet 379:721–728

Munk-Olsen T, Laursen TM, Pedersen CB, Mors O, Mortensen PB (2006) New parents and mental disorders: a population-based register study. JAMA 296:2582–2589

Newport DJ, Viguera AC, Beach AJ, Ritchie JC, Cohen LS, Stowe ZN (2005) Lithium placental passage and obstetrical outcome: implications for clinical management during late pregnancy. Am J Psychiatry 162:2162–2170

NICE (2006) Clinical guideline 38. Bipolar disorder: the management of bipolar disorder in adults, children and adolescents, in primary and secondary care. National Institute for Health and Clinical Excellence, UK-London

Nielsen RE, Damkier P (2012) Pharmacological treatment of unipolar depression during pregnancy and breast-feeding – a clinical overview. Nord J Psychiatry 66:159–166

Nora JJ, Nora AH, Toews WH (1974) Letter: lithium, Ebstein's anomaly, and other congenital heart defects. Lancet 2:594–595

Nordon C, Sutter AL, Verdoux H (2007) Management of women with bipolar disorders from conception through the postpartum period. Presse Med 36:1913–1918

Schou M (1976) What happened later to the lithium babies? A follow-up study of children born without malformations. Acta Psychiatr Scand 54:193–197

Schou M, Goldfield MD, Weinstein MR, Villeneuve A (1973) Lithium and pregnancy. I. Report from the register of lithium babies. Br Med J 2:135–136

Sutter-Dallay AL, Lacaze I, Chazaud C, Rascle N, Dallay D, Rebola M et al (2010) Psychotropes, grossesse et adaptation néonatale du nouveau-né. Étude exploratoire (Unité Mère-Enfant, Bordeaux: 2001–2007). Ann Med Psychol 168:628–631

van der Lugt NM, van de Maat JS, van Kamp IL, Knoppert-van der Klein EA, Hovens JG, Walther FJ (2012) Fetal, neonatal and developmental outcomes of lithium-exposed pregnancies. Early Hum Dev 88:375–378

Viguera AC, Whitfield T, Baldessarini RJ, Newport DJ, Stowe Z, Reminick A et al (2007) Risk of recurrence in women with bipolar disorder during pregnancy: prospective study of mood stabilizer discontinuation. Am J Psychiatry 164:1817–1824

Viguera AC, Tondo L, Koukopoulos AE, Reginaldi D, Lepri B, Baldessarini RJ (2011) Episodes of mood disorders in 2,252 pregnancies and postpartum periods. Am J Psychiatry 168:1179–1185

Weinstein MR, Goldfield M (1975) Cardiovascular malformations with lithium use during pregnancy. Am J Psychiatry 132:529–531

Youngs RM, Chu MS, Meloni EG, Naydenov A, Carlezon WA Jr, Konradi C (2006) Lithium administration to preadolescent rats causes long-lasting increases in anxiety-like behavior and has molecular consequences. J Neurosci 26:6031–6039

Zalzstein E, Koren G, Einarson T, Freedom RM (1990) A case-control study on the association between first trimester exposure to lithium and Ebstein's anomaly. Am J Cardiol 65:817–818

19 Lithium in Children and Adolescents

Nathalie Nanzer and Jean-Michel Aubry

Abstract

In children and adolescents, lithium is mainly used to treat bipolar disorder. From the available data, modest improvement can be anticipated with lithium therapy in youths with acute-phase manic bipolar disorder. Placebo-controlled studies are necessary to determine whether lithium has antidepressant properties in paediatric bipolar disorder. Lithium seems to be an effective long-term treatment for patients with paediatric bipolar disorder who respond to acute treatment with lithium, but close monitoring of side effects during maintenance treatment is necessary. Lithium alone or in combination with an atypical antipsychotic may reduce aggressive behaviours in children and adolescents with conduct disorders.

N. Nanzer, MD
Child Guidance Unit, Division of Child and Adolescent Psychiatry, University Hospitals of Geneva, Chemin des Crêts-de-Champel 41, CH-1206 Geneva, Switzerland
e-mail: nathalie.nanzer@hcuge.ch

J.-M. Aubry, MD (✉)
Mood Disorders Unit, Division of Psychiatric Specialties, Department of Mental Health and Psychiatry, University Hospitals of Geneva, and Faculty of Medicine, University of Geneva, Rue de Lausanne 20 bis, CH-1201 Geneva, Switzerland
e-mail: jean-michel.aubry@hcuge.ch

G.S. Malhi et al. (eds.), *The Science and Practice of Lithium Therapy*,
DOI 10.1007/978-3-319-45923-3_19

Key Points

- The efficacy of lithium for the treatment of mania in youths has not been documented in double-blind studies.
- Lithium in monotherapy seems to have some efficacy in bipolar depression, but double-blind controlled studies are necessary to confirm preliminary results obtained with open-label trials.
- For maintenance treatment, lithium may be an effective longer-term treatment for patients with paediatric bipolar disorder who respond to acute treatment with lithium.
- Regarding safety and tolerability, lithium-related adverse events are frequently reported. Close monitoring of side effects during maintenance treatment is necessary.
- Lithium alone or in combination with an atypical antipsychotic may reduce aggressive behaviours in children and adolescents with conduct disorders, but contradictory results have been reported in available studies.

19.1 Introduction

In paediatric populations, lithium is mainly used to treat bipolar disorder. Most of the studies address the treatment of acute mania in monotherapy or in association with second-generation antipsychotics or other medications. Little is known about the treatment of bipolar depression and maintenance treatment in youth. In this population, lithium has also been studied in conduct disorder or severe mood dysregulation.

Lithium was the first medication to be approved by the US Federal Drug Agency (FDA) for the treatment of mania in bipolar disorder for youths aged 12–17 years. Despite this, few studies have specifically evaluated the efficacy of lithium in the treatment of paediatric bipolar disorder. Indeed, the long history of lithium use with adult bipolar disorder likely influenced the decision to approve it for adolescents (Washburn et al. 2011).

19.2 Lithium for the Treatment of Acute Mania

The pharmacologic agents used to treat mania in youths include lithium, anticonvulsant drugs with mood-stabilizing effects and second-generation antipsychotic medications. Because few large-scale, prospective studies have examined pharmacologic treatment for paediatric bipolar disorder, many of these medications are used without specific FDA approval for this condition (Washburn et al. 2011).

Published randomized controlled trials (RCTs) of traditional mood stabilizers are scarce, but the best available evidence suggests that second-generation antipsychotics (SGAs) as a group are more effective in reducing manic symptoms than are lithium and anticonvulsants (Peruzzolo et al. 2013).

Liu et al. (2011) conducted a systematic review of the available literature on the effectiveness of agents for the treatment of mania, depression and attention deficit hyperactivity disorder (ADHD) in children and adolescents. The review included 29 open-label studies and 17 RCTs covering 2,666 individuals and evaluated mood stabilizers, SGAs and naturopathic compounds (flax oil and omega-3). Modest effects were found for traditional antimanic agents such as lithium, when used as monotherapy. The SGAs as a group were significantly more effective than mood stabilizers and naturopathic compounds on meta-analytic regression of RCTs. On the negative side, SGAs were also associated with increased rates of weight gain and somnolence.

Weller and colleagues (2002) have suggested that lithium could be more efficient when bipolar disorder starts at an early age. However, treatment response may not be as good as in adult populations because of mixed and/or dysphoric manic phases. Frequent psychotic symptoms could also limit lithium response.

In their naturalistic study with 266 children, Masi and colleagues (2010) report that lithium is rarely used in prepubertal children and that it could be less efficient than valproic acid when used in monotherapy, which is in agreement with results from Strober et al. (1990) and Geller et al. (1998). Masi and colleagues (2010) also suggest that lithium may be more efficient than valproic acid for the treatment of bipolar I disorder in youths.

Some clinical factors have been reported to predict poor response to lithium treatment in paediatric bipolar disorder, including prepubertal illness onset and the presence of comorbid ADHD, substance abuse, conduct disorder and personality disorder. Furthermore, mixed manic episodes and rapid cycling, which are commonly seen in adolescents, may underlie poor response to lithium treatment (Kowatch et al. 2005).

Four open-label clinical trials involving lithium carbonate in the treatment of mania in paediatric bipolar disorder have been published (Kowatch et al. 2000; reviewed in Liu et al. 2011; Pavuluri et al. 2006; Kafantaris et al. 2001, 2003). The response rate for manic symptoms ranged from 23 % to 55 % (with an average response of 40 %). Of these four open-label clinical trials, only one assessed the acute effects of lithium as monotherapy and reported a response rate of 38 % (Kowatch et al. 2000). In this study (n = 42. 6–18 years), lithium, divalproex sodium or carbamazepine were compared for 6 weeks. All three drugs were found to be effective, and there was no significant difference between the three drugs for an improvement of at least 50 % on the Young mania scale and of 1 or 2 on the *Clinical Global Impression Improvement Scale* (CGIIS). With these criteria, response rate was 53 and 40 % with divalproex, 38 and 46 % with lithium and 38 and 31 % for carbamazepine.

Several RCTs evaluating the treatment of mania have also been reported in the past 15 years (reviewed in Liu et al. 2011; Peruzzolo et al. 2013). Geller and colleagues (1998) assessed the impact of lithium (6-week, n = 25, age = 16.3 years) on abstinence in substance-dependent adolescents with bipolar disorder type I or II. Results with both intent to treat and completers (n = 21) showed significant differences for improving functioning scores and weekly random urine assays. The authors found that lithium was more effective than placebo in improving functioning scores (Children's Global Assessment Scale: CGAS) but there was no between-group difference in measures of mood symptoms according to the schedule for affective disorders and schizophrenia for school-aged children (K-SADS) mood items. The study had several limitations.

Kowatch et al. (2007) conducted an 8-week RCT of lithium, divalproex or placebo in 153 subjects aged 7–17 years with bipolar I in manic or mixed episode. Lithium showed only a trend toward efficacy but did not clearly separate from placebo (52 % for divalproex, 42 % for lithium, 29 % for placebo). The effect size for lithium was moderate.

Another 8-week RCT was conducted with 279 children and adolescents aged 6–15 years. Subjects were randomly assigned to receive risperidone, lithium (1.1–1.3 mM/L) or divalproex sodium (Geller et al. 2012). The primary outcome measure was the Children Global Impressions for Bipolar Illness Improvement-Mania (CGI-BP-IM). Patients were also assessed with the Modified Side Effects Form for Children and Adolescent and the K-SADS Mania Rating Scale (KMRS). Patients in the risperidone group had a significantly higher response rate than those treated with lithium (68.5 vs 35.6 %; $p<0.001$) or those treated with divalproex sodium (68.5 vs 24.0 %; $p<0.001$). There was no significant difference in response rates between the lithium and the divalproex groups. Mean weight gain was significantly greater with risperidone than lithium (3.31 vs 1.42 kg; $p<0.001$), and greater increase in body mass index and prolactin levels was detected in the risperidone group.

In a double-blind study, Kafantaris et al. (2004) failed to demonstrate separation from placebo. In this study, adolescents responding to lithium treatment were randomized for 2 weeks to a placebo group or to a group continuing lithium treatment. The augmentation of symptoms occurred in both groups without significant differences, but one of the limitations of this study was the very short period of assessment.

Another double-blinded maintenance study (Findling et al. 2005) showed that lithium and divalproex had a similar long-term stabilizing effect in bipolar youth who had been previously stabilized on combination treatment with lithium plus divalproex. This study had a high dropout rate, with only three of 30 patients completing maintenance treatment.

Lithium, like other antimanic drugs, should be prescribed for a sufficient time and at a sufficient dosage to determine the effectiveness of the agent. Generally, a 6–8-week course, at appropriate doses, is recommended before adding another drug or substituting it altogether (Peruzzolo et al. 2013).

In conclusion, modest improvement can be anticipated with lithium therapy in youths with acute-phase manic bipolar disorder, but its efficacy has not been established in double-blind studies.

Combined medication treatment of two mood stabilizers has been studied in two open-label trials. In a study with 90 youths with bipolar I and II, Findling et al. (2003) evaluated in an open-label study lithium in combination with divalproex.

Remission was defined as no affective cycling and no additional treatment with another medication for four consecutive weeks. Primary outcome measures included the Young Mania Rating Scale (YMRS), the Child Depression Rating Scale (CDRS) and the CGI. Nearly half (47 %) of the subjects achieved remission at the end of the 11-week study. At week 8, subjects showed a 70.6 % response rate (50 % improvement in the YMRS). Fifteen patients withdrew from the study due to medication intolerance or side effects (ataxia, proteinuria, emesis, dysphoria, worsening mania).

In another prospective, 8-week open-label combination trial of lithium and divalproex sodium, Findling et al. (2006) assessed 38 children, with a mean age of 10.5 years who relapsed after treatment with either lithium monotherapy or divalproex sodium monotherapy. Thirty-four of the 38 subjects (89 %) responded to the combination treatment, although nearly two-thirds (65 %) of the patients also received a stimulant medication and four received adjunctive antipsychotic treatment.

19.3 Lithium in Bipolar Depression

Adults with bipolar disorder have been shown to spend most of their symptomatic time in the depressed phase of the illness (Judd et al. 2002, 2003). However, although children and adolescents with bipolar disorder experience depressive symptoms and episodes, less is known about the course and prevalence of depression in paediatric bipolar disorder (Chang 2009). Evidence-based treatment options for young people with depression and bipolar illness are still limited (Thomas et al. 2011).

One 6-week open-label study (Patel et al. 2006) examined the effectiveness and tolerability of lithium for the treatment of depression in 27 adolescents aged 12–18 years with bipolar I disorder.

The subjects received lithium 30 mg/kg, which was adjusted to achieve a therapeutic serum level of 1.0–1.2 mM/L. Response (defined as a reduction in CDRS-R score ≥50 %) occurred in 48 % of the subjects, suggesting plausible effectiveness in paediatric bipolar depression, and 30 % of the patients achieved remission (CDRS-R score 28 and a CGI-BP Improvement score of 1 or 2) with lithium monotherapy. Side effects were moderate and did not lead to abnormally high attrition, and the most common of these were headache (74 %), nausea/vomiting (67 %), stomach ache (30 %) and abdominal cramps (19 %).

In summary, there are very few studies regarding bipolar depression in children and adolescents, and all the available trials have limitations, such as small sample sizes and lack of a placebo group. Based on current data, lithium monotherapy seems to be effective.

However, at this stage, we can only agree with Cosgrove et al. (2013) who conclude that placebo-controlled studies would be necessary to conclusively determine whether lithium's antidepressant properties are robust in paediatric bipolar disorder but also that clinicians should currently consider using lithium to treat paediatric bipolar depression.

19.4 Lithium in Maintenance Treatment

Paediatric bipolar disorder has a high relapse rate, chronicity and poor treatment compliance. Approximately a third to two-thirds (35–70 %) of patients relapse within 6 months to 4 years after hospital discharge (DelBello et al. 2007). Strober and colleagues (1990) have reported in a prospective study that 90 % of adolescents with bipolar disorder have a new mood episode when they stop their treatment versus 37.5 % of those continuing lithium treatment.

Based on available studies and although knowledge is still limited, recommendations suggest that a mood stabilizer should be maintained for 12–24 months after a manic episode. Long-term treatment may be necessary, but this has to be evaluated and decided depending on the risk/benefit ratio for each patient (Kowatch et al. 2005).

In one study, adolescents who initially responded to lithium were randomly assigned to lithium or placebo for 2 weeks (Kafantaris et al. 2004). The results suggested that both lithium and placebo had similar rates of symptomatic relapse (52.6 % for lithium, 61.9 % for placebo). As stated by Peruzzolo et al. (2013), despite promising results in the open-label study phase, a large treatment effect for lithium is not evident in the maintenance phase.

In an 18-month double-blind trial, Findling and colleagues (2005) compared lithium and divalproex as maintenance monotherapy. Sixty children and adolescents (5–17 years) who had achieved four consecutive weeks of symptom remission during open-label combination therapy with divalproex and lithium were randomized to receive either drug as monotherapy in a double-blind fashion. The median survival time before symptoms returned was 114 days for lithium and 112 days for divalproex, suggesting that these agents are equally effective in maintaining mood stability. Concomitant treatment of comorbid ADHD with stimulant medication was not associated with earlier time to relapse (Thomas et al. 2011).

Findling and colleagues (2006) conducted an 8-week prospective, open-label trial with 38 patients aged 5–17 years with bipolar I or II who remitted with combination therapy consisting of lithium and divalproex. The main issue raised by the authors was whether patients who achieved stability on combination drug therapy would benefit from receiving more than one drug during long-term treatment. For this purpose, patients subsequently discontinued one of the agents. Those who became symptomatic during maintenance monotherapy with lithium or divalproex presented an 89.5 % remission rate when treated with the same combination, according to the YMRS, CDRS-R, CGAS and CGI-S.

In another randomized double-blind study, Findling and colleagues (2007) compared the efficacy of divalproex and lithium in young patients who had responded to a combination of both medications for 20 weeks. They were then randomized to receive a monotherapy as maintenance treatment up to 76 weeks. Both medications had similar efficacy.

Finally, Findling et al. (2013) examined the long-term effectiveness of lithium for the treatment of paediatric bipolar disorder within the context of combination mood stabilizer therapy for refractory mania and pharmacological treatment of comorbid psychiatric conditions. Forty-one outpatients, aged 7–17 years, with a diagnosis of bipolar I disorder, who demonstrated at least partial response to an 8-week trial of open-label treatment with lithium received open-label lithium for an additional 16 weeks. Up to two adjunctive medications could be prescribed to patients experiencing residual symptoms of mania or comorbid psychiatric conditions, following a standardized algorithm. Twenty-five of the 41 patients (60.9 %) were prescribed adjunctive psychotropic medication for residual symptoms: 13 (31.7 %) for refractory mania and 15 (36.6 %) for attention deficit hyperactivity disorder.

At the end of this phase, 28 patients (68.3 %) met criteria for response (≥50 % reduction from phase I baseline in YMRS, summary score and a CGI-I score of 1 or 2), with 22 (53.7 %) considered to be in remission (YMRS score ≤12 and CGISeverity score of 1 or 2).

These data suggest that patients who initially responded to lithium maintained mood stabilization during continuation treatment, but partial responders did not experience further improvement during phase II, despite the opportunity to receive adjunctive medications.

The most commonly reported (≥20 %) adverse events associated with lithium treatment were vomiting, headache, abdominal pain and tremor.

Regarding safety and tolerability across studies, the most frequent side effects were gastrointestinal disturbances, including nausea, vomiting and diarrhoea. Weight gain, cognitive dulling, tremor and fatigue were also frequently reported (Diaz-Caneja et al. 2014). Therefore, close monitoring of side effects during maintenance treatment is necessary.

In summary, lithium may be an effective longer-term treatment for patients with paediatric bipolar disorder who respond to acute treatment with lithium. Partial responders to acute lithium treatment did not appear to experience substantial symptom improvement during the continuation phase, despite the possibility that adjunctive medication could be prescribed.

19.5 Lithium in Other Psychopathologies

19.5.1 Conduct Disorder

Masi et al. (2009) studied the effectiveness of lithium in children and adolescents with conduct disorder in a retrospective naturalistic study of 60 patients, aged 8–17 years (mean age 14.2). All patients were initially treated with lithium, and the follow-up period was 6–12 months. Co-medication such as second-generation antipsychotics could be added if necessary to achieve satisfactory control of symptoms. At the end of the study, 29/60 patients (48.3 %) were classified as responders, ten receiving lithium monotherapy and 19 receiving lithium plus atypical antipsychotic therapy.

Verbal and physical aggression scores improved significantly both in patients taking lithium as monotherapy and those also taking an add-on atypical antipsychotic. Improvement in self-aggression score was found only when an atypical antipsychotic was added to lithium therapy. Predictors of a positive response to treatment were less severe disease at baseline, lower aggression scores and an impulsive type of aggression (affective, nonpredatory). Adverse effects were gastrointestinal, polydipsia, increased urinary frequency, tremor and increased thyroid stimulating hormone levels. Two patients discontinued treatment because of adverse effects (vomiting and thyroid dysfunction).

In summary, lithium alone or in combination with second-generation antipsychotics may reduce aggressive behaviours in children and adolescents with CD.

Levy and Bloch (2012) conclude that the use of drugs for this indication is broad despite the lack of systematic knowledge on this subject. Trials of lithium yield contradictory results. The most proven efficacy is for the atypical antipsychotics. Valproate or lithium may be possible second or third alternatives, based on some supporting evidence.

19.5.2 Severe Mood Dysregulation (SMD)

Dickstein et al. (2009) led the first RCT of lithium in youths with severe mood dysregulation. Lithium was chosen on the basis of its potential in treating irritability and aggression and neuro-metabolic effects. Youths aged 7–17 years and with SMD were tapered off their medications. Those who continued to meet SMD criteria after a 2-week, single-blind, placebo run-in were randomized to a 6-week double-blind trial of either lithium ($n=14$) or placebo ($n=11$). Forty-five percent ($n=20/45$) of SMD youths were not randomized due to significant clinical improvement during the placebo run-in.

Among randomized patients, there were no significant between-group differences in either clinical (CGI-I and PANSS factor 4 score) or magnetic resonance spectroscopy (based on myoinositol, N-acetyl-aspartate and combined glutamate/glutamine, all referenced to creatine) outcome measures.

19.6 Monitoring and Main Side Effects

While lithium is an US FDA-approved medication for 12–18-year-olds with bipolar disorder, it requires diligent monitoring. Lithium's small therapeutic window explains significant adverse effects, particularly at toxic levels. To achieve an accurate lithium level, blood should be drawn 12 h after a dose (Hamrin and Iennaco 2010).

The same therapeutic levels of lithium used for treating adult patients have been recommended for treating children and adolescents, because definitive youth-specific data to guide dosing are lacking. It is generally now accepted that concentrations should be maintained between 0.6 and 0.8 mmol/L (Grandjean and Aubry 2009a; Malhi et al. 2011), although some authors still favour 0.8–1.2 mmol/L (Thomas et al. 2011).

One approach to reaching the target blood level is to start lithium at a low dose of 20 mg/kg/day or 900 mg/day, divided into two or three daily doses, followed by gradual increases moderated by consideration of the patient's clinical response to the medication and reported side effects (Findling et al. 2008). In an exploratory study with outpatients aged 7–17 years with a diagnosis of bipolar I disorder, Findling and colleagues (2011) propose beginning lithium at a dosage of 300 mg thrice daily, with a 300 mg augmentation during the first week, and subsequent 300 mg weekly increases until therapeutic dosage is met.

Before the start of lithium therapy, a physical examination should be completed, and renal problems must be excluded. Baseline blood work should include a lithium level, blood urea nitrogen, creatinine concentration, thyroid function tests, electrolytes and complete blood count and should be repeated every 2–3 months during the first 6 months of treatment and every 6 months thereafter (Thomas et al. 2011). Once a therapeutic level has been achieved on a stable dose for three consecutive months, lithium levels should be checked every 3 months (Findling et al. 2008). Pregnancy tests should be obtained for adolescents of childbearing age, and education on appropriate measures to prevent pregnancy should be discussed.

Although serious lithium-induced cardiac side effects appear to be rare in youths without preexisting cardiac disease, it is generally recommended that an electrocardiogram is obtained at baseline after a few months of treatment and yearly thereafter (Findling et al. 2008).

Due to lithium's potential to cause birth defects (cardiac malformations, including Ebstein anomaly, in infants exposed during the first trimester, a reliable contraception method should be used (Grandjean and Aubry 2009b). (For more detail, see Chap. 18.) Weight should be monitored because lithium has demonstrated significant weight gain.

It has been asserted that younger children may be more vulnerable to lithium-related side effects than older adolescents, but this assertion has not been definitively confirmed (Campbell et al. 1991; Hagino et al. 1995). Adverse effects of lithium include nausea, vomiting, diarrhoea, increased appetite, weight gain, headache, hypothyroidism, renal function abnormalities, polyuria, polydipsia, leukocytosis, tremors and acne. The dose should be either discontinued or lowered if lithium toxicity occurs. Symptoms of lithium toxicity include ataxia, dysarthria, reduced motor coordination, diarrhoea, vomiting, anorexia, weakness, blurred vision, tinnitus, polyuria, coarse tremor, muscle twitching, irritability and agitation.

Several drugs have been found to increase lithium levels, including carbamazepine, nonsteroidal anti-inflammatory drugs, tetracyclines and thiazide diuretics (Scahill et al. 2001). Theophylline and caffeine promote lithium excretion, resulting in lower serum levels of lithium at the same oral dose. Clinicians should discuss these potential interactions with patients and their families.

19.7 Summary

Lithium is the oldest mood stabilizer and has been used in adult patients with bipolar disorder for more than 50 years. In children and adolescents with mood disorders, empirical evidence is much more limited. For the treatment of mania, available studies suggest that second-generation antipsychotics are more efficient than lithium. Modest improvement at best can be expected with lithium treatment of mania in youths.

Encouraging results have been reported with lithium in the treatment of depression in youths, but there are few open studies, and these have small sample size. Therefore, lithium cannot be recommended for the treatment of depression in young populations based on this level of evidence.

Regarding maintenance treatment, based on the available limited evidence, recommendations suggest that a mood stabilizer should be maintained for 12–24 months after a manic episode. Maintenance of treatment for the longer term has to be evaluated on a case-by-case basis, depending for instance on the severity of the episode, comorbidity and familial history of mood disorders. Long-term lithium treatment could be especially effective for patients who have responded to acute treatment with lithium.

Lithium treatment requires regular monitoring. The same therapeutic levels (0.6–0.8 mmol/L) as in the adult population have been recommended for children and adolescents aged 12–18 years.

Apart from bipolar disorders, lithium monotherapy or in combination with SGAs may reduce aggressive behaviours in children and adolescents with conduct disorder, but contradictory results have been reported.

References

Campbell M, Silva RR, Kafantaris V, Locascio JJ, Gonzalez NM, Lee D, Lynch NS (1991) Predictors of side effects associated with lithium administration in children. Psychopharmacol Bull 27(3):373–380

Chang K (2009) Challenges in the diagnosis and treatment of pediatric bipolar depression. Dialogues Clin Neurosci 11(1):73–80

Cosgrove VE, Roybal D, Chang KD (2013) Bipolar depression in pediatric populations : epidemiology and management. Paediatr Drugs 15(2):83–91. doi:10.1007/s40272-013-0022-8

DelBello MP, Hanseman D, Adler CM, Fleck DE, Strakowski SM (2007) Twelve-month outcome of adolescents with bipolar disorder following first hospitalization for a manic or mixed episode. Am J Psychiatry 164(4):582–590. doi:10.1176/appi.ajp.164.4.582

Diaz-Caneja CM, Moreno C, Llorente C, Espliego A, Arango C, Moreno D (2014) Practitioner review: long-term pharmacological treatment of pediatric bipolar disorder. J Child Psychol Psychiatry 55:59. doi:10.1111/jcpp.12271

Dickstein DP, Towbin KE, Van Der Veen JW, Rich BA, Brotman MA, Knopf L, Onelio L, Pine DS, Leibenluft E (2009) Randomized double-blind placebo-controlled trial of lithium in youths with severe mood dysregulation. J Child Adolesc Psychopharmacol 19(1):61–73. doi:10.1089/cap.2008.044

Findling RL, McNamara NK, Gracious BL, Youngstrom EA, Stansbrey RJ, Reed MD, Demeter CA, Branicky LA, Fisher KE, Calabrese JR (2003) Combination lithium and divalproex sodium in pediatric bipolarity. J Am Acad Child Adolesc Psychiatry 42(8):895–901. doi:10.1097/01.CHI.0000046893.27264.53

Findling RL, McNamara NK, Youngstrom EA, Stansbrey R, Gracious BL, Reed MD, Calabrese JR (2005) Double-blind 18-month trial of lithium versus divalproex maintenance treatment in pediatric bipolar disorder. J Am Acad Child Adolesc Psychiatry 44(5):409–417. doi:10.1097/01.chi.0000155981.83865.ea

Findling RL, McNamara NK, Stansbrey R, Gracious BL, Whipkey RE, Demeter CA, Reed MD, Youngstrom EA, Calabrese JR (2006) Combination lithium and divalproex sodium in pediatric bipolar symptom re-stabilization. J Am Acad Child Adolesc Psychiatry 45(2):142–148. doi:10.1097/01.chi.0000189135.05060.8a

Findling RL, Frazier TW, Youngstrom EA, McNamara NK, Stansbrey RJ, Gracious BL, Reed MD, Demeter CA, Calabrese JR (2007) Double-blind, placebo-controlled trial of divalproex monotherapy in the treatment of symptomatic youth at high risk for developing bipolar disorder. J Clin Psychiatry 68(5):781–788

Findling RL, Frazier JA, Kafantaris V, Kowatch R, McClellan J, Pavuluri M, Sikich L, Hlastala S, Hooper SR, Demeter CA, Bedoya D, Brownstein B, Taylor-Zapata P (2008) The collaborative lithium trials (CoLT): specific aims, methods, and implementation. Child Adolesc Psychiatry Ment Health 2(1):21. doi:10.1186/1753-2000-2-21

Findling RL, Kafantaris V, Pavuluri M, McNamara NK, McClellan J, Frazier JA, Sikich L, Kowatch R, Lingler J, Faber J, Rowles BM, Clemons TE, Taylor-Zapata P (2011) Dosing strategies for lithium monotherapy in children and adolescents with bipolar I disorder. J Child Adolesc Psychopharmacol 21(3):195–205. doi:10.1089/cap.2010.0084

Findling RL, Kafantaris V, Pavuluri M, McNamara NK, Frazier JA, Sikich L, Kowatch R, Rowles BM, Clemons TE, Taylor-Zapata P (2013) Post-acute effectiveness of lithium in pediatric bipolar I disorder. J Child Adolesc Psychopharmacol 23(2):80–90. doi:10.1089/cap.2012.0063

Geller B, Cooper TB, Sun K, Zimerman B, Frazier J, Williams M, Heath J (1998) Double-blind and placebo-controlled study of lithium for adolescent bipolar disorders with secondary substance dependency. J Am Acad Child Adolesc Psychiatry 37(2):171–178. doi:10.1097/00004583-199802000-00009

Geller B, Luby JL, Joshi P, Wagner KD, Emslie G, Walkup JT, Axelson DA, Bolhofner K, Robb A, Wolf DV, Riddle MA, Birmaher B, Nusrat N, Ryan ND, Vitiello B, Tillman R, Lavori P (2012) A randomized controlled trial of risperidone, lithium, or divalproex sodium for initial treatment of bipolar I disorder, manic or mixed phase, in children and adolescents. Arch Gen Psychiatry 69(5):515–528. doi:10.1001/archgenpsychiatry.2011.1508

Grandjean EM, Aubry JM (2009a) Lithium: updated human knowledge using an evidence-based approach. Part II: clinical pharmacology and therapeutic monitoring. CNS Drugs 23(4):331–349

Grandjean EM, Aubry JM (2009b) Lithium: updated human knowledge using an evidence-based approach. Part III: clinical safety. CNS Drugs 23(5):397–418

Hagino OR, Weller EB, Weller RA, Washing D, Fristad MA, Kontras SB (1995) Untoward effects of lithium treatment in children aged four through six years. J Am Acad Child Adolesc Psychiatry 34(12):1584–1590. doi:10.1097/00004583-199512000-00008

Hamrin V, Iennaco JD (2010) Psychopharmacology of pediatric bipolar disorder. Expert Rev Neurother 10(7):1053–1088. doi:10.1586/ern.10.86

Judd LL, Akiskal HS, Schettler PJ, Endicott J, Maser J, Solomon DA, Leon AC, Rice JA, Keller MB (2002) The long-term natural history of the weekly symptomatic status of bipolar I disorder. Arch Gen Psychiatry 59(6):530–537, doi:yoa20425 [pii]

Judd LL, Akiskal HS, Schettler PJ, Coryell W, Endicott J, Maser JD, Solomon DA, Leon AC, Keller MB (2003) A prospective investigation of the natural history of the long-term weekly symptomatic status of bipolar II disorder. Arch Gen Psychiatry 60(3):261–269, doi:yoa20225 [pii]

Kafantaris V, Coletti DJ, Dicker R, Padula G, Kane JM (2001) Adjunctive antipsychotic treatment of adolescents with bipolar psychosis. J Am Acad Child Adolesc Psychiatry 40(12):1448–1456. doi:10.1097/00004583-200112000-00016

Kafantaris V, Coletti D, Dicker R, Padula G, Kane JM (2003) Lithium treatment of acute mania in adolescents: a large open trial. J Am Acad Child Adolesc Psychiatry 42(9):1038–1045. doi:10.1097/01.CHI.0000070247.24125.24

Kafantaris V, Coletti DJ, Dicker R, Padula G, Pleak RR, Alvir JM (2004) Lithium treatment of acute mania in adolescents: a placebo-controlled discontinuation study. J Am Acad Child Adolesc Psychiatry 43(8):984–993. doi:10.1097/01.chi.0000129223.89433.74

Kowatch RA, Suppes T, Carmody TJ, Bucci JP, Hume JH, Kromelis M, Emslie GJ, Weinberg WA, Rush AJ (2000) Effect size of lithium, divalproex sodium, and carbamazepine in children and adolescents with bipolar disorder. J Am Acad Child Adolesc Psychiatry 39(6):713–720. doi:10.1097/00004583-200006000-00009

Kowatch RA, Fristad M, Birmaher B, Wagner KD, Findling RL, Hellander M (2005) Treatment guidelines for children and adolescents with bipolar disorder. J Am Acad Child Adolesc Psychiatry 44(3):213–235, doi:00004583-200503000-00006 [pii]

Kowatch RA, Findling RF, Scheffer RE, et al (2007) Placebo controlled trial of divalproex versus lithium for bipolar disorder. Paper presented at the National meeting of the American Academy of Child and adolescent Psychiatry, Boston, USA

Levy T, Bloch Y (2012) Pharmacotherapy for conduct disorder in children and adolescents. Harefuah 151(7):421–426, 435, 434

Liu HY, Potter MP, Woodworth KY, Yorks DM, Petty CR, Wozniak JR, Faraone SV, Biederman J (2011) Pharmacologic treatments for pediatric bipolar disorder: a review and meta-analysis. J Am Acad Child Adolesc Psychiatry 50(8):749–762. doi:10.1016/j.jaac.2011.05.011, e739

Malhi GS, Tanious M, Gershon S (2011) The lithiumeter: a measured approach. Bipolar Disord 13(3):219–226. doi:10.1111/j.1399-5618.2011.00918.x

Masi G, Milone A, Manfredi A, Pari C, Paziente A, Millepiedi S (2009) Effectiveness of lithium in children and adolescents with conduct disorder: a retrospective naturalistic study. CNS Drugs 23(1):59–69. doi:10.2165/0023210-200923010-00004

Masi G, Perugi G, Millepiedi S, Mucci M, Pfanner C, Berloffa S, Pari C, Gagliano A, D'Amico F, Akiskal HS (2010) Pharmacological response in juvenile bipolar disorder subtypes: a naturalistic retrospective examination. Psychiatry Res 177(1–2):192–198. doi:10.1016/j.psychres.2009.01.022

Patel NC, DelBello MP, Bryan HS, Adler CM, Kowatch RA, Stanford K, Strakowski SM (2006) Open-label lithium for the treatment of adolescents with bipolar depression. J Am Acad Child Adolesc Psychiatry 45(3):289–297. doi:10.1097/01.chi.0000194569.70912.a7

Pavuluri MN, Henry DB, Carbray JA, Sampson GA, Naylor MW, Janicak PG (2006) A one-year open-label trial of risperidone augmentation in lithium nonresponder youth with preschool-onset bipolar disorder. J Child Adolesc Psychopharmacol 16(3):336–350. doi:10.1089/cap.2006.16.336

Peruzzolo TL, Tramontina S, Rohde LA, Zeni CP (2013) Pharmacotherapy of bipolar disorder in children and adolescents: an update. Rev Bras Psiquiatr 35(4):393–405. doi:10.1590/1516-4446-2012-0999

Scahill L, Farkas L, Hamrin V (2001) Lithium in children and adolescents. J Child Adolesc Psychiatr Nurs 14(2):89–93

Strober M, Morrell W, Lampert C, Burroughs J (1990) Relapse following discontinuation of lithium maintenance therapy in adolescents with bipolar I illness: a naturalistic study. Am J Psychiatry 147(4):457–461

Thomas T, Stansifer L, Findling RL (2011) Psychopharmacology of pediatric bipolar disorders in children and adolescents. Pediatr Clin North Am 58(1):173–187. doi:10.1016/j.pcl.2010.10.001, xii

Washburn JJ, West AE, Heil JA (2011) Treatment of pediatric bipolar disorder: a review. Minerva Psichiatr 52(1):21–35

Weller EB, Danielyan AK, Weller RA (2002) Somatic treatment of bipolar disorder in children and adolescents. Child Adolesc Psychiatr Clin N Am 11(3):595–617

Lithium Use in Elderly Patients with Bipolar Disorder

20

Frédéric Limosin and Jean-Pierre Schuster

Abstract

Bipolar disorder is associated with significant burden in elderly patients, including morbidity and mortality, especially by suicide, and impairment of quality of life and functional autonomy. Although lithium constitutes a first-line evidence-based choice in patients with bipolar disorder, it is prescribed less often in the elderly than in younger adults with bipolar disorder. Controlled studies on the efficacy of lithium in elderly patients are sparse, but the available data do not support the idea that side effects, especially on renal function, justify its under-utilisation. Moreover, the anti-inflammatory and neuroprotective effects of lithium may be of particular interest in the older population with bipolar disorder, as they may play a role in preventing and/or controlling the occurrence of certain neurodegenerative processes.

F. Limosin, MD, PhD (✉)
Service de Psychiatrie de l'adulte et du sujet âgé, hôpital Corentin-Celton, groupe hospitalier Hôpitaux Universitaires Paris Ouest, Assistance Publique–Hôpitaux de Paris (AP–HP), 4, parvis Corentin Celton, Issy-les-Moulineaux 92130, France

Université Paris Descartes, Sorbonne Paris-Cité, Paris 75006, France

Inserm, U894, centre de psychiatrie et neurosciences, Paris 75014, France
e-mail: frederic.limosin@aphp.fr

J.-P. Schuster, MD
Service de Psychiatrie de l'adulte et du sujet âgé, hôpital Corentin-Celton, groupe hospitalier Hôpitaux Universitaires Paris Ouest, Assistance Publique–Hôpitaux de Paris (AP–HP), 4, parvis Corentin Celton, Issy-les-Moulineaux 92130, France
e-mail: jean-pierre.schuster@aphp.fr

G.S. Malhi et al. (eds.), *The Science and Practice of Lithium Therapy*,
DOI 10.1007/978-3-319-45923-3_20

Key Points

- The prevalence of bipolar disorder is estimated at 4 % in the general population and ranges from 0.25 to 1.03 % in adults older than 65.
- Traditionally, early-onset bipolar (EOB) and late-onset bipolar (LOB) disorders are distinguished by the cut-off of 50 years.
- Decreasingly prescribed in the elderly, lithium remains prominent among the therapeutic strategies recommended for bipolar disorder in this age group.
- Lower therapeutic plasma concentrations for the elderly, from 0.4 to 0.7 mEq/L, are recommended by some.
- The most frequently reported undesirable effect among elderly patients with a mood disorder being treated with lithium is confusion.
- It is advisable to be particularly vigilant and to carry out regular monitoring of plasma calcium levels for elderly patients being treated with lithium.
- The neuroprotective properties of lithium are promising, but remain a point of contention.

20.1 Introduction

Mental disorders in the elderly represent a growing concern for public health. Psychiatric disorders are prevalent in elderly patients; major depression, in particular, has an estimated prevalence of 40 % among institutionalised patients aged 75 and over. Mental disorders constitute one of the first causes of morbidity and premature mortality in this population. In Europe, elderly people have the highest rate of suicide (in 2005, 16.4–22.9 per 100,000). Suicide in people aged 60 and over accounts for 30 % of the total number of suicide cases.

20.2 Prevalence and Incidence of Bipolar Disorder in Older People

Even though the majority of patients with bipolar disorder are diagnosed before the age of 35, 15–20 % are diagnosed after the age of 55 (Almeida and Fenner 2002). The prevalence of bipolar disorder is lower in older than in younger people, estimated at 4 % in the general population (Hirschfeld et al. 2003) and ranging between 0.25 % and 1.03 % (Unutzer et al. 1998; Ritchie et al. 2004; Goldstein et al. 2006) in adults older than 65. The principal assumptions advanced to account for this low prevalence are the premature high death rate of individuals with a bipolar disorder, a certain degree of symptomatic amendment of the disorder with ageing and greater difficulty in diagnosing bipolar disorder in older adults. The latter arises in part from a degree of inadequacy of the diagnostic criteria used to evaluate adult subjects to sufficiently take account of evolutionary characteristics related to age. It should also be noted that the gender ratio among elderly patients with bipolar disorder suggests a female preponderance of 2:1 (Depp and Jeste 2004).

In a cohort of 8,012 elderly subjects aged 60 years and over, the 3-year incidence rate of bipolar I disorder and bipolar II disorder was 0.54 % (SE = 0.09) and 0.34 % (SE = 0.06), respectively (Chou et al. 2011).

20.3 Clinical Characteristics of Bipolar Disorder in Elderly Subjects

The clinical characteristics of bipolar disorder differ in the elderly from those in younger patients, with more cases of rapid cycling and fewer suicide attempts (Oostervink et al. 2009). Several authors propose a phenotypical dismemberment of bipolar disorder according to age at onset of the disorder (Bellivier et al. 2003; Manchia et al. 2008).

Classically, we distinguished early-onset bipolar (EOB) disorder from late-onset bipolar (LOB) disorder; the age threshold retained being generally 50 years. However, this separation lacks consensus (Lala and Sajatovic 2012) even though the two 'subtypes' show distinct clinical and evolutionary characteristics. In 2004, a review of the literature found an equivalent sex ratio between EOB and LOB. According to the same study, the LOB was associated with a reduced family concentration of the disorder and with greater neurological comorbidities (Depp and Jeste 2004). Clinical specificities according to age at onset remain a point of controversy (Lala and Sajatovic 2012). For Sajatovic and colleagues, manic episodes may be less frequent and of lower severity among patients presenting a LOB compared to those with EOB, with, moreover, a mood that is more often irritable than inflamed (Sajatovic 2002). Some prior studies (Snowdon 1991), but not all (Schurhoff et al. 2000), suggest greater rates of mixed episodes among patients with LOB.

A recent exploratory study over 2 years compared patients presenting with a manic episode and showed that the LOB subjects reached remission and left hospital within shorter time periods than EOB patients (Oostervink et al. 2009). Lastly, Schurhoff and colleagues showed that late-onset episodes are associated with a better rate of response to lithium (Schurhoff et al. 2000).

20.4 Evolution of Lithium Prescription in the Elderly

Although lithium has long been regarded as the gold standard of long-term mood disorder treatment, prescription practice has gradually evolved in favour of valproic acid. In Ontario, Canada, between 1993 and 2001, the annual number of new lithium prescriptions for patients aged over 65 fell from 653 to 281, while prescriptions for valproic acid increased (Shulman et al. 2003). However, Van Melick and colleagues (2012) showed, over a period ranging from 1996 to 2008, that there was no higher rate of discontinuation or substitution of lithium among elderly patients on that treatment than for other age groups. Furthermore, current recommendations for good practice continue to recommend the use of lithium in elderly bipolar subjects (Llorca et al. 2010).

20.5 Pharmacokinetics

Reduction in the volume of distribution, concomitant with a reduction in the glomerular filtration rate and thus renal clearance, explains why therapeutic serum lithium concentrations are obtained in older patients with weaker posology than in younger adults (Sproule et al. 2000). In the same way, an increase in the half-life of elimination and possible neurotoxic effects with weaker doses than in younger adults have encouraged some authors to recommend lower therapeutic plasma concentrations in the elderly (0.4–0.7 mEq/L) (Foster 1992).

20.6 Effectiveness and Tolerance

To date there has been no specific randomised controlled study of the effectiveness of lithium in elderly bipolar subjects for mania or for the prevention of relapse. Nevertheless, some retrospective uncontrolled studies, as well as post hoc analysis of data from a double-blind study, do indicate its effectiveness and tolerance among elderly bipolar subjects (Aziz et al. 2006; Department of Veterans Affairs – Department of Defense 2010). A higher incidence of undesirable effects, reported retrospectively, was found in elderly subjects compared to younger subjects ($p < 0.02$) (Smith and Helms 1982). Another retrospective study over 7 years showed that the most frequently reported undesirable effect among 114 elderly people presenting with a mood disorder treated with lithium was confusion (19.3 %) (Holroyd and Rabins 1994).

Rej and colleagues (2012) took a specific interest in renal complications and analysed 96 articles concerning patients aged 65 or older. The annual incidence of acute renal insufficiency was 1.5 %; the prevalence of chronic renal insufficiency was 1.2–34 % (with risk factors being age, lithium overdose, polyuria and antecedents of a deterioration of renal function); finally, the prevalence of insipid diabetes ranged between 1.8 % and 85 % (with risk factors being the duration of treatment, posology and the nature of the illness). However, on balance, the authors conclude that the frequency of undesirable renal effects in elderly subjects is not sufficient to justify avoidance of lithium treatment.

Concerning induced hypercalcaemia and hyperparathyroidism, this undesirable outcome is asymptomatic and its frequency among the elderly is poorly documented. However, among elderly subjects, this side effect is likely to be more serious, especially when it coexists with a deterioration of renal function. In this context, Lehmann and Lee (2013) carried out a systematic analysis of cases reported in the literature. Their results show that 40 % of cases related to subjects aged over 60 and that the serum calcium concentrations were higher in subjects over 60 as compared to younger subjects. It is thus advisable to be particularly vigilant and to perform regular monitoring of plasma calcium levels among the elderly treated with lithium.

20.7 Lithium and Dementia

Kessing and colleagues showed in 2008 that patients with a greater number of lifetime lithium prescriptions had a lower risk of developing Alzheimer's disease (Kessing et al. 2008). Following these results, several teams tried to determine the mechanism(s) by which lithium could have a protective role against this neurodegenerative disease. It was shown in animal models and in models of cellular culture that lithium increases neuronal survival by various mechanisms: inhibition of apoptosis, regulation of autophagy, increase in metabolism and an increase in the synthesis of neurotrophic factors (Forlenza et al. 2012).

In humans, lithium's neuroprotective effects can be explained by any and all of the following: an increase in the expression of anti-apoptosis genes, inhibition of cellular oxidative stress, synthesis of brain-derived neurotrophic factor (BDNF), cortical thickening, an increase in the density of grey matter, an increase in the volume of the hippocampus and inhibition of the glycogen synthase kinase-3 beta (GSK-3) (Forlenza et al. 2012; Malhi et al. 2013).

At the clinical level, highlighting the beneficial role that lithium might play in reducing the risk of Alzheimer's disease is much more complex. Studies carried out to date are contradictory. A study by Hampel and colleagues (2009) found, at 10 weeks of treatment, no effect of lithium on the activity of GSK-3, the biomarkers in the LCR, or on the cognitive score on the Alzheimer's Disease Assessment Scale (ADAS-Cog). By contrast, Forlenza and colleagues (2011), in a double-blind study, tested lithium versus placebo among patients presenting with a mild cognitive impairment (MCI). After 12 months of treatment, lithium slowed down the progression of cognitive and functional deteriorations and decreased hyperphosphorylation of the protein tau. However, this reduction in the hyperphosphorylation of tau is precisely a key process of continuum MCI-AD.

20.8 Anti-inflammatory Drug Effects of Lithium

During ageing, dysregulation of the dendritic cells results in a reduction in the immune response to infections and an increase in chronic immune reactions with pro-inflammatory synthesis of cytokines (CPI) and chemokines (CK). However, CPI and CK exert cellular toxicity and contribute to a vulnerability to various somatic pathologies, such as Alzheimer's disease, arterial hypertension, arteriosclerosis, diabetes and cancer. In adults, interleukin-10 (IL10) has a powerful effect against the production of CPI and CK, but in elderly subjects the production of IL10 decreases.

In addition, it has been shown that the mechanism of action of lithium on the reduction of inflammation differs between young and old subjects. In the elderly, unlike with young subjects, lithium is unable to induce production of IL10 but decreases the secretion of TNF-α and IL6 by dendritic cells (Agrawal et al. 2013).

20.9 Summary

Decreasingly prescribed to elderly patients, lithium nevertheless remains among the therapeutic strategies recommended for the treatment of bipolar disorder in this age group. Even though the occurrence of certain complications demands additional monitoring of elderly subjects, the profile of tolerance of lithium does not justify its underutilisation. However, it must be noted that there is currently a lack of specific controlled studies concerning the effectiveness of lithium among elderly bipolar subjects, and this needs to be addressed. This is particularly important because, via its anti-inflammatory and neuroprotective properties, lithium could provide a therapeutic alternative that is likely to limit the development of certain neurodegenerative processes.

References

Agrawal S, Gollapudi S, Gupta S, Agrawal A (2013) Dendritic cells from the elderly display an intrinsic defect in the production of IL-10 in response to lithium chloride. Exp Gerontol 48:1285–1292

Almeida OP, Fenner S (2002) Bipolar disorder: similarities and differences between patients with illness onset before and after 65 years of age. Int Psychogeriatr 14:311–322

Aziz R, Lorberg B, Tampi RR (2006) Treatments for late-life bipolar disorder. Am J Geriatr Pharmacother 4:347–364

Bellivier F, Golmard JL, Rietschel M et al (2003) Age at onset in bipolar I affective disorder: further evidence for three subgroups. Am J Psychiatry 160:999–1001

Chou KL, Mackenzie CS, Liang K, Sareen J (2011) Three year incidence and predictors of first-onset of DSM-IV mood, anxiety, and substance use disorders in older adults: results from wave 2 of the National Epidemiologic Survey on Alcohol and Related Conditions. J Clin Psychiatry 72:144–155

Department of Veterans Affairs – Department of Defense: VA/DoD clinical practice guideline for management of bipolar disorder in adults (2010)

Depp CA, Jeste DV (2004) Bipolar disorder in older adults: a critical review. Bipolar Disord 6:343–367

Forlenza OV, Diniz BS, Radanovic M et al (2011) Disease-modifying properties of long-term lithium treatment for amnestic mild cognitive impairment: randomised controlled trial. Br J Psychiatry 198:351–356

Forlenza OV, de Paula VJ, Machado-Vieira R et al (2012) Does lithium prevent Alzheimer's disease? Drugs Aging 29:335–342

Foster JR (1992) Use of lithium in elderly psychiatric patients: a review of the literature. Lithium 3:77–93

Goldstein BI, Herrmann N, Shulman KI (2006) Comorbidity in bipolar disorder among the elderly: results from an epidemiological community sample. Am J Psychiatry 163:319–321

Hampel H, Ewers M, Bürger K et al (2009) Lithium trial in Alzheimer's disease: a randomized, single-blind, placebo-controlled, multicenter 10-week study. J Clin Psychiatry 70:922–931

Hirschfeld RM, Calabrese JR, Weissman MM et al (2003) Screening for bipolar disorder in the community. J Clin Psychiatry 64:53–59
Holroyd S, Rabins PV (1994) A retrospective chart review of lithium side effects in a geriatric outpatient population. Am J Geriatr Psychiatry 2:346–351
Kessing LV, Søndergård L, Forman JL, Andersen PK (2008) Lithium treatment and risk of dementia. Arch Gen Psychiatry 65:1331–1335
Lala SV, Sajatovic M (2012) Medical and psychiatric comorbidities among elderly individuals with bipolar disorder: a literature review. J Geriatr Psychiatry Neurol 25:20–25
Lehmann SW, Lee J (2013) Lithium-associated hypercalcemia and hyperparathyroidism in the elderly: what do we know? J Aff Disord 146:151–157
Llorca PM, Courtet P, Martin P et al (2010) Screening and management of bipolar disorders: results. Encéphale 36(Sup 4):S86–S102
Malhi GS, Tanious M, Das P et al (2013) Potential mechanisms of action of lithium in bipolar disorder. Current understanding. CNS Drugs 27:135–153. doi:10.1007/s40263-013-0039-0
Manchia M, Lampus S, Chillotti C et al (2008) Age at onset in Sardinian bipolar I patients: evidence for three subgroups. Bipolar Disord 10:443–446
Oostervink F, Boomsma MM, Nolen WA (2009) Bipolar disorder in the elderly; different effects of age and of age of onset. J Affect Disord 116:176–183
Rej S, Herrmann N, Shulman K (2012) The effects of lithium on renal function in older adults – a systematic review. J Geriatr Psychiatry Neurol 25:51–61
Ritchie K, Artero S, Beluche I et al (2004) Prevalence of DSM-IV psychiatric disorder in the French elderly population. Br J Psychiatry 184:147–152
Sajatovic M (2002) Aging-related issues in bipolar disorder: a health services perspective. J Geriatr Psychiatry Neurol 15:128–133
Schurhoff F, Bellivier F, Jouvent R et al (2000) Early and late onset bipolar disorders: two different forms of manic depressive illness? J Affect Disord 58:215–221
Shulman KI, Rochon P, Sykora K et al (2003) Changing prescription patterns for lithium and valproic acid in old age: shifting practice without evidence. BMJ 326:960–961
Smith RE, Helms PM (1982) Adverse effects of lithium therapy in the acutely ill elderly patient. J Clin Psychiatry 43:94–99
Snowdon J (1991) A retrospective case-note study of bipolar disorder in old age. Br J Psychiatry 158:485–490
Sproule BA, Hardy BG, Shulman KI (2000) Differential pharmacokinetics of lithium in elderly patients. Drugs Aging 16:165–177
Unutzer J, Simon G, Pabiniak C et al (1998) The treated prevalence of bipolar disorder in a large staff-model HMO. Psychiatr Serv 49:1072–1078
Van Melick EJ, Wilting I, Souverein PC, Egberts TC (2012) Differences in lithium use patterns in the Netherlands: comparing middle-aged and older patients in a database study. Am J Geriatr Pharmacother 10:193–200

Lithium in the Twenty-First Century 21

Guy M. Goodwin

Abstract

The discovery of lithium has important lessons for psychiatry at several levels: in relation to strategies for drug discovery, in understanding the bipolar phenotype, in how insight into lithium's mechanism of action and unique pharmacology relates to its positive actions on the one hand and its adverse effects on the other. Lithium's pharmacology remains of interest, and a lithium mimetic is overdue. All these elements continue to place lithium at the heart of future research in bipolar disorder, just as it has occupied this place in its past. Moreover, findings on neuroprotection may herald new future indications and applications.

Key Points

- Lithium's efficacy may hold the key to developing new and effective treatments for bipolar disorder.
- Lithium's unique mechanisms of action may provide novel insights into the neurobiology of bipolar disorder.
- Lithium should not be forgotten in future research.

21.1 Introduction

The discovery of lithium remains one of the most intriguing stories of modern and not so modern medicine. As this volume illustrates, it has important lessons for psychiatry at several levels. This final chapter summarizes snapshots of the key issues for the future.

G.M. Goodwin, FMedSci
University Department of Psychiatry, University of Oxford, Warneford Hospital, Oxford OX3 7JX, UK
e-mail: guy.goodwin@psych.ox.ac.uk

G.S. Malhi et al. (eds.), *The Science and Practice of Lithium Therapy*,
DOI 10.1007/978-3-319-45923-3_21

21.2 Drug Discovery

Lithium illustrates the phenomenon of serendipity in the discovery of novel treatments. This is important because we face an immediate future where major pharmaceutical companies have, with only a few notable exceptions, withdrawn from research and development in psychiatric disorders. That means that there could well be nothing that is really new for a generation. It reflects a failure of the preceding two decades of brain research to generate the concrete scientific understanding that best underpins applied research in our field.

In the future we will continue to need what I like to call 'guided serendipity'. In other words, discovery that is fortuitous, but nevertheless guided by approximate hypotheses that are plausible and science led. The unexpected use of ketamine to treat depression is the most recent really interesting example. Ketamine is a drug with an established indication that is being 'repurposed', to use the jargon buzzword. There will continue to be innovations in neuroscience, targeted mainly at neurodegeneration. This may be primarily as 'biologicals', such as antibodies, aptamers, RNA antagomirs and cell therapy. However, for all we know, these approaches and the development of more familiar types of drug may offer treatments that turn out to work in psychiatric indications. Guided serendipity, having served us well in the past, may yet do so again.

For this to happen, however, we must be free to experiment in the spirit of John Cade. Much modern regulation throttles experiment and hence innovation. It does so by demanding standards of safety or certainty that are impossible, or impossibly expensive, to satisfy. Too often as a Principal Investigator, I have found myself engaged in the pointless exercise of being required to guarantee to prevent from happening something that has never happened before or to collect detailed data on things that we know have happened before and that are bound to happen again, such as adverse effects.

21.3 The Bipolar Phenotype

Quite apart from the generic regulatory barriers to conducting experimental medicine, bipolar disorder presents a particular additional challenge because it is defined to be an episodic disorder. Studying relatively infrequent severe episodes is very challenging. It means either trials in acutely ill patients or relapse-prevention studies that require years rather than weeks of follow-up. There is thus currently no accepted way to do small-scale, short-term, proof-of-concept studies in bipolar patients, and I would argue considerable pessimism about innovation primarily for this reason. This is not a new situation of course. It partly explains why the risk of developing a drug for a bipolar indication is usually only undertaken by industry for products (e.g. antipsychotics and anticonvulsants) that already have indications.

The episodic view of bipolar disorder often assumes that mood returns to normal between episodes. This clinical view is based on retrospective, and often informal, assessment of mood, which has limited reliability. New methods for prospective intensive mood monitoring are revealing a different and more complex picture, in which chronic mood instability is much more typical of most patients. An emerging hypothesis is that mood instability is a core abnormality of bipolar disorder, underlying persistent functional impairments and creating vulnerability to full episodes of depression or mania. This instability is conceded for patients at risk of relapse, and monitoring mood is recommended as a standard of care in many guidelines. But mood instability as a phenomenon is strangely unexplored and poorly understood. We do not know, for example, what the key bandwidths are over which mood homeostasis is mainly regulated. How does the clinical observation of weekly variations in mood relate to daily, or within day, variations, which also clearly occur? Can such timescales in turn be related to much shorter mechanistic loops amenable to analysis with neurophysiological precision?

A more precise understanding of mood stability would give us a potential target on which to test novel treatments in well-designed, short-term experimental studies. Moreover, mood instability may be of trans-diagnostic significance. It is a relatively common complaint in population studies, beyond the formal diagnoses of bipolar disorder and borderline personality disorder, and is predictive of significant morbidity, including suicide. A major goal of future research must be to define the status of mood instability: as a clinical phenomenon, as a possible emergent property of neural networks expressed by defined genotypes, as a pathophysiological construct and for its therapeutic implications.

21.4 Lithium's Mechanism of Action

Lithium has a remarkable efficacy, but continuing safety concerns. It has long been obvious that a lithium mimetic might improve the benefit/harm ratio. A recent suggestion is that an orphan drug approach could work. This is based on the principle that the structure of putative targets for lithium's action (like inositol monophosphatase) may allow the identification of candidate molecules predicted to interact with the target protein. Ebselen is a drug from the National Institutes of Health Clinical Collection, a chemical library of bioavailable drugs, considered clinically safe but without proven use. It represents a partial lithium mimetic that has the potential, on the one hand, to validate inositol monophosphatase inhibition as a treatment for bipolar disorder and on the other to act as a treatment in itself (Singh et al. 2013).

However, those aspects of lithium's unique pharmacology that relate to its positive actions, on the one hand, and its adverse effects on the other, are still poorly distinguished. Quite simply, we have a way to go in understanding lithium's mechanism(s) of action. Studies of human cells offer an emerging way to improve our capacity to understand molecular pathology in functional disorders of the brain. This is being accelerated by the use of pluripotent stem cells. Fibroblasts from patients with known genetic variants can be grown in culture and transformed to different cell types. The properties of these cells may allow us to define the underlying abnormality in bipolar disorder and to understand how, for example, a drug like lithium exerts its unique profile of effect (Haggarty and Perlis 2013). Such assay systems should also allow us to identify novel drug targets and new drugs.

Even if these exciting possibilities are fruitful, the traditional approach to drug development by industry is, unfortunately, widely perceived to be failing. That failure threatens innovation in many disorders. An important solution may come through precompetitive public/private partnership to identify new targets and ligands (Norris et al. 2014). This may be the only way that we will see a lithium replacement in the foreseeable future.

21.5 Lithium's Clinical Efficacy

In purely practical terms, we need to refine and improve the use of lithium in defined indications. It remains a first-line choice for bipolar I disorder, but we still need to improve our prospectively collected safety data. So far, it appears that some risks have been either overstated (e.g. the risk of foetal malformation) or over conservatively interpreted (the risk of impaired renal function) (see Chap. 16), although we need to understand better the risks from higher doses of lithium in both cases.

Database studies have yielded another fascinating finding: the possibility that lithium acts to prevent the onset of dementia (Kessing et al. 2010). This possibility needs replication and strengthening. However, it has a possible explanation because lithium inhibits the enzyme glycogen synthase kinase-3β (GSK-3β), which will have putative downstream effects to reduce both tau protein phosphorylation and amyloid-β42 production (Diniz et al. 2013). Additional neuroprotective effects have also been attributed to reduced pro-inflammatory status and decreased oxidative stress. Whether neuroprotection turns out to be yet another unexpected advantage of lithium treatment remains to be seen.

21.6 Naturalistic Outcomes

Pharmacoepidemiology offers a long-neglected way of using naturalistic data to establish drug efficacy and to investigate societal value. For antipsychotics and mood stabilizers, this has been demonstrated very recently using Swedish databases (Fazel et al. 2014). The outcome was documented violence. This is a good outcome to select for several reasons. Violence is not uncommon in psychiatric populations; it is a proxy for severity of illness and it has obvious clinical and societal relevance. As an outcome, it can be recurrent (e.g. unlike death), which means that patients act as their own controls before, during and after treatment, and observation can be made over long time periods. The reduction in risk with antipsychotics was 50 % and with mood stabilizers 20 %, during active treatment. Lithium had effects that were confined to patients with bipolar diagnoses. So the study demonstrated large effects of treatment with lithium that were specific to the bipolar group. The negative finding in the non-bipolar group is important because mood stabilizers are often used in non-bipolar patients judged to be at risk of violence. This kind of study moves us beyond the artificial world of the randomized clinical trial. We need this kind of innovation and other real-world measures to prove what we see when we treat patients successfully with lithium and other medicines.

21.7 Summary

When I lecture about lithium, I like to spin a corny introductory line: it is an old drug, 14 billion years old, synthesized 20 min after the Big Bang. It is also old in terms of its discovery (actually rediscovery) for use in modern psychopharmacology. But it has not received the investment in research and development that many less interesting compounds have had. Its use in medicine has also never been driven by effective marketing. Perhaps these are the reasons it seems still able to surprise us. I hope it can continue to do so.

References

Diniz BS, Machado-Vieira R, Forlenza OV (2013) Lithium and neuroprotection: translational evidence and implications for the treatment of neuropsychiatric disorders. Neuropsychiatr Dis Treat 9:493–500

Fazel S, Zetterqvist J, Larsson H, Långström N, Lichtenstein P (2014) Antipsychotics, mood stabilisers, and risk of violent crime. Lancet 384(9949):1206–1214

Haggarty SJ, Perlis RH (2013) Translation: screening for novel therapeutics with disease-relevant cell types derived from human stem cell models. Biol Psychiatry 75(12):952–960

Kessing LV, Forman JL, Andersen PK (2010) Does lithium protect against dementia? Bipolar Disord 12(1):87–94

Norris SMP, Pankevich DE, Davis M, Altevogt BM (2014) Opportunities to improve and accelerate the drug development pipeline http://www.ncbi.nlm.nih.gov/books/NBK195041/

Singh N, Halliday AC, Thomas JM, Kuznetsova OV, Baldwin R, Woon EC, Aley PK, Antoniadou I, Sharo T, Vasudevan SR, Churchill GC (2013) A safe lithium mimetic for bipolar disorder. Nat Commun 4:1332–1339

Appendix 1: Lithium Battery–Clinical

In addition to cognitive changes experienced by patients with bipolar disorder, those receiving lithium also report neurocognitive side effects. As such, these have an important impact on quality of life. Therefore, routine neurocognitive assessment should be conducted as part of a comprehensive monitoring schedule, alongside the monitoring of mood stability, treatment adherence, medication tolerability and potential toxicity (Fig. A.1).

We have previously recommended the 'Lithium Battery–Clinical' for routine side effect assessment, to note and manage neurocognitive concerns (Malhi et al. 2016). The 'Lithium Battery–Clinical' can be used in conjunction with a patient data sheet and patient self-report measures to inform the management of side effects, as well as the interpretation of any changes to clinical profile and lithium treatment. Clinician ratings of neurocognitive function can also be analysed in the context of patient perceptions to gain a complete picture of neurocognitive functioning over time. This may, in turn, be used to reframe any misconceptions, as well as to raise awareness of improvements or decrements in neurocognition and inform necessary management strategies, if clinically indicated.

Where a clinically significant pre-existing or treatment-related neurocognitive deficit is apparent, a referral for further neuropsychological and/or neurological assessment should be considered.

G.S. Malhi et al. (eds.), *The Science and Practice of Lithium Therapy*,
DOI 10.1007/978-3-319-45923-3

Points for clinically evaluating the effects of lithium on neurocognition in bipolar disorder.	**Action points when patients volunteer neurocognitive symptoms and concerns.**
- Allow patients to spontaneously relay concerns regarding their neurocognitive functioning: For example; o Concentration and clarity and speed of thought ("brain fog") o Psychomotor functioning ("slowing", "shakiness") o Production of language ("word finding", "holding thoughts and then being able to verbalise them clearly") o Thinking creatively ("thinking about things in unconventional ways", "producing creative works"). - If patients do not volunteer any complaints, it may be necessary to ask for these specifically. - Actively raise these concerns when initiating lithium treatment or whenever altering any aspect of treatment (e.g., optimising lithium dose as part of strategies to reduce side effects).	- Discuss the various dynamic multifactorial influences (e.g., mood state influence on actual impairment; mood-congruent attribution bias for perceived impairment), and challenge false beliefs (e.g., can use brief cognitive intervention, or at least introduce doubt about causal beliefs). - Suggest adaptations for daily life (e.g., self-monitoring of neurocognitive profile and mood states and episode onset; occupational therapy; behavioural interventions for reinstatement of previous social, leisure, and occupational activities). - Consider longitudinal assessment of the above, cross-referencing clinician and self-report assessments (e.g., assess patient both on and off treatment for comparison, whenever there are changes to treatment with regular review and at least an annual assessment of all of the above for those on long-term lithium treatment).

Fig. A.1 The lithium battery–clinical. Recommended clinical discussion (*left*) and action points (*right*) for probing and managing concerns regarding neurocognition in bipolar disorder and its treatment with lithium treatment. This battery is designed for longitudinal use alongside a patient data sheet and patient self-report measures, such that any concerns can be cross-referenced with changes in clinical profile and lithium therapy (Adapted from Malhi et al. 2016)

Appendix 2: Lithium Battery–Research

In a recent review (Malhi et al. 2016), we concluded that determining the impact of lithium treatment in bipolar disorder on neurocognition is difficult, and further research is required to disentangle the effects of illness versus those of treatment. Hence, we proposed researchers employ a longitudinal research design (e.g. pre-illness, illness and lithium administration time points) along with common neuropsychological assessments. In doing so, greater control over relatively *stable* neurocognitive components will be attained, allowing the components that *change* with lithium treatment to be more evident and malleable. Further, in conducting controlled comparisons in healthy controls and well-characterised patient groups, illness-related effects can be better partitioned. However, there is a vast array of potential neurocognitive domains and assessments that may reveal the effects of lithium from those caused by the illness, which has hindered progress in this area. Thus, we proposed the development of a research measurement framework aimed at partitioning these neurocognitive profiles.

In order to progress, we developed a 'Lithium Battery–Research' assessment that has been limited to a manageable number of components (and associated objective assessments)—derived from those that have shown some initial promise and utility in making these distinctions (Fig. A.2). Like the 'Lithium Battery–Clinical', the 'Lithium Battery–Research' is ideally intended for use alongside a patient data sheet, such that neurocognitive changes can be correlated with clinical changes. The prototype battery was developed with consideration of (limited) strength of evidence, for longitudinal use within a clinical research context (e.g. time and ease of administration), and with the intention to parse out specific neurocognitive domains. For example, comparison of the trail-making task, parts A and B, allows discrimination between psychomotor speed and executive functioning. Furthermore, the stability of components could be distilled with repeated testing, including those that are thought to be unchanged with bipolar disorder illness onset (here, premorbid IQ) or lithium administration per se. For a more detailed discussion of the level of evidence for each domain, confounding factors, research design and the development

Test (Time required in mins)*	Neurocognitive Domain	Anticipated change in lithium-treated bipolar disorder
RAVLT Learning Trials (4)	Verbal learning	Decrease
WAIS Digit Span tests (15)	Executive function and working memory	Decline, then Stable Variable. Serve as Variable Illness Control (covaries with mood variability).
TMT A and B (7)	Psychomotor and Processing speed	Decrease
RAVLT Delayed Recall (1)	Verbal memory	Decrease
ROCFT Learning Trials (5)	Visual learning	Decline then Stable. Serve as stable Illness Control, and to test domain specificity.
WAIS vocabulary/ information test (15)	Premorbid IQ	Stable, Serves as Pre-illness Control.
COWAT (3)	Verbal fluency	Decrease
Purdue pegboard (3)	Motor speed	Decline, then Stable Decrease. Serve as Illness Control, and to test domain specificity.
ROCFT Delayed Recall (5)	Visual memory	Decline then Stable. Serve as stable Illness Control, and to test domain specificity.

Fig. A.2 The lithium battery–research. A prototype battery of tests that can be used to examine neurocognitive changes specific to lithium treatment across identified domains (in *yellow*)—thus enabling disentanglement from relatively stable domains (in *green*). The battery is intended for use alongside a patient data sheet, such that neurocognitive changes can be correlated with changes in the clinical symptom profile and changes to lithium treatment (Adapted from Malhi et al. 2016. *Abbreviations*: *COWAT* Controlled Oral Word Association Test, *RAVLT* Rey Auditory Learning Verbal Test, *ROCFT* Rey Osterreith Complex Figure Test, TMT A and B Trail-Making Task Form A and Form B, WAIS Wechsler Adult Intelligence Scale. Note *This is a suggested sequence, which allows for the required delayed recall intervals for the RAVLT and the ROCFT of at least 20 min. Total time for the complete battery is <60 min

of the 'Lithium Battery–Research', please see Malhi et al. (2016). The cognitive domains covered in the 'Lithium Battery–Research' are increasingly assessed with computerised administration, which may be employed alongside face-to-face clinician-administered tests, in order to increase ease of administration and reliability. For example, psychomotor speed is often assessed using computerised trail-making tasks; however, motor speed specifically and verbal fluency are currently more commonly assessed face-to-face. Working within a 'Lithium Battery' framework may assist in validating the instrument so that it can eventually be implemented in clinical practice.

Index

G.S. Malhi et al. (eds.), *The Science and Practice of Lithium Therapy*,
DOI 10.1007/978-3-319-45923-3

D

E

F

G

H

I